The Cambridge Program for The Mathematics Test

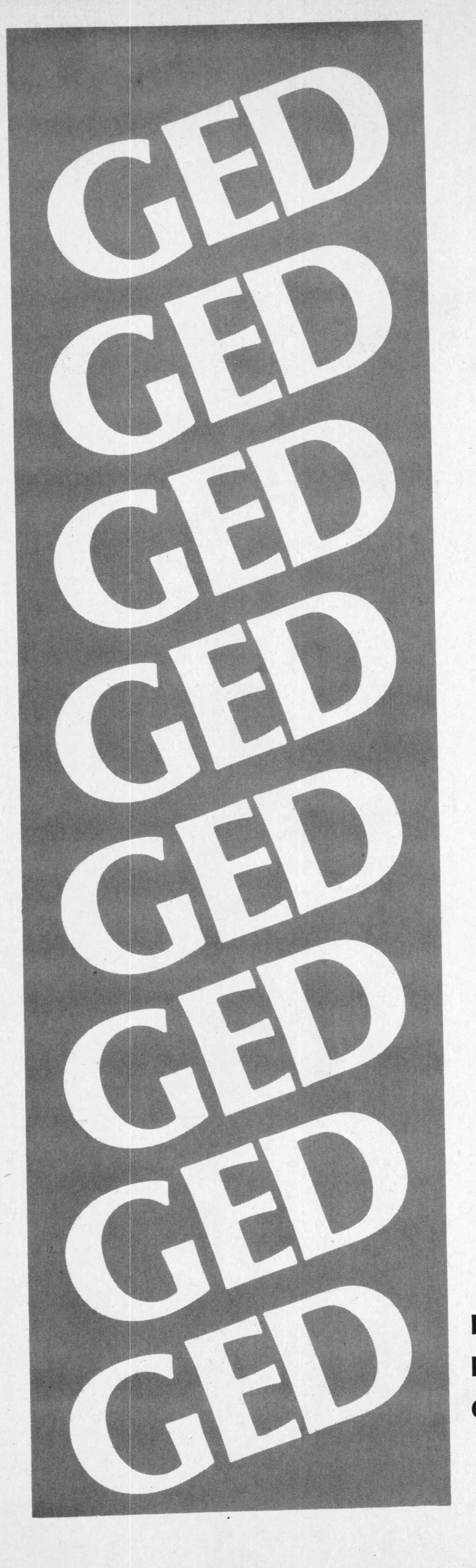

Executive Editor: Brian Schenk
Project Editor: Dennis Mendyk
Contributing Editor: Peter Fondulas

The Cambridge Program for The Mathematics Test

by JERRY HOWETT

CAMBRIDGE
The Adult Education Company
888 Seventh Avenue, New York, New York 10106

ISBN 0-8428-9386-5

4 5 6 7 8 9

CONTENTS

Predictor Test: Basic Math

Before you begin to work with the first part of this book, take the following test. The test will help you to find out how much you already know about basic mathematics. The test also will show you which parts of the book you should study the most.

The test is divided into six parts (Parts A–F). After you complete all six parts, check your answers. In the answer section, circle the numbers of the problems that you missed. After you check your answers, look at the chart that follows. The chart will show you which parts of the book deal with the math skills that gave you the most trouble.

Don't worry if you miss problems on this test. The book will help you to build the math skills that give you trouble.

DIRECTIONS: For each question, find the correct answer.

PART A

1. What is the value of 8 in 38,492?

2. What is the sum of 76, 2,059, and 463?

3. Find the difference between 849 and 7,000.

4. What is the product of 37 and 6,083?

5. How much is 36,432 divided by 72?

6. On Monday, Fred drove 325 miles. Wednesday, he drove 363 miles. Friday, he drove 296 miles. Find the average distance he drove each day.

7. Find the next term in the series 99, 96, 93, 90. . . .

8. Round off 23,518 to the nearest thousand.

9. In November, the Caldwells spent $396 for utilities and mortgage payments, $412 for taxes and Social Security, $183 for car payments and gasoline, $335 for food, and $190 for everything else. How much did they spend altogether that month?

10. The area of the state of Delaware is 2,057 square miles. The area of Pennsylvania is 45,333 square miles. The area of Pennsylvania is how much more than the area of Delaware?

11. Mark made 26 payments of $120 each to pay off his car loan. How much did Mark pay altogether for the loan?

12. Together Manny, Jack, and Jeff caught 42 pounds of fish. They decided to share the fish equally. How many pounds did each person get?

PART B

1. There are 16 ounces in one pound. 7 ounces are what fraction of a pound?

2. Reduce $\frac{60}{400}$ to lowest terms.

3. Find the missing numerator: $\frac{3}{4} = \frac{\ }{48}$.

4. Change $\frac{34}{8}$ to a mixed number and reduce.

5. Change $9\frac{3}{5}$ to an improper fraction.

6. Add $5\frac{9}{16}$ and $2\frac{5}{16}$.

7. Find the sum of $4\frac{3}{4}$, $2\frac{7}{12}$, and $3\frac{4}{9}$.

8. Subtract $4\frac{3}{8}$ from $9\frac{7}{8}$.

9. Take $2\frac{1}{3}$ from $4\frac{4}{5}$.

10. Subtract $9\frac{3}{4}$ from $18\frac{1}{6}$.

11. $\frac{4}{9} \times \frac{4}{5} =$

12. $\frac{8}{15} \times \frac{5}{16} =$

13. $9 \times \frac{11}{12} =$

14. $2\frac{1}{2} \times 2\frac{1}{5} =$

15. $\frac{4}{9} \div \frac{2}{3} =$

16. $5\frac{1}{4} \div \frac{7}{12} =$

17. $7\frac{1}{2} \div 5 =$

18. $1\frac{1}{3} \div 1\frac{7}{9} =$

19. Sandy usually works 40 hours a week. Last week, she worked $1\frac{1}{2}$ hours extra on Tuesday and $2\frac{1}{4}$ hours extra on Thursday. How many hours did she work altogether last week?

20. From a piece of copper tubing 3 yards long, Pete cut a piece $1\frac{3}{4}$ yards long. How long was the piece that was left over?

21. The Pappas family has a monthly budget of $892. They spend $\frac{1}{4}$ of their budget on food. How much do they spend on food each month?

22. Al has a piece of wood 18 inches long. How many $1\frac{1}{2}$-inch long strips can he cut from the piece?

PART C

1. Write fourteen ten-thousandths as a decimal.
2. Which decimal is bigger, .05 or .036?
3. Change .035 to a fraction and reduce.
4. Change $\frac{9}{20}$ to a decimal.
5. Round off 2.4718 to the nearest hundredth.
6. Add 42 + 6.59 + .477.
7. Subtract .807 from 6.3.
8. Multiply 43.8 by .56.
9. $81.6 \div 17 =$

10. $1.044 \div 2.9 =$

11. $232 \div .58 =$

12. In June, the average rainfall in St. Louis is 4.4 inches. In July, the average is 3.7 inches. In August, the average is 2.9 inches. Find the total rainfall for these three months.

13. On Monday morning, the mileage on Jack's car was 14,926.8 miles. Friday evening, the mileage was 15,619.4 miles. How many miles did Jack drive that week?

14. Helen bought 3.75 yards of material at $5.60 a yard. What was the total price of the material?

15. Jeff drove 235 miles in 5.5 hours. To the nearest tenth, what was his average speed in miles per hour?

PART D

1. Change .085 to a percent.

2. Change 275% to a decimal.

3. Change $\frac{5}{16}$ to a percent.

4. Change 64% to a fraction and reduce.

5. Find 35% of 280.

6. Find 6.2% of 720.

7. Find $16\frac{2}{3}$% of 960.

8. 27 is what % of 60?

9. 70% of what number is 560?

10. Find the interest on $1,200 at $4\frac{1}{2}$% annual interest for 1 year and 4 months.

11. Nora and Douglas bought new furniture for $480. They made a down payment of 12%. How much was the down payment?

12. Harry owes $290 on his credit card. Each month, he has to pay a fee of 1.5% of the amount he owes. Find the monthly fee for $290.

13. Paul bought a car radio on sale for $45. The original price of the radio was $60. What was the percent of the discount on the radio?

14. Pete makes $240 a week before deductions. His boss takes out 21% of Pete's pay for taxes and Social Security. How much does Pete take home each week?

15. The Wileys pay $230 a month for rent. Rent is 20% of their monthly budget. What is the Wileys' total monthly budget?

PART E

1. 15 feet = ______ inches.

2. 9,000 pounds = ______ tons.

3. 1 kilogram = ______ grams.

4. 1 kilometer = 0.6 mile. Change 85 kilometers to miles.

5. Add 9 hrs. 55 min. and 8 hrs. 43 min.

6. Subtract 12 lbs. 15 oz. from 14 lbs.

7. Multiply 3 yds. 2 ft. by 5.

8. Divide 10 gals. 2 qts. by 6.

9. The distance from Al's house to the factory where he works is 12 miles. One mile equals 1.6 kilometers. Find the distance in kilometers from Al's house to the factory.

10. Dave worked overtime for 2 hrs. 15 min. on Wednesday, 1 hr. 20 min. on Thursday, and 4 hrs. 50 min. on Saturday. Altogether, how much overtime did Dave work in the week?

11. On Monday, the Ace Aluminum Corporation shipped 12 loads, each weighing 3 tons 400 pounds. Find the total weight of the shipment.

(GO ON TO THE NEXT PAGE)

PART F

Use the table below to answer questions 1 through 4. The table shows Gutenberg High School's passing records for the years 1974 to 1978.

Year	Attempts	Completions	Yards Gained	Touchdowns
1974	167	107	1,185	11
1975	425	273	2,294	25
1976	158	91	1,460	8
1977	361	210	2,620	18
1978	413	231	3,190	25

1. In which year were the most passing attempts made?
2. In which year were the most yards gained?
3. How many more completions were made in 1975 than in 1978?
4. The number of touchdowns in 1976 was what percent of the number in 1978?

Use the bar graph at the right to answer questions 5 through 8. The graph shows the amounts and sources of taxes and fees for New York State in one year.

5. What amount of taxes and fees does New York State collect in one year?
6. How much more does New York State collect in business taxes than in personal income taxes in one year?
7. The amount New York State collects in consumption taxes is what percent of the amount collected in personal income tax?
8. Is gasoline tax income closer to $\frac{1}{2}$, $\frac{1}{3}$, or $\frac{1}{4}$ of the total taxes and fees collected by New York State?

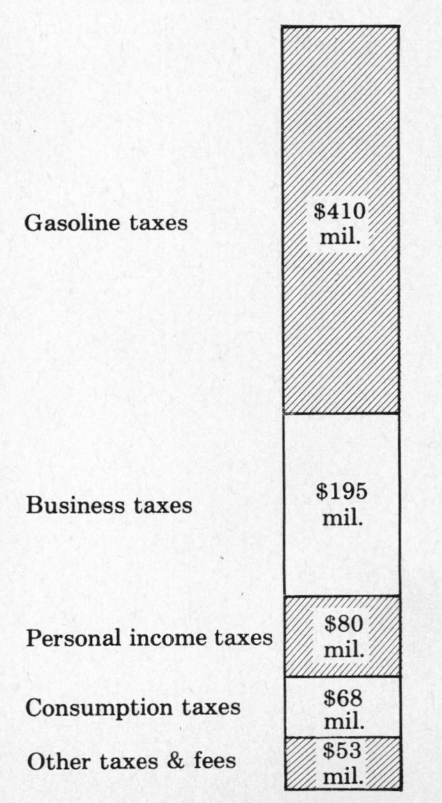

ANSWERS

PART A

1. **8,000** 2. **2,598** 3. **6,151** 4. **225,071**

5. **506** 6. **328 mi.** 7. **87** 8. **24,000**

9. **$1,516** 10. **43,276 sq. mi.** 11. **$3,120** 12. **14 lbs.**

PART B

1. $\frac{7}{16}$ 2. $\frac{3}{20}$ 3. $\frac{36}{48}$ 4. $4\frac{1}{4}$

5. $\frac{48}{5}$ 6. $7\frac{7}{8}$ 7. $10\frac{7}{9}$ 8. $5\frac{1}{2}$

9. $2\frac{7}{15}$ 10. $8\frac{5}{12}$ 11. $\frac{16}{45}$ 12. $\frac{1}{6}$

13. $8\frac{1}{4}$ 14. $5\frac{1}{2}$ 15. $\frac{2}{3}$ 16. **9**

17. $1\frac{1}{2}$ 18. $\frac{3}{4}$ 19. $43\frac{3}{4}$ **hrs.** 20. $1\frac{1}{4}$ **yds.**

21. **$223** 22. **12 strips**

PART C

1. **.0014** 2. **.05** 3. $\frac{7}{200}$ 4. **.45**

5. **2.47** 6. **49.067** 7. **5.493** 8. **24.528**

9. **4.8** 10. **.36** 11. **400** 12. **11 inches**

13. **692.6 mi.** 14. **$21** 15. **42.7 mph**

PART D

1. **8.5%** 2. **2.75** 3. $31\frac{1}{4}\%$ 4. $\frac{16}{25}$

5. **98** 6. **44.64** 7. **160** 8. **45%**

9. **800** 10. **$72** 11. **$57.60** 12. **$4.35**

13. **25%** 14. **$189.60** 15. **$1,150**

PART E

1. **180 in.** 2. **$4\frac{1}{2}$ t.** 3. **1,000 grams** 4. **51 miles**

5. **18 hrs. 38 min.** 6. **1 lb. 1 oz.** 7. **18 yds. 1 ft.**

8. **1 gal. 3 qts.** 9. **19.2 km.** 10. **8 hrs. 25 min.**

11. **38 t. 800 lbs.**

PART F

1. **1975** 2. **1978** 3. **42** 4. **32%**

5. **$806 million** 6. **$115 million** 7. **85%** 8. **$\frac{1}{2}$**

Math Skills Chart

After you check your answers, look at the chart that follows. The chart will show you which parts of this book correspond to the parts of the test that you have just completed. Use the chart to find out which parts of the book you should study the most.

For questions in:	study these pages:	
Part A	17–35	(Whole Numbers)
Part B	36–84	(Fractions)
Part C	85–115	(Decimals)
Part D	116–146	(Percents)
Part E	147–161	(Measurements)
Part F	162–186	(Tables & Graphs)

Make sure that you read the chapter called "Reading in Mathematics" before you do anything else in this book.

Reading in Mathematics

Reading is a basic skill. In order to read, you have to know the letters of the alphabet. You have to know how letters are put together to form words. You have to know the meanings of words and sentences. You have to know the *logic* of the language.

Reading in mathematics is a little different from reading a story or a newspaper. When you read the newspaper, you read normal, everyday language. When you read a math book or a math problem, you have to read two different "languages." First, you have to know the normal, everyday language. Second, you have to know the logic of the math "language."

What is Math "Language"?

If you have ever worked with math before, you know that there are certain signs in math that tell you how to work a problem. For example, look at this problem:

$$13 + 7 = \underline{\qquad}$$

You probably know the two symbols that are in the problem. The first symbol (+) means to add. The second symbol (=) means equal to. If you know what the symbols mean, you can work the problem with little trouble.

$$13 + 7 = \mathbf{20}$$

Such signs as +, −, ×, ÷, and = are the basic signs of math language. However, they are not the only signs. Such signs as %, $>$, $<$, and ∠ also are part of the language. To have a full understanding of math, you also have to know what these signs stand for.

Everyday Language in Math

Read the next problem carefully:

> At the car wash, Jimmy washed 13 cars before lunch. He washed 7 cars after lunch. What was the total number of cars that Jimmy washed?

This problem is written in everyday language. There are no math signs in it. However, if you read the problem carefully, you know that the problem is a math problem.

On the GED Test, most of the problems are written in everyday language. It is your job to figure out which math signs to use to solve the problem. You have to be able to "translate" everyday language into math language.

Read the problem again. What is being asked for in the problem? The problem asks for the total number of cars that Jimmy washed. The word *total* usually means to add. How many cars did Jimmy wash? If you read the problem carefully, you can see that he washed 13 cars before lunch and 7 cars after lunch. To solve the problem, you have to find the *total* of 13 and 7. You have to add.

$$\begin{array}{r} 13 \\ +\ 7 \\ \hline \mathbf{20} \end{array}$$

Whenever you work with a "word" problem in math, you have to translate the everyday language into math language. Sometimes, figuring out what to do in a word problem is hard. However, there are certain steps that you can follow. These steps will make working with word problems easier for you.

Below, you will find the steps that you should use when you work with word problems. After the steps, you will find a word problem. The word problem is worked out with the steps. Study the ways that the steps are used.

Steps for Solving Word Problems

Step 1: Read the problem. Get a mental picture of what is going on.

Step 2: Decide what is being asked in the problem.

Step 3: Decide which facts to use to answer the question.

Step 4: Decide which math *operation* to use.

Step 5: *Estimate* the answer.

Step 6: Figure out the exact answer.

Step 7: Compare the exact answer to the estimate.

Step 8: Read the problem once more to be sure that you answered the right question.

Step 9: Check your answer once more.

(You may have to read the problem three or four times to get all the information. Read the problem as many times as you need to get the information.)

Word Problem:

Bert drives a truck. On Tuesday, he drove 70 miles in 2 hours. On Thursday, he drove 114 miles in 3 hours. On Friday, he drove 44 miles in 1 hour. How many miles did Bert drive in the three days?

(1) 6 (2) 38 (3) 208
(4) 228 (5) 328

Here are the steps that you should use to solve the problem:

Step 1: Read the problem. Get a mental picture of what is going on.

The problem is about Bert's driving a truck.

Step 2: Decide what is being asked in the problem.

The question in the problem is about the number of miles that Bert drove.

Step 3: Decide which facts to use to answer the question.

Since the problem is about total miles, only the numbers that refer to miles should be used. (Notice that the numbers that refer to the hours that Bert drove are not needed to answer the question. Only the miles are needed.)

Step 4: Decide which math *operation* to use.

The question asks: *how many miles?* You have to find the total number of miles. To find the total, you have to add. Addition (+) is the *operation* that you need to answer the question.

Step 5: *Estimate* the answer.

To *estimate* means to make a rough guess. You make an estimate to check yourself. If your estimate and your exact answer are far apart, you should go back and do the problem over. In this problem, there are three numbers that deal with miles: 70, 114, and 44. To estimate the answer to this problem, round off the numbers. The number 70 ends with a zero. It doesn't have to be rounded off. However, the other two numbers do not end in zero. If you round them off, it will be easier to make an estimate.

114 is about	**110**
44 is about	**40**
	+ 70
total mileage is about	**220**

Step 6: Figure out the exact answer.

$$\begin{array}{r} \mathbf{70} \\ \mathbf{114} \\ \underline{+\ \mathbf{44}} \\ \mathbf{228} \end{array} \textbf{ miles}$$

Step 7: Compare the exact answer to the estimate.

228 miles is close to **220** miles.

Step 8: Read the problem once more to be sure that you answered the right question.

The question asks: *how many miles?* Addition (+) is the right operation to use. Only the numbers of miles have to be added.

Step 9: Check your answer once more.

$$\begin{array}{r} 70 \\ 114 \\ \underline{+\ 44} \\ \mathbf{228} \end{array} \textbf{ miles}$$

The correct answer choice is **(4), 228 miles.**

At first, you may think that using these steps to solve word problems is too long. With simple math problems, you may feel that you don't need to work with the steps. Even if you feel this way, you should work with the steps. Practice the steps on problems that you can do easily. Learning the steps will help to build your skills for harder problems.

EXERCISE 1

DIRECTIONS: Work the following word problem. In the space provided, write in the steps that you use to solve the problem. Use all nine steps that were used in the last example.

It takes Ralph 5 minutes to get to work. It takes Terry 8 minutes to get to work. It takes Hal 19 minutes to get to work. How many more minutes does it take Hal to get to work than Ralph and Terry combined?

(1) 3 minutes (2) 6 minutes (3) 11 minutes
(4) 14 minutes (5) 32 minutes

Step 1: Read the problem. Get a mental picture of what is going on.

____

Step 2: Decide what is being asked in the problem.

____

Step 3: Decide which facts to use to answer the question.

____

Step 4: Decide which math operation to use.

____

Step 5: Estimate the answer.

____

Step 6: Figure out the exact answer.

____

Step 7: Compare the exact answer to the estimate.

____

Step 8: Read the problem once more to be sure that you answered the right question.

Step 9: Check your answer once more.

____

Check your answers on page 15.

About the Math Problems on the GED Test

The math part of the GED Test measures two things. First, it measures your ability to use math skills like addition and subtraction to solve problems. Second, the GED Test measures your ability to read and understand math problems that are written in everyday language.

On the GED Test, every problem is followed by five answer choices. You are to choose one of the five answers as the right one for the problem.

Only one of the five answers is right. The rest of the answers are *wrong*.

The steps that you just worked with to answer the word problems were designed for people who are taking the GED. Remember, on the GED Test you will have to answer 50 problems in 90 minutes. You will find it easier to complete the test if you answer each problem in a logical, orderly way. Here is an example of how the steps can help you to answer GED Test questions more easily.

> In one baseball season, the Los Angeles Dodgers won a total of 97 games. Their best pitcher, Sandy Koufax, won 26 games. Don Drysdale won 23 games. Claude Osteen won 15 games. How many games did all the other pitchers on the team win?
>
> (1) 13 (2) 24 (3) 33
> (4) 42 (5) 64

If you do the problem according to the steps, you know that you are being asked to find the number of games that the other pitchers won. In Step 4, you decide which operations to use. To find the number of games the other pitchers won, you have to *subtract* the games that the three pitchers won from the total number of games won. But first, you have to *add* the games that the three pitchers won.

In Step 5, you estimate the answer. First, estimate the total that the three pitchers won. The number 26 is close to 25. The number 23 is close to 20. Add: $25 + 20 + 15 = 60$. Next, subtract the total number of games won. 97 is close to 95. $95 - 60 = 35$. Your estimate is 35.

Now, look at the answer choices again. Which choices are close to 35? Only choices 3 and 4 are close; choice 3 is closer than choice 4. You could make a good guess that answer choice 3 is right.

Find the exact answer. Add: $26 + 23 + 15 = 64$. $97 - 64 = 33$. Answer choice 3 is 33. By reading the problem carefully and by estimating the answer, you are able to make a good guess at the correct answer choice.

Notice that, if you don't read the problem carefully, you can get the wrong answer. If you didn't read the problem all the way through, you might have just added the number of games that the three pitchers won. Your answer (64) would have matched answer choice 5. Even though your answer matched an answer choice, you still would have the wrong answer. It is important for you to make sure you find out what is being asked *before* you try to solve a problem. If you read only part of the problem, you could end up with the wrong answer.

You should never rely totally on guessing when you take the GED Test. However, as you can see, you often can use estimating to narrow the number of possible choices down to two or three. And, with some problems, you might be able to pick the right answer with your estimate and skip working the problem out.

Everyday Words in Math "Language"

Some words that you may use in everyday language also are used in math language. However, the meanings of the words may be different when they are used in math. For example, in everyday language, a *product* is something that you may buy in a store. But in math language, a *product* is the result that you get in a multiplication problem.

Throughout this book, you will find words that are important in math language. The meanings of these words will be explained to you in the book. Study the words of the math language carefully. If you learn the meanings of the words, you will have an easier time when you take the actual GED Test.

The goal of this book is to help you to read, understand, and solve the kinds of problems that appear on the GED Test. Take your time with this book. Even if some of the problems seem easy to you, do all of them. Use the steps that you learned to work all the word problems. By working with easier problems and by going slowly, you will build both your math skills and your confidence. You will increase your chances of doing well on the GED Mathematics Test.

ANSWERS & SOLUTIONS

Step 1: The problem is about the time it takes three people to get to work.

Step 2: The problem asks how many minutes longer it takes Hal to get to work than it takes Ralph and Terry combined.

Step 3: The number of minutes that it takes each person to get to work is needed.

Step 4: First, the minutes that Ralph and Terry take to get to work have to be added. Then, the total should be subtracted from the number of minutes it takes Hal to get to work.

Step 5: 8 minutes is close to 10 minutes. Add: $10 + 5 = 15$. 19 minutes is close to 20 minutes. Subtract: $20 - 15 = 5$. The estimate is 5 minutes.

Step 6: Add: $8 + 5 = 13$. Subtract: $19 - 13 = 6$. The exact answer is 6 minutes.

Step 7: 6 minutes is close to the estimate (5 minutes).

Step 9: Check the answer: $8 + 5 = 13$, and $19 - 13 = 6$. Answer choice 2 (6 minutes) is right.

Whole Numbers

PLACE VALUE

Whole numbers are written with the **digits** 0, 1, 2, 3, 4, 5, 6, 7, 8, and 9. The number 66 has two digits. The number 24,070 has five digits.

The value of a digit depends on its position in the number. Every position has a **place value.** The chart below gives the names of the first ten whole number places.

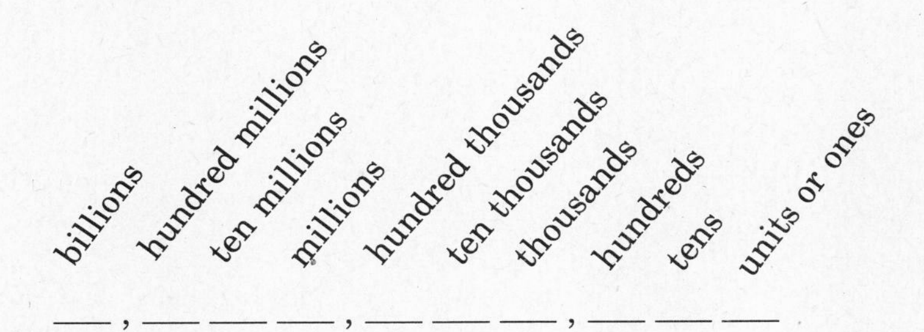

The 6 at the left in 66 ($\underline{6}6$) is in the tens place. It has a value of 6 tens or 60. The 6 at the right ($6\underline{6}$) is in the units or ones place. It has a value of 6 ones or 6. The digit is still 6, but the value is different because of its position.

The 2 in 24,070 is in the ten thousands place. It has a value of 2 ten thousands or 20,000. The 4 is in the thousands place. It has a value of 4 thousands or 4,000. The 7 is in the tens place. It has a value of 7 tens or 70. Notice that the value of the hundreds place and the value of the units place are both 0 in 24,070.

EXAMPLE: Find the value of 8 in 2,840.

Step 1	Step 2
2,840	$\begin{array}{r} 100 \\ \underline{\times 8} \\ \mathbf{800} \end{array}$

Step 1: Tell the name of the place 8 is in. 8 is in the hundreds place.

Step 2: Multiply 8 by the place name.

EXERCISE 1

DIRECTIONS: Find the values of these digits.

1. 4 in 64,520	9 in 1,897	7 in 3,704
2. 5 in 1,580,340	2 in 6,328,139	6 in 463,000,000
3. 8 in 538,927,000	1 in 192,500,000	3 in 28,473

Check your answers on page 31.

Whole Number Operations

You will use the four basic arithmetic operations—addition, subtraction, multiplication, and division—to solve many problems on the GED Test. Here are some hints to help you avoid mistakes with the four basic operations.

Addition—The answer to an addition problem is called the **sum** or **total.** When you set up an addition problem, be sure the numbers are lined up correctly. Put the units under the units, the tens under the tens, the hundreds under the hundreds, and so on.

EXAMPLE: Find the sum of 85, 6,423, and 9.

Step 1	Step 2
85 6,423 + 9	(carries 1 1) 85 6,423 + 9 **6,517**

Step 1: Line up the digits. Put units under units, tens under tens, and so on.
Step 2: Add each column.

Subtraction—The answer to a subtraction problem is called the **difference.** When you take one number from another to find the difference, be sure to put the larger number on top. Line up the numbers with units under units, tens under tens, and so on. Then borrow carefully. It is a good idea to check subtraction problems. To check a subtraction problem, add the answer to the lower number. The sum should equal the top number of the subtraction problem.

EXAMPLE: Take 4,523 from 30,700.

Step 1	Step 2	Step 3
30,700 **− 4,523**	9 2 10 6 1010 **~~3~~ ~~0~~,7 ~~0~~ ~~0~~** −4,523 **26,177**	1 11 4,523 +26,177 **30,700**

Step 1: Line up the problem with the larger number on top. Put units under units, tens under tens, and so on.

Step 2: Borrow and subtract.

Step 3: Check by adding the answer to the lower number. The sum should be the top number in the problem.

Multiplication—The answer to a multiplication problem is called the **product.** To find a product, put the number with more digits on top. When you multiply by the digit in the units place, be sure to start the **partial product** under the units. When you multiply by the digit in the tens column, be sure to start the partial product in the tens column, and so on. Then add the partial products.

EXAMPLE: Multiply 234 and 8,076.

Step 1	Step 2	Step 3	Step 4	Step 5
8,076 **× 234**	8,076 × 234 **32,304**	8,076 × 234 32,304 **242,28**	8,076 × 234 32,304 242,28 **1,615,2**	8,076 × 234 32,304 242,28 1,615,2 **1,889,784**

Step 1: Set up the problem. Put the number with more digits on top.

Step 2: Multiply 8,076 by 4. Start the partial product in the units column.

Step 3: Multiply 8,076 by 3. Start the partial product in the tens column.

Step 4: Multiply 8,076 by 2. Start the partial product in the hundreds column.

Step 5: Add the partial products.

Division—The answer to a division problem is called the **quotient.** Remember to change the order of the numbers when you change from the ÷ sign to the $\overline{)\quad}$ sign. For example, 20 ÷ 4 becomes $4\overline{)20}$.

To find a quotient, repeat these four steps until you finish the problem.

Step 1: Divide.
Step 2: Multiply.
Step 3: Subtract and compare.
Step 4: Bring down the next number.

It is important to line up division problems carefully. When you begin to find a quotient, you must have a digit in the answer for each remaining digit in the problem.

To check a division problem, multiply the answer by the number you divided by. The product should equal the number you divided into.

Be sure you understand every step of the next example.

EXAMPLE: Divide 920 by 23.

Step 1	Step 2	Step 3	Step 4	Step 5	Step 6
23)920	**4** **23)920**	**4** **23)**920 **92**	4 **23)920** −**92** **0**	4 23)920 −92 00	**40** **23)**920 −92 **00**

Step 1: Set up the problem. Put the number you are dividing inside the $\overline{)}$ sign.
Step 2: Divide: $92 \div 23 = 4$. Write 4 above the tens place.
Step 3: Multiply: $4 \times 23 = 92$. Write 92 under 92.
Step 4: Subtract: $92 - 92 = 0$. Compare: 0 is less than 23.
Step 5: Bring down the next number: 0.
Step 6: Divide: 0 cannot be divided. Remember to write 0 above the units place.

EXERCISE 2

DIRECTIONS: Solve each problem.

1. $319 + 73 + 684 + 5 =$

2. $7{,}806 + 82 + 93{,}265 =$

3. $4{,}020 - 384 =$

4. $6{,}000 - 2{,}059 =$

5. $20{,}006 - 12{,}783 =$

6. $14{,}236 - 12{,}855 =$

7. $287 \times 34 =$

8. $6 \times 5{,}039 =$

9. $30{,}460 \times 18 =$

10. $926 \times 700 =$

11. $2{,}454 \div 6 =$

12. $4{,}486 \div 8 =$

13. $2{,}784 \div 32 =$

14. $1{,}454 \div 418 =$

15. Find the sum of 246 and 8,059.

16. Find the difference between 246 and 8,059.

17. Take 1,624 from 4,080.

18. From 33,005 take 9,258.

19. Find the product of 29 and 407.

20. Find the quotient of 2,214 divided by 82.

21. What is the sum of 85, 2,056, and 437?

22. What is the difference between 9,000 and 783?

23. What is the product of 120 and 58?

24. How much is 4,332 divided by 228?

Check your answers on page 31.

SPECIAL KINDS OF WHOLE NUMBER PROBLEMS

Averages

An **average** is a total divided by the number of numbers in the total. An average represents a "middle" or "mean" value for a group of numbers.

EXAMPLE: Sue works as a waitress. She made $24 in tips on Thursday, $42 in tips on Friday, and $48 in tips on Saturday. What was the average total of tips for the three days?

Step 1	Step 2
	$ 38
$24	3)$114
42	− 9
+48	24
$114	− 24
	0

Step 1: Find the total. Add all the tips.
Step 2: Divide the total by the number of days that she worked. The answer is the average.

EXERCISE 3

DIRECTIONS: Find each average.

1. In July, Ann's electricity bill was $14. In August, it was $22. In September, it was $18. What was her average electricity bill for the three months?

2. George Johnson weighs 185 pounds. His wife June weighs 138 pounds. Their daughter Jan weighs 97 pounds. Their son Joe weighs 88 pounds. What is the average weight of the members of the Johnson family?

3. Deborah took five math tests. Her scores were 65, 86, 79, 92, and 88. Find her average score.

4. Paul is a traveling salesman. Monday, he drove 284 miles; Tuesday, 191 miles; Wednesday, 297 miles; Thursday, 162 miles; and Friday, 256 miles. Find the average distance he drove each day.

5. Sam works on commission in a clothing store. The first week in December, he made $146 in commission; the second week, $192; the third week, $233; and the fourth week, $265. Find his average weekly commission for December.

6. On Monday, Don bought 11 gallons of gas. On Wednesday, he bought 9 gallons of gas; on Thursday, he bought 13 gallons of gas. Find the average number of gallons he bought each time.

7. 213 people went to the Pine Bush horse show on Wednesday. 191 people went to it on Thursday, 288 people on Friday, and 304 people on Saturday. Find the average number of people who went to the show each night.

8. In 1980, Mr. Munro made $14,800. Mrs. Munro made $12,500. Their daughter Lee made $6,000, and their son Nick made $4,200. What was the average income for each person in the Munro family?

Check your answers on page 32.

Number Series

A **number series** is a group of numbers written in a special order. 2, 4, 6, 8 . . . is a number series in which each number is 2 more than the number before it. Some series go up by addition or down by subtraction. Other series go up by multiplication or down by division. Each number in a series is called a *term.*

To find a term in a number series, find out how each step in the series is made. Then apply that step to set the term you want. Remember that the steps in a series must follow a regular pattern.

EXAMPLE: Find the next term in the series 98, 95, 92, 89 . . . .

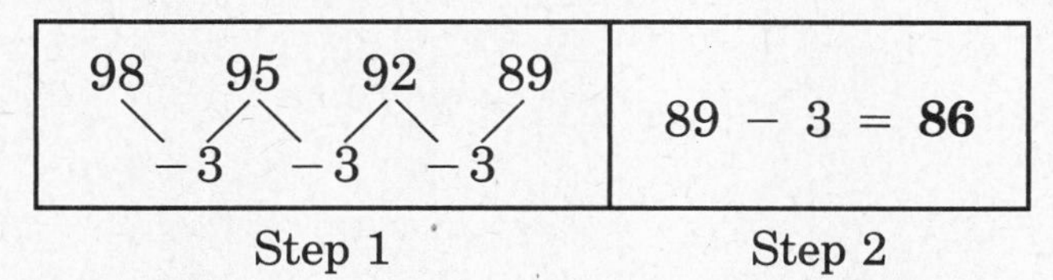

Step 1 Step 2

Step 1: Find out how each step in the series is made. Each term is 3 less than the term before it.
Step 2: Subtract 3 from the last term to get the answer.

EXAMPLE: Find the sixth term in the series 1, 5, 25, 125 . . . .

1 5 25 125	125	625
×5 ×5 ×5	×5	×5
	625	**3,125**
Step 1	Step 2	Step 3

Step 1: Find out how each step in the series is made. Each term is 5 times as big as the term before it.
Step 2: Multiply 125 by 5 to find the fifth term, 625.
Step 3: Multiply 625 by 5 to find the sixth term, 3,125.

EXERCISE 4

DIRECTIONS: Find the terms.

1. Find the next term in the series 5, 10, 20, 40 . . . .
2. Find the fifth term in the series 74, 70, 66 . . . .
3. Find the sixth term in the series 2, 4, 8, 14 . . . .
4. Find the next term in the series 400, 200, 100, 50 . . . .
5. Find the next term in the series 5, 8, 6, 9, 7, 10 . . . .
6. Find the sixth term in the series 1, 4, 16, 64 . . . .
7. Find the next term in the series 243, 81, 27, 9 . . . .
8. Find the seventh term in the series 1, 5, 2, 6, 3 . . . .

Check your answers on page 32.

Rounding Off

Sometimes, whole numbers are more exact than they need to be. For example, Jack makes $188 a week at his job. He makes about $190 a week. $190 is $188 rounded off to the nearest ten. Rounding off is a way of making numbers easier to read and use.

To round off a number, mark the digit in the place you want to round off to. If the digit at the right of this place is more than 4, add 1 to the digit you marked. If the digit at the right is less than 5, leave the digit you marked as it is. Then replace the digits at the right of the digit you marked with zeros.

EXAMPLE: Round off 487 to the nearest ten.

487	487	490
Step 1	Step 2	Step 3

Step 1: Find the digit in the tens place, 8.
Step 2: Look at the digit to the right of 8. The digit is 7. Since 7 is more than 4, add 1 to the 8.
Step 3: Put a 9 in the tens place. Write 0 in the units place.

EXAMPLE: Round off 682,384 to the nearest thousand.

682,384	682,384	682,000
Step 1	Step 2	Step 3

Step 1: Find the digit in the thousands place, 2.
Step 2: Look at the digit to the right of 2. The digit is 3. Since 3 is less than 4, the 2 stays the same.
Step 3: Write zeros in the places to the right of 2.

EXERCISE 5

DIRECTIONS: Round off these numbers.

1. Round off each number to the nearest ten:

   78 164 3,198 2,433

2. Round off each number to the nearest hundred:

   847 1,273 6,580 351

3. Round off each number to the nearest thousand:

   3,196 41,826 28,752 149,628

4. Round off each number to the nearest hundred thousand:

   777,500 316,450 567,300 3,470,992

5. Round off each number to the nearest million:

   5,648,000 12,387,000 32,479,000 188,750,000

Check your answers on page 33.

Whole Number Word Problems

Most of the problems that test whole number skills on the GED Test are word problems. Here are some hints to help you identify the operations you will need to solve whole number problems.

Addition problems are easy to recognize. **Sum, total, combined,** and **altogether** are key words that usually mean to add. To solve many GED problems, you will use other operations along with addition. You have already practiced one example of these problems—finding averages. To get an average, you must use division along with addition.

Subtraction problems are also easy to recognize. The key word **difference** means to subtract. Words like **how much more** and **how much less** also mean to subtract. Two other words, **gross** and **net,** are often used in subtraction problems. *Gross* refers to a total before any deductions are taken from it. *Net* is the amount that is left after the deductions are taken.

EXAMPLE: Jeff gets a gross salary of $280 a week. His employer takes out $52 for taxes and Social Security. Find Jeff's net weekly salary.

$280	gross salary
−52	deductions
$228	net salary

The key words **product** and **times** are not often used in multiplication word problems. Instead of key words, watch for situations that mean to multiply. For example, you may be told the price of one item. Then you are asked to find the price of several items. Or, you may be told how far someone can drive a car on one gallon of gas. Then you are asked to find how far he can drive the car on several gallons of gas.

EXAMPLE: Collette can drive her car for 18 miles on one gallon of gas. How far can she drive with 12 gallons of gas?

$$\begin{array}{r} 18 \\ \times 12 \\ \hline 36 \\ 18 \\ \hline 216 \end{array} \text{ miles}$$

The key word **quotient** is almost never used in division problems. But the words **split** and **share** usually mean to divide.

EXAMPLE: John picked 85 pounds of apples. He wants to share the apples equally with four friends. How many pounds will each person get?

$$\begin{array}{r} 17 \text{ pounds} \\ 5\overline{)85 \text{ pounds}} \end{array}$$

(John and his four friends are 5 people.)

Other situations may mean to divide. For example, you may be told the price of several items. Then you are asked to find the price of one item.

EXAMPLE: Sally paid $8 for 4 pounds of ground beef. Find the price of one pound.

$$\begin{array}{r} \$2 \\ 4\overline{)\$8} \end{array}$$

In many problems, you will need more than one step and more than one operation to get answers. List for yourself the steps you need to take before you start.

EXERCISE 6 (WORD PROBLEMS)

DIRECTIONS: Choose the correct answer to each problem.

1. It costs about $23 a day, without mileage, to rent a car. How much does it cost, before mileage charges, to rent a car for a month that has 31 days?

   (1) $261 (2) $540 (3) $690 (4) $713 (5) $931

2. Last year, the town of Montvale spent $16,593,650 for police and fire protection. The town spent $1,108,212 for welfare and health. How much more did the town spend for police and fire protection than for welfare and health?

   (1) $15,485,438 (2) $15,495,448 (3) $15,497,442
   (4) $16,485,448 (5) $16,511,530

3. John, Tom, Bill, and Fred started a record store. At the end of one year, they had a profit of $27,936. They decided to share the profit equally. How much did each person get?

   (1) $9,312 (2) $7,982 (3) $6,984
   (4) $5,587 (5) $4,656

4. Ezra Hicks died in 1978 after his 94th birthday. In what year was he born?

   (1) 1864 (2) 1872 (3) 1874 (4) 1878 (5) 1884

5. Cheryl made a gross salary of $12,500 last year. Her employer deducted $2,518 from her salary. What was Cheryl's net income last year?

   (1) $9,981 (2) $9,982 (3) $9,992
   (4) $10,981 (5) $10,982

6. Sara bought 12 towels at a sale. She wanted to divide them equally among herself and her 3 sisters. How many towels did each of the sisters get?

   (1) 2 (2) 3 (3) 4 (4) 5 (5) 6

7. Bert Roberts is an auto mechanic. Last year, he made $16,456. Betty Roberts is a clerk. Last year, she made $11,294. Their son Henry works part-time at a grocery store. He made $3,367 last year. Find the combined income of the Roberts family last year.

   (1) $30,117 (2) $31,117 (3) $31,217
   (4) $31,218 (5) $31,227

8. Sandy paid $63 for three pairs of jeans. What was the price of one pair of jeans?

   (1) $32 (2) $24 (3) $23 (4) $22 (5) $21

9. It costs about $1,850 a year to educate one student at Franklin School. There are 230 students in the school. Find the total cost for educating all the students at Franklin School for a year.

   (1) $92,500 (2) $415,500 (3) $415,550
   (4) $425,500 (5) $435,500

10. In 1950, there were 14,273 people living in Elmford. In 1980, there were 8,498 more people living in Elmford than in 1950. How many people lived in Elmford in 1980?

    (1) 22,771 (2) 23,781 (3) 23,881
    (4) 24,721 (5) 26,701

11. Mr. Schmidt ordered 9 boxes of dress shirts to sell in his store. There were 12 shirts in each box. He sent 2 boxes of shirts back because they were the wrong colors. How many shirts did Mr. Schmidt keep?

    (1) 132 (2) 84 (3) 72 (4) 11 (5) 7

12. In 1975, there were 1,036,000 divorces in the U.S. In 1977, there were 1,091,000 divorces. In 1978, there were 1,128,000 divorces. How many more divorces were there in 1978 than in 1977?

    (1) 92,000 (2) 55,000 (3) 37,000
    (4) 27,000 (5) 22,000

13. On weekends, Karen works as a waitress. On Friday, she made $63. On Saturday, she made $79. On Sunday, she made $68. Find the average amount she made for each day.

    (1) $69 (2) $70 (3) $74 (4) $79 (5) $212

14. In 1960, there were 9,706,397 people living in Ohio. In 1970, there were 10,652,017 people living there. By how much did the number of people increase from 1960 to 1970? Round your answer off to the nearest hundred thousand.

    (1) 800,000 (2) 900,000 (3) 920,000
    (4) 950,000 (5) 1,000,000

15. Find the fifth term in the series 2, 4, 8, 16 . . . .

    (1) 32 (2) 28 (3) 24 (4) 22 (5) 20

16. Fred pays $12 for a gallon of paint. How much does he pay for 13 gallons of paint?

    (1) $132 (2) $149 (3) $150 (4) $156 (5) $159

17. Mount Hood is 11,239 feet high. Mount Whitney is 14,494 feet high. How much higher is Mount Whitney than Mount Hood?

    (1) 4,265 ft. (2) 3,365 ft. (3) 3,255 ft.
    (4) 2,265 ft. (5) 1,745 ft.

18. Al wants to split a board 102 inches long into 6 equal pieces. How long will each piece be?

    (1) 15 in. (2) 16 in. (3) 17 in.
    (4) 18 in. (5) 19 in.

19. Oregon became a state in 1859. For how many years had Oregon been a state in 1980?

(1) 141 (2) 139 (3) 131 (4) 129 (5) 121

20. The yearly budget for the town of Staunton is $150,000. By the end of June, the town had spent $89,480. How much money was left in the budget for the rest of the year?

(1) $60,520 (2) $61,520 (3) $61,610
(4) $62,510 (5) $70,519

21. Mr. and Mrs. Wiley bought new furniture marked $650. They agreed to pay for the furniture by making 15 equal monthly payments of $52 each. How much more than $650 did the furniture cost them?

(1) $130 (2) $125 (3) $120 (4) $78 (5) $68

22. In one month, the Simpsons spent $462 for utilities and mortgage payments, $436 for income taxes and Social Security, $194 for car payments and gasoline, $323 for food, and $245 for everything else. How much did they spend altogether that month?

(1) $1,650 (2) $1,660 (3) $1,760
(4) $1,860 (5) $1,950

23. George makes $295 for a five-day work week. How much does he make each day?

(1) $68 (2) $59 (3) $58 (4) $56 (5) $54

24. The Fulton County Stadium can hold 52,532 people. One Saturday afternoon, the stadium was full. The average price for a ticket was $4. Find the total price paid for tickets at the stadium that day.

(1) $219,128 (2) $219,028 (3) $210,128
(4) $209,028 (5) $200,028

25. In 1970, there were 2,809,813 people living in Los Angeles, 715,674 in San Francisco, and 697,027 in San Diego. What was the total number of people living in the three cities?

(1) 3,506,804 (2) 3,524,604 (3) 3,525,487
(4) 4,212,414 (5) 4,222,514

26. Mr. Seltzer ordered 3 cases of tomato soup, 4 cases of chicken soup, and 2 cases of bean soup for his store. Each case held 12 cans. How many cans of soup did he order altogether?

(1) 9 (2) 96 (3) 108 (4) 112 (5) 120

27. Of the 432 students who graduated from Central High School, 184 went to college and 93 went into the army. How many of the graduates did not go into the army?

    (1) 55 (2) 148 (3) 249 (4) 248 (5) 339

28. On Thursday night, 1,129 people went to the All-City basketball tournament. On Friday night, 1,347 people went to the tournament. On Saturday, 1,406 people went. What was the average number of people at the tournament each night?

    (1) 1,294 (2) 1,356 (3) 1,438
    (4) 2,476 (5) 3,882

29. The area of Alaska is 586,412 square miles. The area of Rhode Island is 1,214 square miles. To the nearest hundred square miles, how much bigger is Alaska than Rhode Island?

    (1) 575,000 sq. mi. (2) 576,000 sq. mi. (3) 576,198 sq. mi.
    (4) 585,100 sq. mi. (5) 585,200 sq. mi.

30. Find the next term in this series: 1, 2, 4, 7, 11 . . . .

    (1) 12 (2) 13 (3) 14 (4) 15 (5) 16

Check your answers on page 33.

ANSWERS & SOLUTIONS

EXERCISE 1

1. **4,000** **90** **700**
2. **500,000** **20,000** **60,000,000**
3. **8,000,000** **100,000,000** **3**

EXERCISE 2

1. **1,081**

```
  319
   73
  684
+   5
-----
1,081
```

2. **101,153**

```
  7,806
     82
+93,265
-------
101,153
```

3. **3,636**

```
   9 11
 3 10 1 10
 4, 0 2 0
 -  3 8 4
 --------
 3, 6 3 6
```

4. **3,941**

```
   9 9
 5 10 10 10
 6, 0 0 0
-2, 0 5 9
---------
 3, 9 4 1
```

5. **7,223**

```
    9  9
 1 10 10 10
 2 0, 0 0 6
-1 2, 7 8 3
-----------
    7, 2 2 3
```

6. **1,381**

```
      11
  3  1 13
 1 4, 2 3 6
-1 2, 8 5 5
-----------
   1, 3 8 1
```

7. **9,758**

```
  287
 × 34
-----
1 148
8 61
-----
9,758
```

8. **30,234**

```
 5,039
×    6
------
30,234
```

9. **548,280**

```
 30,460
×    18
-------
243 680
304 60
-------
548,280
```

10. **648,200**

```
    926
  × 700
-------
648,200
```

11. **409**

```
   409
6)2,454
  2 4
   05
    0
   54
   54
```

12. **560 r 6**

```
  560 r 6
8)4,486
  4 0
   48
   48
    06
     0
     6
```

13. **87**

```
     87
32)2,784
   2 56
    224
    224
```

14. **3 r 200**

```
      3 r 200
418)1,454
    1 254
      200
```

15. **8,305**

```
  246
+8,059
------
 8,305
```

16. **7,813**

```
 7 10
 8, 0 5 9
-    2 4 6
----------
 7, 8 1 3
```

17. **2,456**

```
 3 10 7 10
 4, 0 8 0
-1, 6 2 4
---------
 2, 4 5 6
```

18. **23,747**

```
   12  9  9
 2 2 10 10 15
 3 3, 0 0 5
-    9, 2 5 8
-------------
 2 3, 7 4 7
```

19. **11,803**

```
   407
 ×  29
------
 3 663
 8 14
------
11,803
```

20. **27**

```
     27
82)2,214
   1 64
    574
    574
```

21. **2,578**

```
    85
 2,056
+  437
 2,578
```

22. **8,217**

```
     9 9
  8 10 10 10
  9,0 0 0
−   7 8 3
  8,2 1 7
```

23. **6,960**

```
  120
 ×58
  960
 6 00
6,960
```

24. **19**

```
       19
228)4,332
    2 28
    2 052
    2 052
```

EXERCISE 3

1. **$18**

```
 $14      $18
  22    3)$54
  18       3
 $54      24
          24
```

2. **127 lbs.**

```
  185 lbs.      127 lbs.
  138        4)508 lbs.
   97
+  88
  508 lbs.
```

3. **82**

```
  65      82
  86    5)410
  79
  92
 +88
 410
```

4. **238 mi.**

```
   284 mi.        238 mi.
   191        5)1,190 mi.
   297
   162
  +256
 1,190 mi.
```

5. **$209**

```
 $146      $209
  192    4)$836
  233
 +265
 $836
```

6. **11 gals.**

```
 11 gals.     11 gals.
  9         3)33 gals.
+13
 33 gals.
```

7. **249**

```
  213      249
  191    4)996
  288
 +304
  996
```

8. **$9,375**

```
 $14,800      $ 9,375
  12,500    4)$37,500
   6,000
   4,200
 $37,500
```

EXERCISE 4

1. **80** 5, 10, 20, 40, **80** (×2 ×2 ×2 ×2)

2. **58** 74, 70, 66, 62, **58** (−4 −4 −4 −4)

3. **32** 2, 4, 8, 14, 22, **32** (+2 +4 +6 +8 +10)

4. **25** 400, 200, 100, 50, **25** (÷2 ÷2 ÷2 ÷2)

5. **8** 5 8 6 9 7 10 **8** (+3 −2 +3 −2 +3 −2)

6. **1,024** 1 4 16 64 256 **1,024** (×4 ×4 ×4 ×4 ×4)

7. **3** 243 81 27 9 **3** (÷3 ÷3 ÷3 ÷3)

8. **4** 1 5 2 6 3 7 **4** (+4 −3 +4 −3 +4 −3)

EXERCISE 5

1. **80** **160** **3,200** **2,430**
2. **800** **1,300** **6,600** **400**
3. **3,000** **42,000** **29,000** **150,000**
4. **800,000** **300,000** **600,000** **3,500,000**
5. **6,000,000** **12,000,000** **32,000,000** **189,000,000**

EXERCISE 6 (WORD PROBLEMS)

1. **(4) $713**

```
  $ 23
  ×31
    23
   69
  $713
```

2. **(1) $15,485,438**

```
  $16,593,650
  −1,108,212
  $15,485,438
```

3. **(3) $6,984**

```
    $6,984
 4)$27,936
   24
    3 9
    3 6
      33
      32
       16
       16
```

4. **(5) 1884**

```
  1978
  − 94
  1884
```

5. **(2) $9,982**

```
  $12,500
  −2,518
  $ 9,982
```

6. **(2) 3**

```
   3 sisters
  +1 Sara
   4
```

```
    3
 4)12
   12
```

7. **(2) $31,117**

```
  $16,456
   11,294
   +3,367
  $31,117
```

8. **(5) $21**

```
    $21
 3)$63
    6
    03
     3
```

9. **(4) $425,500**

```
  $1,850
    ×230
  55 500
 370 0
$425,500
```

10. **(1) 22,771**

```
  14,273
+  8,498
  22,771
```

11. **(2) 84**

9 × 12 = 108
2 × 12 = 24

```
  108
 − 24
   84
```

12. **(3) 37,000**

```
 1,128,000
−1,091,000
    37,000
```

(Notice that the 1975 number has nothing to do with the problem.)

13. **(2) $70**

```
 $63
  79
  68
$210
```

```
   $70
3)$210
```

14. **(2) 900,000**

```
 10,652,017
− 9,706,397
    945,620
```

to the nearest hundred thousand = **900,000**

15. **(1) 32**

2 ×2→ 4 ×2→ 8 ×2→ 16 ×2→ **32**

16. **(4) $156**

```
 $12
 ×13
  36
 12
$156
```

17. **(3) 3,255 ft.**

```
 14,494 ft.
−11,239
  3,255 ft.
```

18. **(3) 17 in.**

```
  17 in.
6)102 in.
```

19. **(5) 121**

```
 1980
−1859
  121
```

20. **(1) $60,520**

```
$150,000
 −89,480
 $60,520
```

21. **(1) $130**

```
 $52
 ×15
 260
 52
$780
```

```
 $780
 −650
 $130
```

22. **(2) $1,660**

```
  $462
   436
   194
   323
  +245
$1,660
```

23. **(2) $59**

```
  $59
5)$295
```

24. **(3) $210,128**

```
  52,532
×      4
$210,128
```

25. **(5) 4,222,514**

$$\begin{array}{r} 2{,}809{,}813 \\ 715{,}674 \\ +697{,}027 \\ \hline \mathbf{4{,}222{,}514} \end{array}$$

26. **(3) 108**

$$\begin{array}{rr} 3 \times 12 = & 36 \\ 4 \times 12 = & 48 \\ 2 \times 12 = & +24 \\ \hline & \mathbf{108} \end{array}$$

27. **(5) 339**

$$\begin{array}{r} 432 \\ -93 \\ \hline \mathbf{339} \end{array}$$

(Notice that 284 has nothing to do with the problem.)

28. **(1) 1,294**

$$\begin{array}{r} 1{,}129 \\ 1{,}347 \\ +1{,}406 \\ \hline 3{,}882 \end{array} \qquad 3\overline{)3{,}882}\ \ \mathbf{1{,}294}$$

29. **(5) 585,200 sq. mi.**

$$\begin{array}{r} 586{,}412 \text{ sq. mi.} \\ -\quad 1{,}214 \\ \hline 585{,}198 \text{ sq. mi.} \end{array}$$

to the nearest hundred = **585,200**

30. **(5) 16**

1 2 4 7 11 **16**

+1 +2 +3 +4 +5

Fractions

A **fraction** shows a part of a whole thing. Three quarters are three of the four equal parts in a dollar. Three quarters are $\frac{3}{4}$, or three-fourths, of a dollar. Five inches are five of the twelve equal parts in a foot. Five inches are $\frac{5}{12}$, or five-twelfths, of a foot.

Writing Fractions

The top number in a fraction is called the **numerator.** It tells you how many parts you have. The bottom number is called the **denominator.** It tells you how many parts there are in the whole thing. In the fraction $\frac{3}{4}$, 3 is the numerator and 4 is the denominator. You have three parts. The whole is 4 parts. In the fraction $\frac{5}{12}$, 5 is the numerator and 12 is the denominator. You have 5 parts. The whole is 12 parts.

EXAMPLE: There are seven days in a week. 5 days are what fraction of a week?

$$\mathbf{\frac{5}{7}}$$

Step 1: Choose the denominator. 7 tells how many parts are in a whole week. 7 is the denominator.

Step 2: Choose the numerator. 5 tells how many parts there are. 5 is the numerator.

EXERCISE 1

DIRECTIONS: Write a fraction for each problem.

1. There are 60 minutes in one hour. 37 minutes are what fraction of an hour?
2. There are 12 inches in one foot. 11 inches are what fraction of a foot?
3. There are 20 nickels in one dollar. 7 nickels are what fraction of a dollar?
4. There are 144 square inches in one square foot. 5 square inches are what fraction of a square foot?

5. There are 1,000 meters in one kilometer. 123 meters are what fraction of a kilometer?
6. There are 4 quarts in one gallon. One quart is what fraction of a gallon?
7. There are 16 ounces in one pound. 3 ounces are what fraction of a pound?
8. There are 365 days in one year. 31 days are what fraction of a year?
9. There are 2,000 pounds in one ton. 3 pounds are what fraction of a ton?
10. There are 100 centimeters in one meter. 99 centimeters are what fraction of a meter?

Check your answers on page 67.

Forms of Fractions

Fractions may be found in one of three forms:

Proper fraction: The numerator is always less than the denominator: $\frac{1}{2}, \frac{5}{24}, \frac{185}{200}$. A proper fraction does not have all the parts of the whole. The value of a proper fraction is always less than one whole.

Improper fraction: The numerator is as big or bigger than the denominator: $\frac{6}{6}, \frac{9}{8}, \frac{45}{5}$. When the numerator and the denominator are the same, the value of the improper fraction is exactly one whole. $\frac{6}{6}$ is equal to one. When the numerator is bigger than the denominator, the value of the improper fraction is more than one whole. $\frac{9}{8}$ and $\frac{45}{5}$ are each more than one.

Mixed number: A whole number and a proper fraction are written next to each other: $2\frac{3}{4}, 1\frac{1}{2}, 9\frac{11}{12}$. In the mixed number $2\frac{3}{4}$, 2 is the whole number and $\frac{3}{4}$ is the fraction. The value of a mixed number is always more than one whole.

EXERCISE 2

DIRECTIONS: Answer the following questions.

1. Circle the proper fractions in this list:

$\frac{5}{3}$ $\frac{6}{6}$ $\frac{2}{9}$ $3\frac{3}{4}$ $\frac{19}{20}$ $\frac{1}{12}$ $\frac{6}{1}$ $1\frac{1}{15}$

2. Circle the improper fractions in this list:

   $2\frac{4}{5}$ $\frac{8}{2}$ $\frac{2}{8}$ $\frac{15}{16}$ $5\frac{1}{2}$ $\frac{7}{7}$ $\frac{3}{9}$ $\frac{7}{2}$

3. Circle the mixed numbers in this list:

   $\frac{11}{12}$ $\frac{5}{8}$ $\frac{9}{5}$ $4\frac{1}{6}$ $\frac{5}{9}$ $\frac{3}{3}$ $1\frac{5}{8}$ $10\frac{2}{15}$

4. Circle the numbers with a value of more than one in this list:

   $\frac{12}{4}$ $\frac{2}{2}$ $\frac{1}{8}$ $6\frac{1}{3}$ $\frac{14}{15}$ $\frac{2}{17}$ $\frac{9}{1}$ $8\frac{1}{6}$

Check your answers on page 67.

Reducing Fractions

Reducing means dividing both the numerator and the denominator by a number that goes into them evenly. Reducing changes the numbers in a fraction, but it does not change the value. Remember that $\frac{1}{2}$ of a dollar or a 50¢ piece is the same amount of money as $\frac{2}{4}$ of a dollar or two 25¢ pieces.

Reducing is one of the most important fraction operations. Answers to fraction problems should always be reduced.

EXAMPLE: Reduce $\frac{10}{12}$.

$\frac{10 \div 2}{12 \div 2} = \frac{5}{6}$	$\frac{10}{12} = \frac{5}{6}$
Step 1	Step 2

Step 1: Divide 10 and 12 by a number that goes evenly into both of them. 2 divides evenly into both 10 and 12.

Step 2: Check to see if any other number besides 1 goes evenly into both 5 and 6. No other number divides evenly into both. $\frac{5}{6}$ is reduced as far as it will go. (The equal sign tells you that $\frac{5}{6}$ has the same value as $\frac{10}{12}$.)

Sometimes a fraction can be reduced more than once.

EXAMPLE: Reduce $\frac{32}{48}$.

$\frac{32 \div 8}{48 \div 8} = \frac{4}{6}$	$\frac{4 \div 2}{6 \div 2} = \frac{2}{3}$	$\frac{32}{48} = \frac{4}{6} = \frac{2}{3}$
Step 1	Step 2	Step 3

Step 1: Divide 32 and 48 by a number that goes evenly into both of them. 8 divides evenly into 32 and 48.
Step 2: Check to see if any other number besides 1 goes evenly into both 4 and 6. 2 divides evenly into both numbers.
Step 3: Check to see if any other number besides 1 goes evenly into both 2 and 3. No other number divides evenly into both. $\frac{2}{3}$ is reduced as far as it will go.

A fraction is in **lowest terms** when it is reduced as far as it will go.

Some fractions are hard to reduce. Here are some hints and examples to help you decide how to reduce fractions:

When the numbers in a fraction end with 5 or 0, reduce the fraction by 5.

EXAMPLE: Reduce $\frac{45}{80}$.

$\frac{45 \div 5}{80 \div 5} = \frac{9}{16}$	$\frac{45}{80} = \frac{9}{16}$
Step 1	Step 2

Step 1: Divide both 45 and 80 by 5.
Step 2: Check to see if any other number besides 1 goes evenly into both 9 and 16. No other number divides evenly into both. $\frac{9}{16}$ is reduced to lowest terms.

When both the numerator and the denominator end with 0, reduce the fraction by 10.

EXAMPLE: Reduce $\frac{60}{400}$.

$\frac{60 \div 10}{400 \div 10} = \frac{6}{40}$	$\frac{6 \div 2}{40 \div 2} = \frac{3}{20}$	$\frac{60}{400} = \frac{6}{40} = \frac{3}{20}$
Step 1	Step 2	Step 3

Step 1: Divide both 60 and 400 by 10.
Step 2: Check to see if any other number besides 1 goes evenly into 6 and 40. 2 divides evenly into both.
Step 3: Check to see if any other number besides 1 goes evenly into both 3 and 20. No other number divides evenly into both. $\frac{3}{20}$ is reduced to lowest terms.

After you reduce a fraction, always check to see if it can be reduced more.

EXERCISE 3

DIRECTIONS: Reduce each fraction to lowest terms.

1. $\frac{4}{20} =$ $\qquad \frac{7}{42} =$ $\qquad \frac{3}{30} =$ $\qquad \frac{6}{18} =$ $\qquad \frac{9}{45} =$ $\qquad \frac{8}{56} =$

2. $\frac{35}{45} =$ $\qquad \frac{40}{65} =$ $\qquad \frac{25}{30} =$ $\qquad \frac{60}{85} =$ $\qquad \frac{15}{20} =$ $\qquad \frac{30}{55} =$

3. $\frac{24}{72} =$ $\qquad \frac{26}{39} =$ $\qquad \frac{14}{84} =$ $\qquad \frac{27}{54} =$ $\qquad \frac{19}{38} =$ $\qquad \frac{45}{75} =$

4. $\frac{60}{200} =$ $\qquad \frac{32}{60} =$ $\qquad \frac{48}{64} =$ $\qquad \frac{80}{300} =$ $\qquad \frac{22}{36} =$ $\qquad \frac{20}{1{,}000} =$

5. $\frac{60}{144} =$ $\qquad \frac{70}{330} =$ $\qquad \frac{33}{55} =$ $\qquad \frac{24}{40} =$ $\qquad \frac{48}{72} =$ $\qquad \frac{14}{70} =$

Check your answers on page 67.

Raising Fractions to Higher Terms

When you add and subtract fractions, you often have to **raise** the fractions **to higher terms.** Raising to higher terms is the opposite of reducing. To raise a fraction to higher terms, multiply both the numerator and the denominator by the same number.

EXAMPLE: Raise the fraction to higher terms. Find the missing numerator: $\frac{3}{4} = \frac{\ }{32}$.

$\begin{array}{r} \mathbf{8} \\ \mathbf{4\overline{)32}} \end{array}$	$\frac{\mathbf{3 \times 8}}{\mathbf{4 \times 8}} = \frac{\mathbf{24}}{\mathbf{32}}$
Step 1	Step 2

Step 1: Divide the bigger denominator by the smaller denominator.

Step 2: Multiply both the first numerator and the first denominator by 8. $\frac{24}{32}$ is $\frac{3}{4}$ raised to higher terms.

Check to see if the fraction is correctly raised to higher terms. Reduce the new fraction. The reduced fraction should be the same as the old fraction. For example, reduce $\frac{24}{32}$ by 8.

$$\frac{\mathbf{24 \div 8}}{\mathbf{32 \div 8}} = \frac{\mathbf{3}}{\mathbf{4}}$$

EXERCISE 4

DIRECTIONS: Raise each fraction to higher terms. Find each missing numerator.

1. $\frac{2}{3} = \frac{}{12}$ $\frac{3}{5} = \frac{}{35}$ $\frac{4}{9} = \frac{}{45}$ $\frac{5}{8} = \frac{}{16}$ $\frac{1}{4} = \frac{}{32}$

2. $\frac{6}{7} = \frac{}{42}$ $\frac{9}{10} = \frac{}{40}$ $\frac{1}{3} = \frac{}{36}$ $\frac{2}{15} = \frac{}{60}$ $\frac{5}{6} = \frac{}{18}$

3. $\frac{13}{20} = \frac{}{100}$ $\frac{7}{12} = \frac{}{48}$ $\frac{19}{25} = \frac{}{75}$ $\frac{1}{16} = \frac{}{32}$ $\frac{11}{24} = \frac{}{96}$

4. $\frac{7}{50} = \frac{}{200}$ $\frac{1}{5} = \frac{}{50}$ $\frac{3}{8} = \frac{}{72}$ $\frac{2}{9} = \frac{}{36}$ $\frac{5}{12} = \frac{}{36}$

Check your answers on page 68.

Changing Improper Fractions to Whole or Mixed Numbers

The answers to many fraction problems are improper fractions. It is easier to read these answers if you change them to whole or mixed numbers. The answer choices to many fraction problems on the GED Test are given as whole or mixed numbers. To change an improper fraction, divide the denominator into the numerator.

EXAMPLE: Change $\frac{46}{8}$ to a whole or mixed number.

Step 1	Step 2	Step 3
$\begin{array}{r} 5 \\ 8\overline{)\,46} \\ -40 \\ \hline 6 \end{array}$	$\frac{46}{8} = 5\frac{6}{8}$	$5\frac{6 \div 2}{8 \div 2} = 5\frac{3}{4}$

Step 1: Divide the denominator into the numerator.
Step 2: Write the answer to the division (5) as a whole number. Write the remainder over the denominator.
Step 3: Reduce $5\frac{6}{8}$ by 2.

EXERCISE 5

DIRECTIONS: Change each improper fraction to a whole or mixed number. Reduce each fraction that is left over.

1. $\frac{16}{3} =$ $\frac{20}{9} =$ $\frac{21}{8} =$ $\frac{9}{2} =$ $\frac{14}{7} =$ $\frac{15}{4} =$ $\frac{35}{5} =$

2. $\frac{27}{6} =$ $\frac{32}{12} =$ $\frac{30}{9} =$ $\frac{45}{10} =$ $\frac{46}{8} =$ $\frac{48}{15} =$ $\frac{26}{4} =$

Check your answers on page 69.

Changing Mixed Numbers to Improper Fractions

When you multiply and divide fractions, you often have to change mixed numbers to improper fractions. To change a mixed number to an improper fraction, follow these steps:

Step 1: Multiply the denominator by the whole number.
Step 2: Add the numerator.
Step 3: Write the total over the denominator.

EXAMPLE: Change $6\frac{3}{8}$ into an improper fraction.

$\begin{array}{r} 8 \\ \times\ 6 \\ \hline 48 \end{array}$	$\begin{array}{r} 48 \\ +\ 3 \\ \hline 51 \end{array}$	$6\frac{3}{8} = \frac{51}{8}$
Step 1	Step 2	Step 3

Step 1: Multiply the denominator by the whole number: $8 \times 6 = 48$.
Step 2: Add the numerator to the product: $3 + 48 = 51$.
Step 3: Write the total over the denominator: $\frac{51}{8}$.

EXERCISE 6

DIRECTIONS: Change each mixed number into an improper fraction.

1. $3\frac{2}{3} =$ $5\frac{1}{2} =$ $1\frac{7}{10} =$ $2\frac{5}{6} =$ $4\frac{3}{8} =$ $1\frac{1}{2} =$ $3\frac{5}{9} =$

2. $6\frac{3}{4} =$ $8\frac{4}{7} =$ $10\frac{5}{6} =$ $5\frac{2}{9} =$ $2\frac{4}{5} =$ $12\frac{7}{8} =$ $16\frac{1}{3} =$

Check your answers on page 69.

ADDING FRACTIONS

Adding Fractions with the Same Denominators

To add fractions when the denominators are alike, first add the numerators. Then put the total over the denominator.

EXAMPLE: Add $\frac{2}{7}$ and $\frac{4}{7}$.

Step 1	Step 2
$\frac{2}{7} + \frac{4}{7} = 6$	$\frac{2}{7} + \frac{4}{7} = \frac{6}{7}$

Step 1: Add the numerators: $2 + 4 = 6$.
Step 2: Write the total over the denominator: $\frac{6}{7}$.

EXAMPLE: Add $\frac{5}{8}$ and $\frac{7}{8}$.

Step 1	Step 2	Step 3	Step 4
$\frac{5}{8} + \frac{7}{8} = 12$	$\frac{5}{8} + \frac{7}{8} = \frac{12}{8}$	$\frac{12}{8} = 1\frac{4}{8}$	$1\frac{4}{8} = 1\frac{1}{2}$

Step 1: Add the numerators: $5 + 7 = 12$.
Step 2: Write the total over the denominator: $\frac{12}{8}$.
Step 3: Change the improper fraction to a mixed number: $1\frac{4}{8}$.
Step 4: Reduce the fraction to lowest terms: $1\frac{4}{8} = 1\frac{1}{2}$.

With mixed numbers, add the fractions and whole numbers separately.

EXAMPLE: Add $3\frac{4}{9} + 2\frac{7}{9} =$

Steps 1-2	Step 3	Step 4	Step 5
$\begin{array}{r} 3\frac{4}{9} \\ +2\frac{7}{9} \\ \hline \frac{11}{9} \end{array}$	$\begin{array}{r} 3\frac{4}{9} \\ +2\frac{7}{9} \\ \hline 5\frac{11}{9} \end{array}$	$\frac{11}{9} = 1\frac{2}{9}$	$\begin{array}{r} 5 \\ +1\frac{2}{9} \\ \hline 6\frac{2}{9} \end{array}$

Step 1: Add the numerators: $4 + 7 = 11$.

Step 2: Write the total over the denominator: $\frac{11}{9}$.

Step 3: Add the whole numbers: $3 + 2 = 5$.

Step 4: Change the improper fraction to a mixed number: $\frac{11}{9} = 1\frac{2}{9}$.

Step 5: Add the mixed number to the whole number: $5 + 1\frac{2}{9} = 6\frac{2}{9}$.

EXERCISE 7

DIRECTIONS: Add and reduce.

1.	$\begin{array}{r} \frac{5}{9} \\ +\frac{2}{9} \\ \hline \end{array}$	$\begin{array}{r} \frac{3}{8} \\ +\frac{7}{8} \\ \hline \end{array}$	$\begin{array}{r} \frac{3}{4} \\ +\frac{3}{4} \\ \hline \end{array}$	$\begin{array}{r} 2\frac{5}{6} \\ +4\frac{1}{6} \\ \hline \end{array}$	$\begin{array}{r} 3\frac{4}{5} \\ +9\frac{3}{5} \\ \hline \end{array}$	$\begin{array}{r} 6\frac{7}{12} \\ +1\frac{11}{12} \\ \hline \end{array}$
2.	$\begin{array}{r} \frac{7}{10} \\ +\frac{9}{10} \\ \hline \end{array}$	$\begin{array}{r} \frac{11}{16} \\ +\frac{3}{16} \\ \hline \end{array}$	$\begin{array}{r} \frac{2}{7} \\ +\frac{4}{7} \\ \hline \end{array}$	$\begin{array}{r} 8\frac{5}{18} \\ +3\frac{11}{18} \\ \hline \end{array}$	$\begin{array}{r} 2\frac{9}{20} \\ +9\frac{17}{20} \\ \hline \end{array}$	$\begin{array}{r} 10\frac{23}{36} \\ +4\frac{35}{36} \\ \hline \end{array}$

Check your answers on page 69.

Adding Fractions with Unlike Denominators

Often, the fractions in addition problems do not have the same denominators. In these problems you must find a **common denominator.** A common denominator is a number that can be divided evenly by every denominator in the problem. The lowest number that can be divided evenly by every denominator in the problem is called the **lowest common denominator,** or **LCD.**

Sometimes the biggest denominator in the problem is the LCD.

EXAMPLE: Add $\frac{2}{5}$ and $\frac{7}{20}$.

Step 1	Step 2	Step 3	Step 4
$5\overline{)20}$ = 4	$\frac{2 \times 4}{5 \times 4} = \frac{8}{20}$	$\frac{8}{20} + \frac{7}{20} = \frac{15}{20}$	$\frac{15}{20} = \frac{3}{4}$

Step 1: Find the LCD. 5 divides evenly into 20. 20 is the LCD.
Step 2: Raise $\frac{2}{5}$ to a fraction with 20 as the denominator.
Step 3: Add the fractions: $\frac{8}{20} + \frac{7}{20} = \frac{15}{20}$.
Step 4: Reduce the answer to lowest terms.

The biggest denominator is not always the common denominator. For example, look at this problem:

EXAMPLE: $\frac{5}{8} + \frac{1}{3} =$

Step 1	Step 2	Step 3
$8 \times 3 = 24$	$\frac{5 \times 3}{8 \times 3} = \frac{15}{24}$ $\frac{1 \times 8}{3 \times 8} = \frac{8}{24}$	$\frac{15}{24} + \frac{8}{24} = \frac{23}{24}$

Step 1: Find the LCD. Multiply the two denominators together.
Step 2: Raise each fraction to higher terms.
Step 3: Add the fractions: $\frac{15}{24} + \frac{8}{24} = \frac{23}{24}$.

Sometimes it is not a good idea to multiply the denominators together. For example, look at this problem: $2\frac{5}{9} + \frac{4}{15}$. If you multiply the denominators together, you get $9 \times 15 = 135$. 9 and 15 both divide evenly into 135, but 135 is not the LCD. It is easy to make mistakes using big denominators.

To find the LCD, go through the multiplication table of the biggest denominator until you find a number the other denominator divides into evenly:

$15 \times 1 = 15$. 9 does not divide evenly into 15.
$15 \times 2 = 30$. 9 does not divide evenly into 30.
$15 \times 3 = 45$. 9 does divide evenly into 45.

45 is the LCD for this problem.

The last method to find the LCD works well when you add three or more fractions.

EXERCISE 8

DIRECTIONS: Find the LCD and add each problem. Be sure to reduce each answer and change improper fractions to whole or mixed numbers.

1. $\frac{3}{4} + \frac{5}{12}$ $\qquad$ $\frac{7}{9} + \frac{1}{3}$ $\qquad$ $\frac{2}{5} + \frac{13}{15}$ $\qquad$ $4\frac{3}{20} + 3\frac{3}{4}$ $\qquad$ $2\frac{5}{24} + 9\frac{3}{8}$

2. $\frac{3}{5} + \frac{1}{4}$ $\qquad$ $\frac{1}{9} + \frac{2}{7}$ $\qquad$ $\frac{2}{3} + \frac{7}{8}$ $\qquad$ $8\frac{4}{15} + 8\frac{1}{2}$ $\qquad$ $5\frac{3}{4} + 6\frac{7}{9}$

3. $\frac{7}{10}$, $\frac{1}{2}$, $+\frac{1}{4}$ $\qquad$ $\frac{5}{12}$, $\frac{5}{8}$, $+\frac{2}{3}$ $\qquad$ $\frac{5}{6}$, $\frac{3}{8}$, $+\frac{9}{16}$ $\qquad$ $6\frac{1}{2}$, $2\frac{5}{9}$, $+4\frac{1}{6}$ $\qquad$ $3\frac{4}{15}$, $1\frac{1}{4}$, $+9\frac{5}{6}$

Check your answers on page 70.

SUBTRACTING FRACTIONS

Subtracting Fractions with the Same Denominators

When fractions in a subtraction problem have the same denominator, subtract the numerators. Then write the difference over the denominator.

EXAMPLE: Subtract $\frac{2}{15}$ from $\frac{8}{15}$.

Step 1	Step 2	Step 3
$\frac{8}{15} - \frac{2}{15} = \mathbf{6}$	$\frac{8}{15} - \frac{2}{15} = \mathbf{\frac{6}{15}}$	$\frac{6 \div 3}{15 \div 3} = \mathbf{\frac{2}{5}}$

Step 1: Subtract the numerators: $8 - 2 = 6$.
Step 2: Write the difference over the denominator: $\frac{6}{15}$.
Step 3: Reduce the fraction to lowest terms.

With mixed numbers, subtract the fractions and the whole numbers separately.

EXAMPLE: $6\frac{5}{8} - 2\frac{3}{8} =$

Step 1	Step 2	Step 3	Step 4
$\begin{array}{r} 6\frac{5}{8} \\ -2\frac{3}{8} \\ \hline 2 \end{array}$	$\begin{array}{r} 6\frac{5}{8} \\ -2\frac{3}{8} \\ \hline \frac{2}{8} \end{array}$	$\begin{array}{r} 6\frac{5}{8} \\ -2\frac{3}{8} \\ \hline 4\frac{2}{8} \end{array}$	$4\frac{2 \div 2}{8 \div 2} = 4\frac{1}{4}$

Step 1: Subtract the numerators: $5 - 3 = 2$.
Step 2: Write the difference over the denominator: $\frac{2}{8}$.
Step 3: Subtract the whole numbers: $6 - 2 = 4$.
Step 4: Reduce the fraction part of the answer to lowest terms.

EXERCISE 9

DIRECTIONS: Subtract and reduce each problem.

1. $\begin{array}{r} \frac{4}{7} \\ -\frac{1}{7} \\ \hline \end{array}$ $\quad \begin{array}{r} \frac{11}{12} \\ -\frac{5}{12} \\ \hline \end{array}$ $\quad \begin{array}{r} \frac{7}{8} \\ -\frac{1}{8} \\ \hline \end{array}$ $\quad \begin{array}{r} 8\frac{5}{6} \\ -2\frac{1}{6} \\ \hline \end{array}$ $\quad \begin{array}{r} 7\frac{9}{10} \\ -1\frac{3}{10} \\ \hline \end{array}$ $\quad \begin{array}{r} 6\frac{13}{16} \\ -2\frac{9}{16} \\ \hline \end{array}$

2. $\begin{array}{r} \frac{5}{9} \\ -\frac{2}{9} \\ \hline \end{array}$ $\quad \begin{array}{r} \frac{4}{5} \\ -\frac{3}{5} \\ \hline \end{array}$ $\quad \begin{array}{r} \frac{17}{18} \\ -\frac{5}{18} \\ \hline \end{array}$ $\quad \begin{array}{r} 5\frac{3}{4} \\ -3\frac{1}{4} \\ \hline \end{array}$ $\quad \begin{array}{r} 4\frac{19}{20} \\ -1\frac{11}{20} \\ \hline \end{array}$ $\quad \begin{array}{r} 8\frac{11}{36} \\ -2\frac{7}{36} \\ \hline \end{array}$

Check your answers on page 71.

Subtracting Fractions with Unlike Denominators

Often, the fractions in subtraction problems do not have the same denominators. In these problems, find a common denominator. Then change each fraction to a new fraction with the common denominator. Subtract the new fractions.

EXAMPLE: Subtract $5\frac{1}{4}$ from $6\frac{2}{3}$.

Step 1	Step 2	Step 3	Step 4
$12 \div 3 = 4$ $12 \div 4 = 3$	$\frac{2 \times 4}{3 \times 4} = \frac{8}{12}$ $\frac{1 \times 3}{4 \times 3} = \frac{3}{12}$	$6\frac{8}{12}$ $-5\frac{3}{12}$ $\frac{5}{12}$	$6\frac{8}{12}$ $-5\frac{3}{12}$ $1\frac{5}{12}$

Step 1: Find the LCD. 3 and 4 both divide evenly into 12. 12 is the LCD.
Step 2: Raise the fractions to higher terms, with 12 as the denominator.
Step 3: Subtract the fractions.
Step 4: Subtract the whole numbers.

EXERCISE 10

DIRECTIONS: Subtract and reduce each problem.

1. $\frac{3}{4} - \frac{1}{8}$ $\quad$ $\frac{11}{12} - \frac{1}{2}$ $\quad$ $\frac{4}{5} - \frac{1}{4}$ $\quad$ $9\frac{6}{7} - 2\frac{1}{2}$ $\quad$ $8\frac{5}{9} - 5\frac{1}{3}$

2. $\frac{1}{2} - \frac{2}{9}$ $\quad$ $\frac{3}{4} - \frac{1}{6}$ $\quad$ $\frac{7}{8} - \frac{5}{12}$ $\quad$ $7\frac{11}{15} - 4\frac{1}{6}$ $\quad$ $5\frac{4}{5} - 1\frac{3}{8}$

Check your answers on page 71.

Borrowing

In some subtraction problems, there is no top fraction to take the bottom fraction from. To get a fraction in the top number, **borrow.** To borrow means to take 1 from the whole number at the top. Then rewrite the 1 as a fraction. Remember that any fraction with the same numerator and denominator is equal to 1. The numerator and denominator should be the same as the denominator of the other fraction in the problem.

EXAMPLE: $8 - 3\frac{4}{5} =$

Step 1	Step 2	Step 3
$8 = \mathbf{7\frac{5}{5}}$	$\overset{7}{\cancel{8}}\frac{5}{5}$ $-3\frac{4}{5}$ $\mathbf{\frac{1}{5}}$	$\overset{7}{\cancel{8}}\frac{5}{5}$ $-3\frac{4}{5}$ $\mathbf{4\frac{1}{5}}$

Step 1: Borrow 1 from 8. Change the 1 to $\frac{5}{5}$, since 5 is the LCD.
Step 2: Subtract the fractions.
Step 3: Subtract the whole numbers.

In other subtraction problems, the top fraction is not big enough to subtract the bottom fraction. To get a bigger fraction on top, borrow 1 from the whole number at the top. Rewrite the 1 as a fraction. Then add this fraction to the old top fraction. Study the next example carefully.

EXAMPLE: Subtract $3\frac{4}{9}$ from $5\frac{2}{9}$.

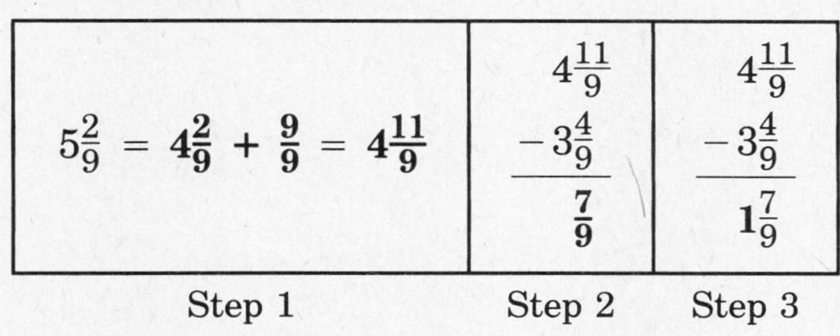

Step 1: Borrow 1 from 5. Change the 1 to $\frac{9}{9}$, and add the $\frac{9}{9}$ to the other fraction, $\frac{2}{9}$.
Step 2: Subtract the fractions.
Step 3: Subtract the whole numbers.

EXERCISE 11

DIRECTIONS: Subtract and reduce.

1. $5 - 2\frac{3}{8}$ $4 - 3\frac{5}{7}$ $9 - 7\frac{2}{3}$ $12 - 4\frac{7}{9}$ $6 - 2\frac{3}{10}$

2. $11 - 3\frac{5}{12}$ $8 - 6\frac{11}{16}$ $6 - 3\frac{2}{5}$ $7 - 4\frac{1}{2}$ $15 - 9\frac{3}{4}$

3.	$5\frac{3}{8}$ $-2\frac{7}{8}$	$7\frac{1}{6}$ $-3\frac{5}{6}$	$10\frac{2}{5}$ $-4\frac{4}{5}$	$6\frac{4}{9}$ $-5\frac{8}{9}$	$2\frac{3}{10}$ $-1\frac{9}{10}$
4.	$8\frac{2}{7}$ $-6\frac{6}{7}$	$6\frac{5}{12}$ $-2\frac{11}{12}$	$11\frac{1}{3}$ $-4\frac{2}{3}$	$12\frac{1}{4}$ $-3\frac{3}{4}$	$4\frac{7}{16}$ $-1\frac{15}{16}$

Check your answers on page 72.

When the denominators in a subtraction problem are different, first change each fraction to a new fraction with a common denominator. Then borrow if you need to.

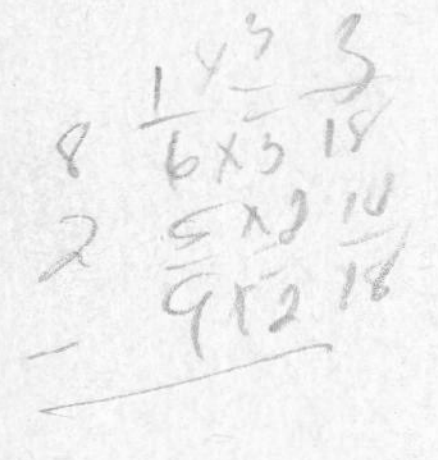

EXAMPLE: $8\frac{1}{6} - 2\frac{5}{9} =$

Step 1	Step 2	Steps 3-4	Step 5
$\mathbf{18 \div 6 = 3}$ $\mathbf{18 \div 9 = 2}$	$\frac{1 \times 3}{6 \times 3} = \mathbf{\frac{3}{18}}$ $\frac{5 \times 2}{9 \times 2} = \mathbf{\frac{10}{18}}$	$1 = \frac{18}{18}$ $\mathbf{\frac{18}{18} + \frac{3}{18} = \frac{21}{18}}$ $8\frac{3}{18} = \mathbf{7\frac{21}{18}}$	$7\frac{21}{18}$ $-2\frac{10}{18}$ $\mathbf{5\frac{11}{18}}$

Step 1: Find the LCD. Both 6 and 9 divide evenly into 18. 18 is the LCD.

Step 2: Raise both fractions to higher terms, with 18 as the denominator.

Step 3: $\frac{3}{18}$ is less than $\frac{10}{18}$. To subtract, you must borrow 1 from the 8 and change it to $\frac{18}{18}$.

Step 4: Add the $\frac{18}{18}$ to $\frac{3}{18}$.

Step 5: Subtract the fractions and the whole numbers.

EXERCISE 12

DIRECTIONS: Subtract and reduce.

1.	$8\frac{2}{5}$ $-3\frac{3}{4}$	$5\frac{1}{2}$ $-1\frac{7}{8}$	$6\frac{3}{7}$ $-2\frac{2}{3}$	$10\frac{2}{9}$ $-3\frac{5}{6}$	$4\frac{1}{3}$ $-2\frac{3}{4}$
2.	$9\frac{1}{2}$ $-7\frac{11}{15}$	$11\frac{1}{2}$ $-5\frac{2}{3}$	$7\frac{3}{10}$ $-4\frac{2}{3}$	$12\frac{1}{6}$ $-8\frac{7}{8}$	$15\frac{4}{9}$ $-6\frac{2}{3}$

Check your answers on page 73.

MULTIPLYING FRACTIONS

Multiplying Fractions by Fractions

When you multiply whole numbers (except 1 and 0), the answer is bigger than the two numbers you multiply. When you multiply fractions, the answer is smaller than either of the two fractions.

When you multiply fractions, you find *a part of a part.* For example, if you multiply $\frac{1}{2}$ by $\frac{1}{2}$, you find $\frac{1}{2}$ *of* $\frac{1}{2}$. You know that $\frac{1}{2}$ of $\frac{1}{2}$ dollar is $\frac{1}{4}$ dollar. The answer is smaller than either of the fractions you multiplied.

You do not need a common denominator when you multiply fractions. Multiply the numerators together. Then multiply the denominators together. Reduce the answer.

EXAMPLE: Multiply $\frac{7}{9}$ by $\frac{5}{8}$.

$\frac{7}{9} \times \frac{5}{8} = \frac{\mathbf{35}}{}$	$\frac{7}{9} \times \frac{5}{8} = \frac{\mathbf{35}}{\mathbf{72}}$
Step 1	Step 2

Step 1: Multiply the numerators: $7 \times 5 = 35$.

Step 2: Multiply the denominators: $9 \times 8 = 72$. $\frac{35}{72}$ is reduced to lowest terms.

EXERCISE 13

DIRECTIONS: Multiply and reduce.

1. $\frac{4}{5} \times \frac{3}{5} =$ $\qquad \frac{5}{7} \times \frac{2}{3} =$ $\qquad \frac{3}{8} \times \frac{7}{8} =$ $\qquad \frac{1}{2} \times \frac{9}{10} =$ $\qquad \frac{2}{3} \times \frac{2}{9} =$

2. $\frac{3}{7} \times \frac{6}{11} =$ $\qquad \frac{3}{4} \times \frac{5}{8} =$ $\qquad \frac{4}{9} \times \frac{2}{5} =$ $\qquad \frac{8}{9} \times \frac{1}{3} =$ $\qquad \frac{5}{7} \times \frac{4}{9} =$

Check your answers on page 74.

Canceling

Canceling is a way of making many fraction multiplication problems easier. It is like reducing. To cancel, divide a numerator and a denominator by a number that goes evenly into both of them.

EXAMPLE: $\frac{4}{9} \times \frac{5}{6} =$

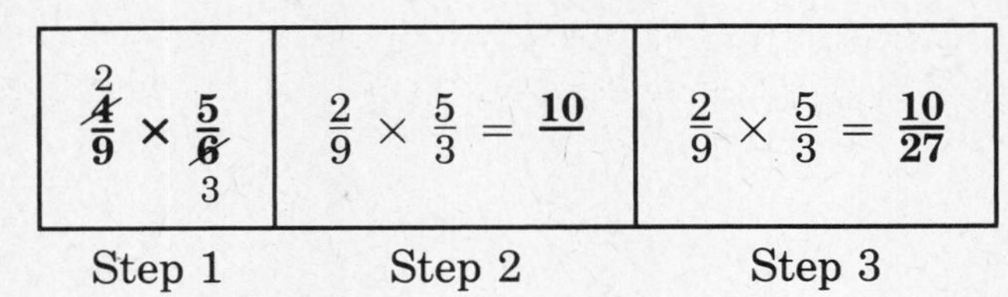

Step 1: Cancel 4 and 6. Divide 4 and 6 by 2. $4 \div 2 = 2$. $6 \div 2 = 3$. (IMPORTANT: You cannot cancel two numerators or two denominators together. In this problem, you cannot cancel the 9 and the 3.)

Step 2: Multiply the numerators: $2 \times 5 = 10$.

Step 3: Multiply the denominators: $9 \times 3 = 27$. $\frac{10}{27}$ is reduced to lowest terms.

Sometimes you can cancel more than once.

EXAMPLE: Multiply $\frac{9}{10}$ by $\frac{5}{12}$.

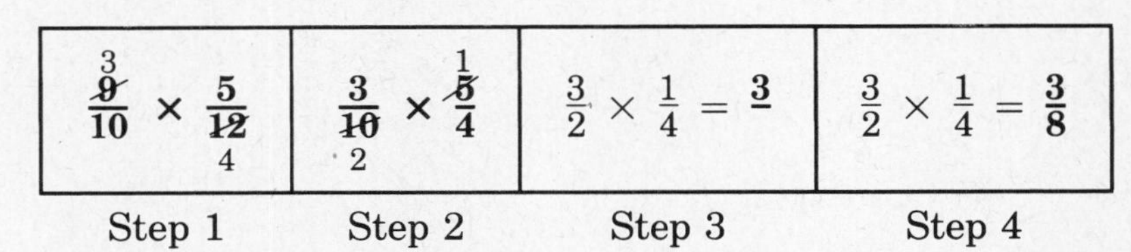

Step 1: Cancel 9 and 12. Divide 9 and 12 by 3. $9 \div 3 = 3$. $12 \div 3 = 4$.

Step 2: Cancel 5 and 10. Divide 5 and 10 by 5. $5 \div 5 = 1$. $10 \div 5 = 2$.

Step 3: Multiply the numerators: $3 \times 1 = 3$.

Step 4: Multiply the denominators: $2 \times 4 = 8$. $\frac{3}{8}$ is reduced to lowest terms.

Canceling also makes multiplying three fractions easier. The numerator and denominator you cancel do not have to be next to each other. You also can cancel several times in one problem. Study this example carefully.

EXAMPLE: $\frac{3}{4} \times \frac{16}{25} \times \frac{5}{8} =$

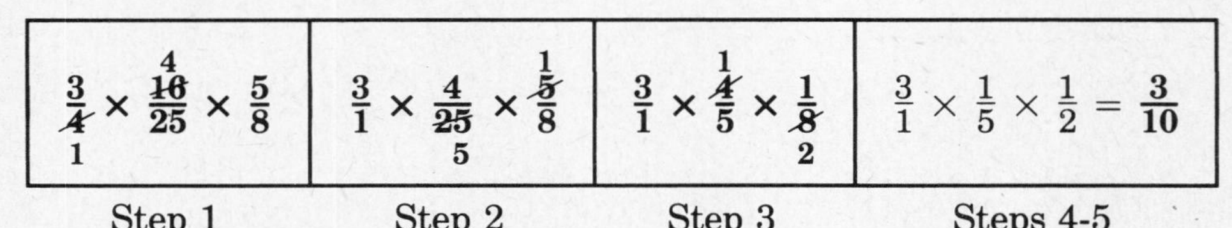

Step 1: Cancel 16 and 4. Divide 16 and 4 by 4. $16 \div 4 = 4$. $4 \div 4 = 1$.

Step 2: Cancel 5 and 25. Divide 5 and 25 by 5. $5 \div 5 = 1$. $25 \div 5 = 5$.

Step 3: Cancel 4 and 8. Divide 4 and 8 by 4. $4 \div 4 = 1$. $8 \div 4 = 2$.
Step 4: Multiply the numerators: $3 \times 1 \times 1 = 3$.
Step 5: Multiply the denominators: $1 \times 5 \times 2 = 10$. $\frac{3}{10}$ is reduced to lowest terms.

EXERCISE 14

DIRECTIONS: Cancel, multiply, and reduce.

1. $\frac{5}{8} \times \frac{11}{15} =$ $\quad \frac{4}{9} \times \frac{15}{16} =$ $\quad \frac{2}{3} \times \frac{9}{10} =$ $\quad \frac{3}{10} \times \frac{20}{21} =$ $\quad \frac{9}{16} \times \frac{8}{15} =$

2. $\frac{7}{24} \times \frac{3}{14} =$ $\quad \frac{9}{20} \times \frac{4}{5} =$ $\quad \frac{5}{12} \times \frac{3}{20} =$ $\quad \frac{7}{18} \times \frac{16}{21} =$ $\quad \frac{8}{15} \times \frac{3}{4} =$

3. $\frac{5}{9} \times \frac{6}{7} \times \frac{3}{4} =$ $\quad \frac{1}{2} \times \frac{9}{10} \times \frac{4}{15} =$ $\quad \frac{5}{6} \times \frac{4}{9} \times \frac{3}{4} =$ $\quad \frac{3}{16} \times \frac{8}{9} \times \frac{4}{5} =$

4. $\frac{3}{4} \times \frac{1}{2} \times \frac{16}{21} =$ $\quad \frac{8}{15} \times \frac{1}{4} \times \frac{5}{8} =$ $\quad \frac{3}{8} \times \frac{2}{9} \times \frac{4}{5} =$ $\quad \frac{2}{3} \times \frac{9}{20} \times \frac{5}{6} =$

Check your answers on page 74.

Multiplying Fractions by Whole Numbers and Mixed Numbers

To multiply a whole number and a fraction, first write the whole number as a fraction. Write the whole number as the numerator and 1 as the denominator. Then cancel, multiply, and reduce.

EXAMPLE: Multiply $\frac{7}{9}$ by 12.

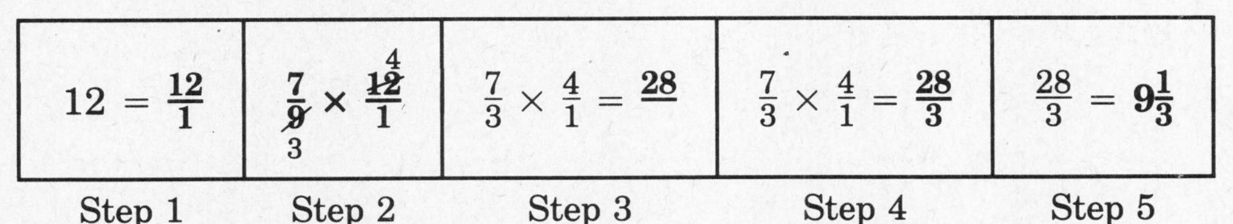

$12 = \frac{12}{1}$	$\frac{7}{\cancel{9}_3} \times \frac{\cancel{12}^4}{1}$	$\frac{7}{3} \times \frac{4}{1} = \frac{28}{}$	$\frac{7}{3} \times \frac{4}{1} = \frac{28}{3}$	$\frac{28}{3} = 9\frac{1}{3}$
Step 1	Step 2	Step 3	Step 4	Step 5

Step 1: Write 12 as a fraction. Use 1 as the denominator.
Step 2: Cancel 12 and 9. Divide 12 and 9 by 3. $12 \div 3 = 4$. $9 \div 3 = 3$.
Step 3: Multiply the numerators: $7 \times 4 = 28$.
Step 4: Multiply the denominators: $3 \times 1 = 3$.
Step 5: Change the improper fraction to a mixed number: $\frac{28}{3} = 9\frac{1}{3}$.

To multiply with mixed numbers, first change the mixed numbers to improper fractions. Then cancel and multiply the fractions.

EXAMPLE: $2\frac{1}{4} \times \frac{2}{15} =$

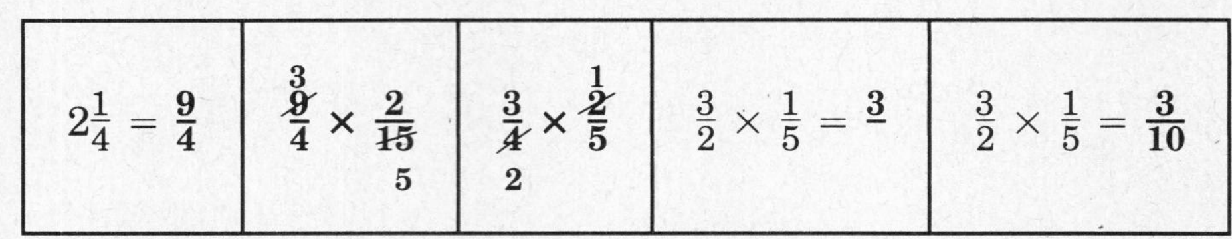

$2\frac{1}{4} = \frac{9}{4}$	$\frac{\not{9}^{\,3}}{4} \times \frac{2}{\not{15}_{\,5}}$	$\frac{3}{\not{4}_{\,2}} \times \frac{\not{2}^{\,1}}{5}$	$\frac{3}{2} \times \frac{1}{5} = \mathbf{3}$	$\frac{3}{2} \times \frac{1}{5} = \frac{\mathbf{3}}{\mathbf{10}}$
Step 1	Step 2	Step 3	Step 4	Step 5

Step 1: Change the mixed number to an improper fraction.
Step 2: Cancel 9 and 15. Divide 9 and 15 by 3. $9 \div 3 = 3$. $15 \div 3 = 5$.
Step 3: Cancel 2 and 4. Divide 2 and 4 by 2. $2 \div 2 = 1$. $4 \div 2 = 2$.
Step 4: Multiply the numerators: $3 \times 1 = 3$.
Step 5: Multiply the denominators: $2 \times 5 = 10$.

EXERCISE 15

DIRECTIONS: Cancel, multiply, and reduce. Change each improper fraction answer to a whole or mixed number.

1. $\frac{2}{3} \times 6 =$ $\quad \frac{5}{6} \times 9 =$ $\quad \frac{3}{10} \times 5 =$ $\quad \frac{1}{4} \times 3 =$ $\quad \frac{5}{9} \times 12 =$

2. $4 \times \frac{5}{6} =$ $\quad 2 \times \frac{3}{5} =$ $\quad 8 \times \frac{7}{12} =$ $\quad 11 \times \frac{2}{3} =$ $\quad 7 \times \frac{9}{14} =$

3. $2\frac{2}{3} \times \frac{5}{8} =$ $\quad 1\frac{5}{9} \times \frac{3}{4} =$ $\quad 2\frac{6}{7} \times \frac{7}{10} =$ $\quad \frac{2}{9} \times 2\frac{1}{2} =$ $\quad \frac{8}{15} \times 2\frac{1}{4} =$

4. $2\frac{1}{4} \times 1\frac{2}{3} =$ $\quad 1\frac{3}{4} \times 1\frac{5}{7} =$ $\quad 3\frac{3}{5} \times 1\frac{1}{9} =$ $\quad 1\frac{5}{16} \times 6\frac{2}{3} =$

5. $2\frac{2}{5} \times \frac{3}{8} \times 1\frac{1}{9} =$ $\quad 2\frac{2}{3} \times \frac{5}{6} \times 3\frac{3}{5} =$ $\quad 1\frac{1}{6} \times 1\frac{1}{2} \times 2\frac{4}{7} =$

Check your answers on page 75.

DIVIDING FRACTIONS

Dividing by Fractions

Dividing by a fraction means finding how many times a fraction goes into another number. For example, if you divide $\frac{1}{2}$ by $\frac{1}{4}$, you find out how many times $\frac{1}{4}$ goes into $\frac{1}{2}$. You know that $\frac{1}{4}$ dollar goes into $\frac{1}{2}$ dollar exactly twice. Write this problem as $\frac{1}{2} \div \frac{1}{4}$. Read the ÷ sign as the words "divided by."

To divide by a fraction, **invert** the fraction at the right, and follow the rules for multiplying fractions. To invert means to rewrite a fraction with the numerator on the bottom and the denominator on top. For example, when you invert $\frac{3}{5}$, you get $\frac{5}{3}$.

EXAMPLE: Divide $\frac{1}{2}$ by $\frac{1}{4}$.

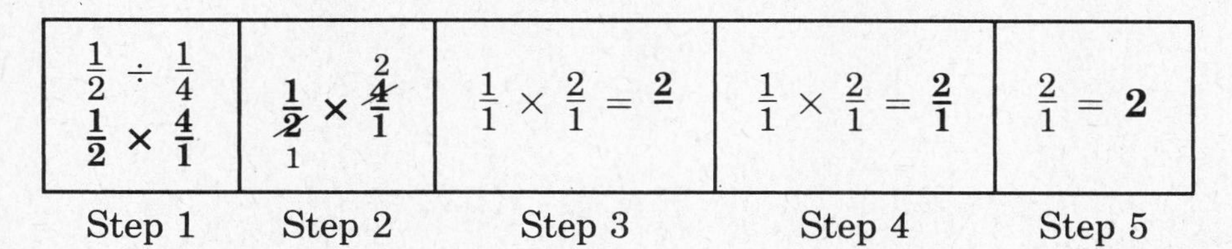

Step 1	Step 2	Step 3	Step 4	Step 5
$\frac{1}{2} \div \frac{1}{4}$ $\frac{1}{2} \times \frac{4}{1}$	$\frac{1}{\cancel{2}_1} \times \frac{\cancel{4}^2}{1}$	$\frac{1}{1} \times \frac{2}{1} = \mathbf{2}$	$\frac{1}{1} \times \frac{2}{1} = \frac{2}{1}$	$\frac{2}{1} = \mathbf{2}$

Step 1: Invert the fraction you are dividing by. $\frac{1}{4}$ becomes $\frac{4}{1}$. Change the ÷ sign to a × sign.
Step 2: Cancel 4 and 2. Divide 4 and 2 by 2. $4 \div 2 = 2$. $2 \div 2 = 1$.
Step 3: Multiply the numerators: $1 \times 2 = 2$.
Step 4: Multiply the denominators: $1 \times 1 = 1$.
Step 5: Change $\frac{2}{1}$ to a whole number: $\frac{2}{1} = 2$.

EXERCISE 16

DIRECTIONS: Divide, change each improper fraction answer to a whole or mixed number, and reduce.

1. $\frac{3}{5} \div \frac{9}{10} =$ $\quad \frac{4}{9} \div \frac{8}{15} =$ $\quad \frac{2}{3} \div \frac{4}{9} =$ $\quad \frac{5}{12} \div \frac{3}{8} =$ $\quad \frac{9}{10} \div \frac{2}{5} =$

2. $\frac{7}{8} \div \frac{7}{12} =$ $\quad \frac{8}{15} \div \frac{4}{25} =$ $\quad \frac{3}{4} \div \frac{6}{11} =$ $\quad \frac{2}{7} \div \frac{4}{5} =$ $\quad \frac{4}{5} \div \frac{12}{13} =$

Check your answers on page 76.

Dividing Whole Numbers and Mixed Numbers by Fractions

When you divide a whole number by a fraction, you find out how many times the fraction goes into the whole number. To divide a whole number by a fraction, first write the whole number as a fraction with 1 as the denominator. Then invert the fraction at the right and follow the rules for multiplying fractions.

EXAMPLE: Divide 10 by $\frac{8}{9}$.

Step 1	Step 2	Step 3	Steps 4-5	Step 6
$10 = \mathbf{\frac{10}{1}}$	$\frac{10}{1} \div \frac{8}{9}$ $\mathbf{\frac{10}{1} \times \frac{9}{8}}$	$\mathbf{\frac{\overset{5}{\cancel{10}}}{1} \times \frac{9}{\underset{4}{\cancel{8}}}}$	$\frac{5}{1} \times \frac{9}{4} = \mathbf{\frac{45}{4}}$	$\frac{45}{4} = \mathbf{11\frac{1}{4}}$

Step 1: Write 10 as a fraction. Use 1 as the denominator.
Step 2: Invert the fraction you are dividing by. $\frac{8}{9}$ becomes $\frac{9}{8}$. Change the $\div$ sign to a $\times$ sign.
Step 3: Cancel 10 and 8. Divide 10 and 8 by 2. $10 \div 2 = 5$. $8 \div 2 = 4$.
Step 4: Multiply the numerators: $5 \times 9 = 45$.
Step 5: Multiply the denominators: $1 \times 4 = 4$.
Step 6: Change the improper fraction to a mixed number: $\frac{45}{4} = 11\frac{1}{4}$.

To divide a mixed number by a fraction, first change the mixed number to an improper fraction. Then invert the fraction at the right and follow the rules for multiplying fractions.

EXAMPLE: $3\frac{1}{5} \div \frac{8}{9} =$

Step 1	Step 2	Step 3	Steps 4-5	Step 6
$3\frac{1}{5} = \mathbf{\frac{16}{5}}$	$\frac{16}{5} \div \frac{8}{9}$ $\mathbf{\frac{16}{5} \times \frac{9}{8}}$	$\mathbf{\frac{\overset{2}{\cancel{16}}}{5} \times \frac{9}{\underset{1}{\cancel{8}}}}$	$\frac{2}{5} \times \frac{9}{1} = \mathbf{\frac{18}{5}}$	$\frac{18}{5} = \mathbf{3\frac{3}{5}}$

Step 1: Change the mixed number to an improper fraction.
Step 2: Invert the fraction you are dividing by. $\frac{8}{9}$ becomes $\frac{9}{8}$. Change the $\div$ sign to a $\times$ sign.
Step 3: Cancel 16 and 8. Divide 16 and 8 by 8. $16 \div 8 = 2$. $8 \div 8 = 1$.
Step 4: Multiply the numerators: $2 \times 9 = 18$.
Step 5: Multiply the denominators: $5 \times 1 = 5$.
Step 6: Change the improper fraction to a mixed number: $\frac{18}{5} = 3\frac{3}{5}$.

EXERCISE 17

DIRECTIONS: Divide, change every improper fraction answer to a whole or mixed number, and reduce.

1.	$8 \div \frac{4}{5} =$	$3 \div \frac{6}{7} =$	$6 \div \frac{2}{5} =$	$10 \div \frac{5}{8} =$	$2 \div \frac{3}{4} =$
2.	$5 \div \frac{15}{16} =$	$9 \div \frac{6}{7} =$	$7 \div \frac{7}{10} =$	$4 \div \frac{14}{19} =$	$3 \div \frac{6}{13} =$
3.	$1\frac{7}{8} \div \frac{3}{4} =$	$2\frac{2}{9} \div \frac{5}{6} =$	$2\frac{2}{3} \div \frac{1}{2} =$	$1\frac{5}{7} \div \frac{4}{7} =$	$6\frac{1}{4} \div \frac{5}{8} =$
4.	$2\frac{5}{8} \div \frac{7}{12} =$	$1\frac{5}{16} \div \frac{3}{20} =$	$1\frac{7}{9} \div \frac{2}{3} =$	$3\frac{3}{8} \div \frac{1}{4} =$	$2\frac{2}{15} \div \frac{2}{5} =$

Check your answers on page 77.

Dividing Fractions and Mixed Numbers by Whole Numbers

When you divide a fraction by a whole number, you "split" the fraction into smaller parts. For example, if you divide a fraction by 3, you split the fraction into 3 equal parts. To divide a fraction or a mixed number by a whole number, first write the whole number as a fraction with a denominator of 1. Then invert the fraction at the right and follow the rules for multiplying fractions.

EXAMPLE: Divide $\frac{4}{5}$ by 6.

$6 = \mathbf{\frac{6}{1}}$	$\frac{4}{5} \div \frac{6}{1}$ $\mathbf{\frac{4}{5} \times \frac{1}{6}}$	$\frac{\overset{2}{\cancel{4}}}{5} \times \frac{1}{\underset{3}{\cancel{6}}}$	$\frac{2}{5} \times \frac{1}{3} = \mathbf{2}$	$\frac{2}{5} \times \frac{1}{3} = \mathbf{\frac{2}{15}}$
Step 1	Step 2	Step 3	Step 4	Step 5

Step 1: Write 6 as a fraction. Use 1 as the denominator.
Step 2: Invert the fraction you are dividing by. $\frac{6}{1}$ becomes $\frac{1}{6}$. Change the $\div$ sign to a $\times$ sign.
Step 3: Cancel 4 and 6. Divide 4 and 6 by 2. $4 \div 2 = 2$. $6 \div 2 = 3$.
Step 4: Multiply the numerators: $2 \times 1 = 2$.
Step 5: Multiply the denominators: $5 \times 3 = 15$. $\frac{2}{15}$ is reduced to lowest terms.

When you divide a mixed number by a whole number, remember to change the mixed number to an improper fraction.

EXERCISE 18

DIRECTIONS: Divide, change every improper fraction to a whole or mixed number, and reduce.

1. $\frac{2}{3} \div 8 =$ $\quad$ $\frac{6}{7} \div 4 =$ $\quad$ $\frac{3}{8} \div 9 =$ $\quad$ $\frac{4}{5} \div 12 =$ $\quad$ $\frac{9}{10} \div 3 =$

2. $1\frac{1}{5} \div 2 =$ $\quad$ $3\frac{1}{3} \div 5 =$ $\quad$ $3\frac{1}{2} \div 7 =$ $\quad$ $2\frac{2}{5} \div 6 =$ $\quad$ $1\frac{2}{3} \div 15 =$

Check your answers on page 78.

Dividing by Mixed Numbers

To divide by a mixed number, change the mixed number to an improper fraction. Then invert the fraction at the right and follow the rules for multiplying fractions.

If the number you are dividing is a whole number or a mixed number, remember to rewrite it as a fraction.

EXAMPLE: $4\frac{1}{2} \div 1\frac{1}{5} =$

$4\frac{1}{2} = \mathbf{\frac{9}{2}}$ $1\frac{1}{5} = \mathbf{\frac{6}{5}}$	$\frac{9}{2} \div \frac{6}{5}$ $\mathbf{\frac{9}{2} \times \frac{5}{6}}$	$\frac{\overset{3}{\cancel{9}}}{2} \times \frac{5}{\underset{2}{\cancel{6}}}$	$\frac{3}{2} \times \frac{5}{2} = \mathbf{\frac{15}{4}}$	$\frac{15}{4} = \mathbf{3\frac{3}{4}}$
Step 1	Step 2	Step 3	Steps 4-5	Step 6

Step 1: Change the mixed numbers to improper fractions.

Step 2: Invert the fraction you are dividing by. $\frac{6}{5}$ becomes $\frac{5}{6}$. Change the $\div$ sign to a $\times$ sign.

Step 3: Cancel 9 and 6. Divide 9 and 6 by 3. $9 \div 3 = 3$. $6 \div 3 = 2$.

Step 4: Multiply the numerators: $3 \times 5 = 15$.

Step 5: Multiply the denominators: $2 \times 2 = 4$.

Step 6: Change the improper fraction to a mixed number. $\frac{15}{4} = 3\frac{3}{4}$.

EXERCISE 19

DIRECTIONS: Divide, change every improper fraction answer to a whole or mixed number, and reduce.

1. $\frac{5}{9} \div 1\frac{2}{3} =$ $\quad$ $\frac{7}{8} \div 4\frac{1}{2} =$ $\quad$ $\frac{3}{20} \div 2\frac{2}{5} =$ $\quad$ $\frac{4}{7} \div 2\frac{2}{7} =$ $\quad$ $\frac{5}{6} \div 3\frac{1}{3} =$

2. $14 \div 3\frac{1}{2} =$ $\quad$ $4 \div 1\frac{5}{6} =$ $\quad$ $6 \div 1\frac{1}{8} =$ $\quad$ $10 \div 3\frac{3}{4} =$ $\quad$ $9 \div 3\frac{3}{5} =$

3. $7\frac{1}{2} \div 3\frac{3}{7} =$ $\quad 2\frac{1}{10} \div 6\frac{3}{4} =$ $\quad 3\frac{8}{9} \div 4\frac{2}{3} =$ $\quad 1\frac{1}{4} \div 2\frac{1}{12} =$ $\quad 1\frac{7}{8} \div 4\frac{3}{8} =$

Check your answers on page 79.

Fraction Word Problems

Most fraction problems on the GED Test are word problems. Here are some hints to help you identify the operations you will need to solve fraction problems.

Addition and subtraction problems are easy to recognize. You have already learned some of the key words for addition problems—**sum, total, altogether,** and **combined.** You have also learned some of the key words for subtraction—**difference, how much more,** and **how much less.** Also, the words **gross** and **net** in a problem often are clues to subtract.

Multiplication problems are sometimes harder to recognize. The word **of** immediately followed by a fraction usually means to multiply. But *of* is a very common word. It does not always mean to multiply. Compare the next two examples.

EXAMPLE: Jim bought a used car for $2,480. He paid $\frac{3}{10}$ of the price as a down payment. How much was his down payment?

Here, *of* means to multiply:

$$\frac{3}{10} \text{ of } \$2{,}480 = \frac{3}{\cancel{10}_{1}} \times \frac{\cancel{2480}^{248}}{1} = \frac{744}{1} = \$744$$

EXAMPLE: In the town of Midvale, $\frac{1}{2}$ of the budget goes for education, $\frac{1}{5}$ of the budget for police and fire protection, and $\frac{1}{20}$ of the budget for parks and recreation. These items make up what fraction of the whole budget?

Here *of* does not help. This is an addition problem.

$$\begin{array}{rcl} \frac{1}{2} & = & \frac{10}{20} \\ \frac{1}{5} & = & \frac{4}{20} \\ +\ \frac{1}{20} & = & \frac{1}{20} \\ \hline & & \frac{15}{20} = \frac{3}{4} \end{array}$$

Watch for situations that mean to multiply. For example, you may be told the price of one item. Then you are asked to find the price of several items.

EXAMPLE: One pound of steak costs \$4. Find the cost of a $2\frac{1}{4}$-pound steak.

Multiply the price of one pound by $2\frac{1}{4}$.

$$2\frac{1}{4} \times \$4 =$$

$$\frac{9}{\cancel{4}_1} \times \frac{\cancel{4}^1}{1} = \frac{9}{1} = \$9$$

Division problems are the hardest to set up correctly. Remember that the thing being divided must come first in the problem. The words **cut, share,** and **split** usually mean to divide.

EXAMPLE: Jeff wants to cut a board $5\frac{1}{4}$ feet long into 6 equal pieces. How long will each piece be?

Divide the length of the board by 6. Remember that the thing being divided must come first. The board $5\frac{1}{4}$ feet long is being divided. Put it first.

$$5\frac{1}{4} \div 6 = \frac{21}{4} \div \frac{6}{1} = \frac{\cancel{21}^7}{4} \times \frac{1}{\cancel{6}_2} = \frac{7}{8} \text{ ft.}$$

The word **per** in miles *per* gallon and miles *per* hour means to divide.

EXAMPLE: Alice drove 95 miles in $2\frac{1}{2}$ hours. Find her average speed in miles per hour.

$$95 \div 2\frac{1}{2} = \frac{95}{1} \div \frac{5}{2} = \frac{\cancel{95}^{19}}{1} \times \frac{2}{\cancel{5}_1} = 38 \text{ miles per hour}$$

Watch for situations that mean to divide. For example, you may be told the cost of several items. Then you are asked to find the cost of one item.

EXAMPLE: Jill paid \$15 for $3\frac{3}{4}$ yards of material. Find the cost of one yard of material.

Here the money is being divided into smaller amounts. Put the money first.

$$\$15 \div 3\frac{3}{4} = \frac{15}{1} \div \frac{15}{4} = \frac{\cancel{15}^1}{1} \times \frac{4}{\cancel{15}_1} = \frac{4}{1} = \$4$$

In many problems you will need more than one step and more than one operation to get answers. List for yourself the steps you need to take before you start.

EXERCISE 20 (WORD PROBLEMS)

DIRECTIONS: Choose the correct answer to each problem.

1. The distance from Joe's house to the factory where he works is 15 miles. On his way to work, Joe stopped for gas $8\frac{3}{10}$ miles from his house. How much farther did Joe have to drive to get to work?

   (1) $6\frac{3}{10}$ mi. (2) $6\frac{7}{10}$ mi. (3) $7\frac{3}{10}$ mi.
   (4) $7\frac{7}{10}$ mi. (5) $12\frac{7}{10}$ mi.

2. Last year, Phil made $13,500. He spent $\frac{1}{4}$ of his income on rent. How much did Phil spend on rent last year?

   (1) $281 (2) $1,350 (3) $2,700
   (4) $3,275 (5) $3,375

3. Rachel drove 159 miles on $10\frac{3}{5}$ gallons of gasoline. How far did she drive on one gallon of gasoline?

   (1) 12 mi. (2) 15 mi. (3) 17 mi.
   (4) 23 mi. (5) 55 mi.

4. The Potters spend $\frac{3}{8}$ of their income for food, $\frac{1}{4}$ of their income for rent, and $\frac{3}{16}$ of their income for electricity. Together these items are what fraction of the Potters' income?

   (1) $\frac{7}{16}$ (2) $\frac{9}{16}$ (3) $\frac{11}{16}$ (4) $\frac{13}{16}$ (5) $\frac{7}{8}$

5. Dave usually works 35 hours a week. One week in November, he was sick and missed $9\frac{3}{4}$ hours of work. How many hours did Dave work that week?

   (1) $24\frac{1}{4}$ hrs. (2) $25\frac{1}{4}$ hrs. (3) $25\frac{3}{4}$ hrs.
   (4) $26\frac{1}{4}$ hrs. (5) $26\frac{3}{4}$ hrs.

6. A recipe for corn bread calls for $\frac{1}{2}$ teaspoon of salt, 2 cups of sifted flour, and $\frac{2}{3}$ cup of cornmeal. Jane wants to make twice the amount of corn bread in the recipe. How much cornmeal should she use?

   (1) $\frac{2}{3}$ cup (2) 1 cup (3) $1\frac{1}{3}$ cups
   (4) $1\frac{2}{3}$ cups (5) 2 cups

7. How many $\frac{3}{4}$-inch strips of wood can be cut from a piece of wood 72 inches long?

   (1) 18 (2) 24 (3) 42 (4) 54 (5) 96

8. Eddie drove 140 miles in $3\frac{1}{2}$ hours. What was his average speed?

   (1) 32 mph (2) 38 mph (3) 40 mph
   (4) 44 mph (5) 60 mph

9. Colin bought $2\frac{1}{2}$ pounds of box nails, $3\frac{7}{16}$ pounds of masonry nails, and $4\frac{3}{8}$ pounds of roofing nails. Find the total weight of the items Colin bought.

   (1) $9\frac{11}{16}$ lbs. (2) $10\frac{1}{8}$ lbs. (3) $10\frac{5}{16}$ lbs.
   (4) $10\frac{7}{16}$ lbs. (5) $10\frac{11}{16}$ lbs.

10. Jack bought a coat that regularly sold for $69. It was on sale for $\frac{1}{3}$ off. What was the sale price of the coat?

    (1) $23 (2) $46 (3) $60 (4) $63 (5) $66

11. A share of Apex Steel stock sold Wednesday for $24\frac{3}{8}$. Thursday, it was down $\frac{1}{2}$. Friday, it was up $\frac{5}{8}$. Monday, it was down $\frac{1}{4}$. What was the closing price of the stock on Monday?

    (1) $24\frac{1}{4}$ (2) $24\frac{3}{8}$ (3) $24\frac{3}{4}$ (4) $25\frac{1}{4}$ (5) $25\frac{3}{4}$

12. Four friends went fishing. They agreed to share the fish equally. Together they caught $17\frac{3}{4}$ pounds of fish. How many pounds did each person get?

    (1) $4\frac{3}{16}$ lbs. (2) $4\frac{1}{4}$ lbs. (3) $4\frac{3}{8}$ lbs.
    (4) $4\frac{7}{16}$ lbs. (5) $4\frac{15}{16}$ lbs.

13. The yearly budget for the town of Staunton is $285,000. $\frac{1}{3}$ of the town's budget goes for education. $\frac{3}{5}$ of the education budget goes for teachers' salaries. Find the total amount of teachers' salaries in Staunton for a year.

    (1) $17,000 (2) $19,000 (3) $57,000
    (4) $95,000 (5) $114,000

14. Dorothy bought $6\frac{1}{2}$ yards of material. She used $3\frac{3}{4}$ yards to make a pair of curtains. How much material did she have left?

    (1) $2\frac{3}{4}$ yds. (2) $3\frac{1}{4}$ yds. (3) $3\frac{3}{4}$ yds.
    (4) $4\frac{1}{4}$ yds. (5) $4\frac{1}{2}$ yds.

15. Sam works part-time at the lumber yard. Monday, he worked $2\frac{1}{2}$ hours. Thursday, he worked $1\frac{3}{4}$ hours. Saturday, he worked $6\frac{2}{3}$ hours. How many hours did he work altogether that week?

(1) $9\frac{1}{12}$ hrs. (2) $9\frac{1}{2}$ hrs. (3) $10\frac{1}{12}$ hrs.
(4) $10\frac{11}{12}$ hrs. (5) $12\frac{5}{12}$ hrs.

16. Heather is filling cans with $\frac{1}{2}$ pound of cooked peaches in each can. How many cans does she need for 9 pounds of cooked peaches?

(1) $4\frac{1}{2}$ (2) 9 (3) 12 (4) 15 (5) 18

17. Nick bought a piece of lumber 72 inches long to make a wall shelf for the kitchen. He sawed two $10\frac{1}{4}$-inch long pieces for the sides and three $15\frac{3}{4}$-inch long pieces for the shelves. How long was the original piece after he finished sawing?

(1) $2\frac{3}{4}$ in. (2) $3\frac{3}{4}$ in. (3) $4\frac{1}{4}$ in.
(4) $14\frac{1}{2}$ in. (5) 19 in.

18. In 1960, there were $2\frac{1}{2}$ million people in Oakwood County. In 1980, there were $4\frac{1}{3}$ million people. By how much did the number of people in Oakwood County grow from 1960 to 1980?

(1) $1\frac{5}{6}$ million (2) $2\frac{1}{3}$ million (3) $2\frac{1}{2}$ million
(4) $2\frac{5}{6}$ million (5) $6\frac{5}{6}$ million

19. Joe Twyman bought $2\frac{3}{4}$ tons of scrap metal at $216 a ton. How much did he pay for the scrap metal?

(1) $378 (2) $432 (3) $486 (4) $540 (5) $594

20. On Monday morning, John drove $2\frac{3}{10}$ miles to his son's school, $5\frac{9}{10}$ miles to a gas station, and $6\frac{7}{10}$ miles to his job. How many miles did John drive altogether on Monday morning?

(1) $14\frac{1}{5}$ mi. (2) $14\frac{9}{10}$ mi. (3) $15\frac{3}{5}$ mi.
(4) $15\frac{9}{10}$ mi. (5) 16 mi.

21. Felix needed some $\frac{3}{4}$-inch pieces of wood. How many $\frac{3}{4}$-inch pieces could he get from a 5-foot board?

(1) 6 (2) 12 (3) 45 (4) 60 (5) 80

22. The price of a sweater in December was \$28. In January, the sweater was on sale for $\frac{1}{4}$ off the December price. How much did the sweater cost in January?

(1) \$7 (2) \$14 (3) \$21 (4) \$28 (5) \$35

23. In June, George weighed 216 pounds. He went on a diet and lost $33\frac{1}{2}$ pounds by September. How much did George weigh in September?

(1) $180\frac{1}{2}$ lbs. (2) $181\frac{1}{2}$ lbs. (3) $182\frac{1}{2}$ lbs.
(4) $183\frac{1}{2}$ lbs. (5) $184\frac{1}{2}$ lbs.

24. Manny worked 7 hours on Thursday for \$6 an hour. He worked overtime that day for $2\frac{1}{3}$ hours at \$9 an hour. How much did he make altogether on Thursday?

(1) \$57 (2) \$63 (3) \$64 (4) \$77 (5) \$78

25. Mr. Meyer is $71\frac{3}{4}$ inches tall. His son James is $2\frac{1}{2}$ inches taller than Mr. Meyer. How tall is James?

(1) $69\frac{1}{4}$ in. (2) $73\frac{1}{4}$ in. (3) $73\frac{3}{4}$ in.
(4) $74\frac{1}{4}$ in. (5) $79\frac{3}{4}$ in.

26. Ellen's weekly gross salary is \$282. Her employer takes out $\frac{1}{6}$ of her salary for taxes and Social Security. How much does Ellen take home each week?

(1) \$47 (2) \$235 (3) \$243 (4) \$245 (5) \$329

27. There are 28,640 registered voters in Berkley. In the last election, $\frac{5}{8}$ of the them voted. How many people voted in Berkley's last election?

(1) 10,740 (2) 17,900 (3) 21,480
(4) 35,800 (5) 44,320

28. The Suburban Development Company bought 36 acres of farmland from Jane Miller. The company plans to build houses on the land. Each house will be on a $1\frac{1}{2}$-acre lot. How many houses will the company build on the land?

    (1) 24 (2) 27 (3) 36 (4) 45 (5) 54

29. The Uptown Community Association wants to raise $\$2\frac{1}{2}$ million to build a new sports center. So far, they have raised $\$1\frac{7}{8}$ million for the center. How much more do they need?

    (1) $\$\frac{3}{8}$ million (2) $\$\frac{5}{8}$ million (3) $\$1\frac{3}{8}$ million
    (4) $\$1\frac{3}{4}$ million (5) $\$2\frac{5}{8}$ million

30. Robert wants to cut a pipe $76\frac{1}{2}$ inches long into $4\frac{1}{2}$-inch long pieces. How many pieces can he cut from the pipe?

    (1) 14 (2) 15 (3) 16 (4) 17 (5) 18

Check your answers on page 80.

Fractions Review

These problems will give you a chance to practice the skills you learned in the fractions section of this book. With each answer is the name of the section in which the skills needed for the problem are explained. For every problem you miss, review the section in which the skills for that problem are explained. Then try the problem again.

DIRECTIONS: Solve each problem.

1. There are 36 inches in one yard. 17 inches are what fraction of a yard?

2. Reduce $\frac{16}{44}$ to lowest terms.

3. Find the missing numerator: $\frac{9}{16} = \frac{}{48}$

4. Change $\frac{60}{9}$ to a mixed number and reduce.

5. Change $8\frac{5}{9}$ to an improper fraction.

6. Add $2\frac{7}{15}$ and $4\frac{11}{15}$.

7. Find the sum of $3\frac{5}{8}$, $2\frac{1}{3}$, and $4\frac{5}{6}$.

8. Take $4\frac{7}{20}$ from $9\frac{13}{20}$.

9. Subtract $3\frac{9}{16}$ from $5\frac{7}{8}$.

10. Take $2\frac{2}{3}$ from $8\frac{3}{8}$.

11. $\frac{5}{8} \times \frac{3}{7} =$

12. $\frac{7}{12} \times \frac{8}{35} =$

13. $\frac{3}{8} \times 24 =$

14. $3\frac{3}{4} \times 5\frac{3}{5} =$

15. $\frac{7}{10} \div \frac{3}{4} =$

16. $12 \div \frac{5}{6} =$

17. $4\frac{1}{6} \div \frac{5}{8} =$

18. $8\frac{2}{5} \div 7 =$

19. $4\frac{4}{9} \div 6\frac{2}{3} =$

20. Maxine bought $2\frac{1}{2}$ pounds of chicken, $1\frac{5}{16}$ pounds of fish, and $2\frac{1}{4}$ pounds of ground beef. Find the total weight of the things Maxine bought.

21. A share of Acme Copper stock sold for $62\frac{1}{2}$ on Thursday. Friday, it was down $\frac{1}{4}$. Monday, it was up $\frac{3}{8}$. Find the price of the stock on Monday.

22. The annual budget for Southville is \$3,360,000. $\frac{5}{16}$ of the budget goes for police and fire protection. What amount goes for police and fire protection?

Check your answers on page 82.

ANSWERS & SOLUTIONS

EXERCISE 1

1. $\frac{37}{60}$ 2. $\frac{11}{12}$ 3. $\frac{7}{20}$ 4. $\frac{5}{144}$ 5. $\frac{123}{1000}$ 6. $\frac{1}{4}$ 7. $\frac{3}{16}$

8. $\frac{31}{365}$ 9. $\frac{3}{2000}$ 10. $\frac{99}{100}$

EXERCISE 2

1. $\frac{2}{9}$ $\frac{19}{20}$ $\frac{1}{12}$ 2. $\frac{8}{2}$ $\frac{7}{7}$ $\frac{7}{2}$ 3. $4\frac{1}{6}$ $1\frac{5}{8}$ $10\frac{2}{15}$

4. $\frac{12}{4}$ $6\frac{1}{3}$ $\frac{9}{1}$ $8\frac{1}{6}$

EXERCISE 3

1. $\frac{1}{5}$ $\frac{1}{6}$ $\frac{1}{10}$ $\frac{1}{3}$ $\frac{1}{5}$

$\frac{4 \div 4}{20 \div 4} = \frac{1}{5}$ $\frac{7 \div 7}{42 \div 7} = \frac{1}{6}$ $\frac{3 \div 3}{30 \div 3} = \frac{1}{10}$ $\frac{6 \div 6}{18 \div 6} = \frac{1}{3}$ $\frac{9 \div 9}{45 \div 9} = \frac{1}{5}$

$\frac{1}{7}$

$\frac{8 \div 8}{56 \div 8} = \frac{1}{7}$

2. $\frac{7}{9}$ $\frac{8}{13}$ $\frac{5}{6}$ $\frac{12}{17}$ $\frac{3}{4}$

$\frac{35 \div 5}{45 \div 5} = \frac{7}{9}$ $\frac{40 \div 5}{65 \div 5} = \frac{8}{13}$ $\frac{25 \div 5}{30 \div 5} = \frac{5}{6}$ $\frac{60 \div 5}{85 \div 5} = \frac{12}{17}$ $\frac{15 \div 5}{20 \div 5} = \frac{3}{4}$

$\frac{6}{11}$

$\frac{30 \div 5}{55 \div 5} = \frac{6}{11}$

3. $\frac{1}{3}$ $\frac{2}{3}$ $\frac{1}{6}$ $\frac{1}{2}$ $\frac{1}{2}$

$\frac{24 \div 24}{72 \div 24} = \frac{1}{3}$ $\frac{26 \div 13}{39 \div 13} = \frac{2}{3}$ $\frac{14 \div 14}{84 \div 14} = \frac{1}{6}$ $\frac{27 \div 27}{54 \div 27} = \frac{1}{2}$ $\frac{19 \div 19}{38 \div 19} = \frac{1}{2}$

$\frac{3}{5}$

$\frac{45 \div 15}{75 \div 15} = \frac{3}{5}$

4. $\frac{3}{10}$ $\frac{8}{15}$ $\frac{3}{4}$ $\frac{4}{15}$

$\frac{60 \div 20}{200 \div 20} = \frac{3}{10}$ $\frac{32 \div 4}{60 \div 4} = \frac{8}{15}$ $\frac{48 \div 16}{64 \div 16} = \frac{3}{4}$ $\frac{80 \div 20}{300 \div 20} = \frac{4}{15}$

$\frac{11}{18}$ $\frac{1}{50}$

$\frac{22 \div 2}{36 \div 2} = \frac{11}{18}$ $\frac{20 \div 20}{1{,}000 \div 20} = \frac{1}{50}$

5. $\frac{5}{12}$ $\frac{7}{33}$ $\frac{3}{5}$ $\frac{3}{5}$

$\frac{60 \div 12}{144 \div 12} = \frac{5}{12}$ $\frac{70 \div 10}{330 \div 10} = \frac{7}{33}$ $\frac{33 \div 11}{55 \div 11} = \frac{3}{5}$ $\frac{24 \div 8}{40 \div 8} = \frac{3}{5}$

$\frac{2}{3}$ $\frac{1}{5}$

$\frac{48 \div 24}{72 \div 24} = \frac{2}{3}$ $\frac{14 \div 14}{70 \div 14} = \frac{1}{5}$

EXERCISE 4

1. $\frac{8}{12}$ $\frac{21}{35}$ $\frac{20}{45}$ $\frac{10}{16}$ $\frac{8}{32}$

$\frac{2 \times 4}{3 \times 4} = \frac{8}{12}$ $\frac{3 \times 7}{5 \times 7} = \frac{21}{35}$ $\frac{4 \times 5}{9 \times 5} = \frac{20}{45}$ $\frac{5 \times 2}{8 \times 2} = \frac{10}{16}$ $\frac{1 \times 8}{4 \times 8} = \frac{8}{32}$

2. $\frac{36}{42}$ $\frac{36}{40}$ $\frac{12}{36}$ $\frac{8}{60}$ $\frac{15}{18}$

$\frac{6 \times 6}{7 \times 6} = \frac{36}{42}$ $\frac{9 \times 4}{10 \times 4} = \frac{36}{40}$ $\frac{1 \times 12}{3 \times 12} = \frac{12}{36}$ $\frac{2 \times 4}{15 \times 4} = \frac{8}{60}$ $\frac{5 \times 3}{6 \times 3} = \frac{15}{18}$

3. $\frac{65}{100}$ $\frac{28}{48}$ $\frac{57}{75}$ $\frac{2}{32}$ $\frac{44}{96}$

$\frac{13 \times 5}{20 \times 5} = \frac{65}{100}$ $\frac{7 \times 4}{12 \times 4} = \frac{28}{48}$ $\frac{19 \times 3}{25 \times 3} = \frac{57}{75}$ $\frac{1 \times 2}{16 \times 2} = \frac{2}{32}$ $\frac{11 \times 4}{24 \times 4} = \frac{44}{96}$

4. $\frac{28}{200}$ $\quad$ $\frac{10}{50}$ $\quad$ $\frac{27}{72}$ $\quad$ $\frac{8}{36}$ $\quad$ $\frac{15}{36}$

$\frac{7 \times 4}{50 \times 4} = \frac{28}{200}$ $\quad$ $\frac{1 \times 10}{5 \times 10} = \frac{10}{50}$ $\quad$ $\frac{3 \times 9}{8 \times 9} = \frac{27}{72}$ $\quad$ $\frac{2 \times 4}{9 \times 4} = \frac{8}{36}$ $\quad$ $\frac{5 \times 3}{12 \times 3} = \frac{15}{36}$

EXERCISE 5

1. $5\frac{1}{3}$ $\quad$ $2\frac{2}{9}$ $\quad$ $2\frac{5}{8}$ $\quad$ $4\frac{1}{2}$ $\quad$ 2 $\quad$ $3\frac{3}{4}$ $\quad$ 7

$$\begin{array}{r} 5\frac{1}{3} \\ 3\overline{)16} \\ \underline{15} \\ 1 \end{array} \quad \begin{array}{r} 2\frac{2}{9} \\ 9\overline{)20} \\ \underline{18} \\ 2 \end{array} \quad \begin{array}{r} 2\frac{5}{8} \\ 8\overline{)21} \\ \underline{16} \\ 5 \end{array} \quad \begin{array}{r} 4\frac{1}{2} \\ 2\overline{)9} \\ \underline{8} \\ 1 \end{array} \quad \begin{array}{r} 2 \\ 7\overline{)14} \\ \underline{14} \end{array} \quad \begin{array}{r} 3\frac{3}{4} \\ 4\overline{)15} \\ \underline{12} \\ 3 \end{array} \quad \begin{array}{r} 7 \\ 5\overline{)35} \\ \underline{35} \end{array}$$

2. $4\frac{1}{2}$ $\quad$ $2\frac{2}{3}$ $\quad$ $3\frac{1}{3}$ $\quad$ $4\frac{1}{2}$

$$\begin{array}{r} 4\frac{3}{6} = 4\frac{1}{2} \\ 6\overline{)27} \\ \underline{24} \\ 3 \end{array} \quad \begin{array}{r} 2\frac{8}{12} = 2\frac{2}{3} \\ 12\overline{)32} \\ \underline{24} \\ 8 \end{array} \quad \begin{array}{r} 3\frac{3}{9} = 3\frac{1}{3} \\ 9\overline{)30} \\ \underline{27} \\ 3 \end{array} \quad \begin{array}{r} 4\frac{5}{10} = 4\frac{1}{2} \\ 10\overline{)45} \\ \underline{40} \\ 5 \end{array}$$

$5\frac{3}{4}$ $\quad$ $3\frac{1}{5}$ $\quad$ $6\frac{1}{2}$

$$\begin{array}{r} 5\frac{6}{8} = 5\frac{3}{4} \\ 8\overline{)46} \\ \underline{40} \\ 6 \end{array} \quad \begin{array}{r} 3\frac{3}{15} = 3\frac{1}{5} \\ 15\overline{)48} \\ \underline{45} \\ 3 \end{array} \quad \begin{array}{r} 6\frac{2}{4} = 6\frac{1}{2} \\ 4\overline{)26} \\ \underline{24} \\ 2 \end{array}$$

EXERCISE 6

1. $3\frac{2}{3} = \frac{11}{3}$ $\quad$ $5\frac{1}{2} = \frac{11}{2}$ $\quad$ $1\frac{7}{10} = \frac{17}{10}$ $\quad$ $2\frac{5}{6} = \frac{17}{6}$ $\quad$ $4\frac{3}{8} = \frac{35}{8}$ $\quad$ $1\frac{1}{2} = \frac{3}{2}$ $\quad$ $3\frac{5}{9} = \frac{32}{9}$

2. $6\frac{3}{4} = \frac{27}{4}$ $\quad$ $8\frac{4}{7} = \frac{60}{7}$ $\quad$ $10\frac{5}{6} = \frac{65}{6}$ $\quad$ $5\frac{2}{9} = \frac{47}{9}$ $\quad$ $2\frac{4}{5} = \frac{14}{5}$ $\quad$ $12\frac{7}{8} = \frac{103}{8}$ $\quad$ $16\frac{1}{3} = \frac{49}{3}$

EXERCISE 7

1. $\frac{7}{9}$ $\quad$ $1\frac{1}{4}$ $\quad$ $1\frac{1}{2}$ $\quad$ 7 $\quad$ $13\frac{2}{5}$

$$\begin{array}{r} \frac{5}{9} \\ +\frac{2}{9} \\ \hline \frac{7}{9} \end{array} \quad \begin{array}{r} \frac{3}{8} \\ +\frac{7}{8} \\ \hline \frac{10}{8} = 1\frac{2}{8} = 1\frac{1}{4} \end{array} \quad \begin{array}{r} \frac{3}{4} \\ +\frac{3}{4} \\ \hline \frac{6}{4} = 1\frac{2}{4} = 1\frac{1}{2} \end{array} \quad \begin{array}{r} 2\frac{5}{6} \\ +4\frac{1}{6} \\ \hline 6\frac{6}{6} = 7 \end{array} \quad \begin{array}{r} 3\frac{4}{5} \\ +9\frac{3}{5} \\ \hline 12\frac{7}{5} = 13\frac{2}{5} \end{array}$$

$\mathbf{8\frac{1}{2}}$

$$\begin{array}{r} 6\frac{7}{12} \\ +1\frac{11}{12} \\ \hline 7\frac{18}{12} = 8\frac{6}{12} = \mathbf{8\frac{1}{2}} \end{array}$$

2. $\mathbf{1\frac{3}{5}}$

$$\begin{array}{r} \frac{7}{10} \\ +\frac{9}{10} \\ \hline \frac{16}{10} = 1\frac{6}{10} = \mathbf{1\frac{3}{5}} \end{array}$$

$\mathbf{\frac{7}{8}}$

$$\begin{array}{r} \frac{11}{16} \\ +\frac{3}{16} \\ \hline \frac{14}{16} = \mathbf{\frac{7}{8}} \end{array}$$

$\mathbf{\frac{6}{7}}$

$$\begin{array}{r} \frac{2}{7} \\ +\frac{4}{7} \\ \hline \mathbf{\frac{6}{7}} \end{array}$$

$\mathbf{11\frac{8}{9}}$

$$\begin{array}{r} 8\frac{5}{18} \\ +3\frac{11}{18} \\ \hline 11\frac{16}{18} = \mathbf{11\frac{8}{9}} \end{array}$$

$\mathbf{12\frac{3}{10}}$

$$\begin{array}{r} 2\frac{9}{20} \\ +9\frac{17}{20} \\ \hline 11\frac{26}{20} = 12\frac{6}{20} = \mathbf{12\frac{3}{10}} \end{array}$$

$\mathbf{15\frac{11}{18}}$

$$\begin{array}{r} 10\frac{23}{36} \\ +\ 4\frac{35}{36} \\ \hline 14\frac{58}{36} = 15\frac{22}{36} = \mathbf{15\frac{11}{18}} \end{array}$$

EXERCISE 8

1. $\mathbf{1\frac{1}{6}}$

$$\begin{array}{r} \frac{3}{4} = \frac{9}{12} \\ +\frac{5}{12} = \frac{5}{12} \\ \hline \frac{14}{12} = 1\frac{2}{12} = \mathbf{1\frac{1}{6}} \end{array}$$

$\mathbf{1\frac{1}{9}}$

$$\begin{array}{r} \frac{7}{9} = \frac{7}{9} \\ +\frac{1}{3} = \frac{3}{9} \\ \hline \frac{10}{9} = \mathbf{1\frac{1}{9}} \end{array}$$

$\mathbf{1\frac{4}{15}}$

$$\begin{array}{r} \frac{2}{5} = \frac{6}{15} \\ +\frac{13}{15} = \frac{13}{15} \\ \hline \frac{19}{15} = \mathbf{1\frac{4}{15}} \end{array}$$

$\mathbf{7\frac{9}{10}}$

$$\begin{array}{r} 4\frac{3}{20} = 4\frac{3}{20} \\ +3\frac{3}{4} = 3\frac{15}{20} \\ \hline 7\frac{18}{20} = \mathbf{7\frac{9}{10}} \end{array}$$

$\mathbf{11\frac{7}{12}}$

$$\begin{array}{r} 2\frac{5}{24} = 2\frac{5}{24} \\ +9\frac{3}{8} = 9\frac{9}{24} \\ \hline 11\frac{14}{24} = \mathbf{11\frac{7}{12}} \end{array}$$

2. $\mathbf{\frac{17}{20}}$

$$\begin{array}{r} \frac{3}{5} = \frac{12}{20} \\ +\frac{1}{4} = \frac{5}{20} \\ \hline \mathbf{\frac{17}{20}} \end{array}$$

$\mathbf{\frac{25}{63}}$

$$\begin{array}{r} \frac{1}{9} = \frac{7}{63} \\ +\frac{2}{7} = \frac{18}{63} \\ \hline \mathbf{\frac{25}{63}} \end{array}$$

$\mathbf{1\frac{13}{24}}$

$$\begin{array}{r} \frac{2}{3} = \frac{16}{24} \\ +\frac{7}{8} = \frac{21}{24} \\ \hline \frac{37}{24} = \mathbf{1\frac{13}{24}} \end{array}$$

$\mathbf{16\frac{23}{30}}$

$$\begin{array}{r} 8\frac{4}{15} = 8\frac{8}{30} \\ +8\frac{1}{2} = 8\frac{15}{30} \\ \hline \mathbf{16\frac{23}{30}} \end{array}$$

$\mathbf{12\frac{19}{36}}$

$$\begin{array}{r} 5\frac{3}{4} = 5\frac{27}{36} \\ +6\frac{7}{9} = 6\frac{28}{36} \\ \hline 11\frac{55}{36} = \mathbf{12\frac{19}{36}} \end{array}$$

3. $\mathbf{1\frac{9}{20}}$ $\qquad$ $\mathbf{1\frac{17}{24}}$ $\qquad$ $\mathbf{1\frac{37}{48}}$

$$\begin{array}{r} \frac{7}{10} = \frac{14}{20} \\ \frac{1}{2} = \frac{10}{20} \\ +\frac{1}{4} = \frac{5}{20} \\ \hline \frac{29}{20} = \mathbf{1\frac{9}{20}} \end{array} \qquad \begin{array}{r} \frac{5}{12} = \frac{10}{24} \\ \frac{5}{8} = \frac{15}{24} \\ +\frac{2}{3} = \frac{16}{24} \\ \hline \frac{41}{24} = \mathbf{1\frac{17}{24}} \end{array} \qquad \begin{array}{r} \frac{5}{6} = \frac{40}{48} \\ \frac{3}{8} = \frac{18}{48} \\ +\frac{9}{16} = \frac{27}{48} \\ \hline \frac{85}{48} = \mathbf{1\frac{37}{48}} \end{array}$$

$\mathbf{13\frac{2}{9}}$ $\qquad$ $\mathbf{14\frac{7}{20}}$

$$\begin{array}{r} 6\frac{1}{2} = 6\frac{9}{18} \\ 2\frac{5}{9} = 2\frac{10}{18} \\ +4\frac{1}{6} = 4\frac{3}{18} \\ \hline 12\frac{22}{18} = 13\frac{4}{18} = \mathbf{13\frac{2}{9}} \end{array} \qquad \begin{array}{r} 3\frac{4}{15} = 3\frac{16}{60} \\ 1\frac{1}{4} = 1\frac{15}{60} \\ +9\frac{5}{6} = 9\frac{50}{60} \\ \hline 13\frac{81}{60} = 14\frac{21}{60} = \mathbf{14\frac{7}{20}} \end{array}$$

EXERCISE 9

1. $\mathbf{\frac{3}{7}}$ $\qquad$ $\mathbf{\frac{1}{2}}$ $\qquad$ $\mathbf{\frac{3}{4}}$ $\qquad$ $\mathbf{6\frac{2}{3}}$ $\qquad$ $\mathbf{6\frac{3}{5}}$ $\qquad$ $\mathbf{4\frac{1}{4}}$

$$\begin{array}{r} \frac{4}{7} \\ -\frac{1}{7} \\ \hline \mathbf{\frac{3}{7}} \end{array} \qquad \begin{array}{r} \frac{11}{12} \\ -\frac{5}{12} \\ \hline \frac{6}{12} = \mathbf{\frac{1}{2}} \end{array} \qquad \begin{array}{r} \frac{7}{8} \\ -\frac{1}{8} \\ \hline \frac{6}{8} = \mathbf{\frac{3}{4}} \end{array} \qquad \begin{array}{r} 8\frac{5}{6} \\ -2\frac{1}{6} \\ \hline 6\frac{4}{6} = \mathbf{6\frac{2}{3}} \end{array} \qquad \begin{array}{r} 7\frac{9}{10} \\ -1\frac{3}{10} \\ \hline 6\frac{6}{10} = \mathbf{6\frac{3}{5}} \end{array} \qquad \begin{array}{r} 6\frac{13}{16} \\ -2\frac{9}{16} \\ \hline 4\frac{4}{16} = \mathbf{4\frac{1}{4}} \end{array}$$

2. $\mathbf{\frac{1}{3}}$ $\qquad$ $\mathbf{\frac{1}{5}}$ $\qquad$ $\mathbf{\frac{2}{3}}$ $\qquad$ $\mathbf{2\frac{1}{2}}$ $\qquad$ $\mathbf{3\frac{2}{5}}$ $\qquad$ $\mathbf{6\frac{1}{9}}$

$$\begin{array}{r} \frac{5}{9} \\ -\frac{2}{9} \\ \hline \frac{3}{9} = \mathbf{\frac{1}{3}} \end{array} \qquad \begin{array}{r} \frac{4}{5} \\ -\frac{3}{5} \\ \hline \mathbf{\frac{1}{5}} \end{array} \qquad \begin{array}{r} \frac{17}{18} \\ -\frac{5}{18} \\ \hline \frac{12}{18} = \mathbf{\frac{2}{3}} \end{array} \qquad \begin{array}{r} 5\frac{3}{4} \\ -3\frac{1}{4} \\ \hline 2\frac{2}{4} = \mathbf{2\frac{1}{2}} \end{array} \qquad \begin{array}{r} 4\frac{19}{20} \\ -1\frac{11}{20} \\ \hline 3\frac{8}{20} = \mathbf{3\frac{2}{5}} \end{array} \qquad \begin{array}{r} 8\frac{11}{36} \\ -2\frac{7}{36} \\ \hline 6\frac{4}{36} = \mathbf{6\frac{1}{9}} \end{array}$$

EXERCISE 10

1. $\mathbf{\frac{5}{8}}$ $\qquad$ $\mathbf{\frac{5}{12}}$ $\qquad$ $\mathbf{\frac{11}{20}}$ $\qquad$ $\mathbf{7\frac{5}{14}}$ $\qquad$ $\mathbf{3\frac{2}{9}}$

$$\begin{array}{r} \frac{3}{4} = \frac{6}{8} \\ -\frac{1}{8} = \frac{1}{8} \\ \hline \mathbf{\frac{5}{8}} \end{array} \qquad \begin{array}{r} \frac{11}{12} = \frac{11}{12} \\ -\frac{1}{2} = \frac{6}{12} \\ \hline \mathbf{\frac{5}{12}} \end{array} \qquad \begin{array}{r} \frac{4}{5} = \frac{16}{20} \\ -\frac{1}{4} = \frac{5}{20} \\ \hline \mathbf{\frac{11}{20}} \end{array} \qquad \begin{array}{r} 9\frac{6}{7} = 9\frac{12}{14} \\ -2\frac{1}{2} = 2\frac{7}{14} \\ \hline \mathbf{7\frac{5}{14}} \end{array} \qquad \begin{array}{r} 8\frac{5}{9} = 8\frac{5}{9} \\ -5\frac{1}{3} = 5\frac{3}{9} \\ \hline \mathbf{3\frac{2}{9}} \end{array}$$

2. $\mathbf{\frac{5}{18}}$ $\quad$ $\mathbf{\frac{7}{12}}$ $\quad$ $\mathbf{\frac{11}{24}}$ $\quad$ $\mathbf{3\frac{17}{30}}$ $\quad$ $\mathbf{4\frac{17}{40}}$

$$\begin{array}{r} \frac{1}{2} = \frac{9}{18} \\ -\frac{2}{9} = \frac{4}{18} \\ \hline \mathbf{\frac{5}{18}} \end{array} \qquad \begin{array}{r} \frac{3}{4} = \frac{9}{12} \\ -\frac{1}{6} = \frac{2}{12} \\ \hline \mathbf{\frac{7}{12}} \end{array} \qquad \begin{array}{r} \frac{7}{8} = \frac{21}{24} \\ -\frac{5}{12} = \frac{10}{24} \\ \hline \mathbf{\frac{11}{24}} \end{array} \qquad \begin{array}{r} 7\frac{11}{15} = 7\frac{22}{30} \\ -4\frac{1}{6} = 4\frac{5}{30} \\ \hline \mathbf{3\frac{17}{30}} \end{array} \qquad \begin{array}{r} 5\frac{4}{5} = 5\frac{32}{40} \\ -1\frac{3}{8} = 1\frac{15}{40} \\ \hline \mathbf{4\frac{17}{40}} \end{array}$$

EXERCISE 11

1. $\mathbf{2\frac{5}{8}}$ $\quad$ $\mathbf{\frac{2}{7}}$ $\quad$ $\mathbf{1\frac{1}{3}}$ $\quad$ $\mathbf{7\frac{2}{9}}$ $\quad$ $\mathbf{3\frac{7}{10}}$

$$\begin{array}{r} \overset{4}{\cancel{5}}\frac{8}{8} \\ -2\frac{3}{8} \\ \hline \mathbf{2\frac{5}{8}} \end{array} \qquad \begin{array}{r} \overset{3}{\cancel{4}}\frac{7}{7} \\ -3\frac{5}{7} \\ \hline \mathbf{\frac{2}{7}} \end{array} \qquad \begin{array}{r} \overset{8}{\cancel{9}}\frac{3}{3} \\ -7\frac{2}{3} \\ \hline \mathbf{1\frac{1}{3}} \end{array} \qquad \begin{array}{r} \overset{11}{\cancel{12}}\frac{9}{9} \\ -\ 4\frac{7}{9} \\ \hline \mathbf{7\frac{2}{9}} \end{array} \qquad \begin{array}{r} \overset{5}{\cancel{6}}\frac{10}{10} \\ -2\frac{3}{10} \\ \hline \mathbf{3\frac{7}{10}} \end{array}$$

2. $\mathbf{7\frac{7}{12}}$ $\quad$ $\mathbf{1\frac{5}{16}}$ $\quad$ $\mathbf{2\frac{3}{5}}$ $\quad$ $\mathbf{2\frac{1}{2}}$ $\quad$ $\mathbf{5\frac{1}{4}}$

$$\begin{array}{r} \overset{10}{\cancel{11}}\frac{12}{12} \\ -\ 3\frac{5}{12} \\ \hline \mathbf{7\frac{7}{12}} \end{array} \qquad \begin{array}{r} \overset{7}{\cancel{8}}\frac{16}{16} \\ -6\frac{11}{16} \\ \hline \mathbf{1\frac{5}{16}} \end{array} \qquad \begin{array}{r} \overset{5}{\cancel{6}}\frac{5}{5} \\ -3\frac{2}{5} \\ \hline \mathbf{2\frac{3}{5}} \end{array} \qquad \begin{array}{r} \overset{6}{\cancel{7}}\frac{2}{2} \\ -4\frac{1}{2} \\ \hline \mathbf{2\frac{1}{2}} \end{array} \qquad \begin{array}{r} \overset{14}{\cancel{15}}\frac{4}{4} \\ -\ 9\frac{3}{4} \\ \hline \mathbf{5\frac{1}{4}} \end{array}$$

3. $\mathbf{2\frac{1}{2}}$ $\qquad$ $\mathbf{3\frac{1}{3}}$

$$\begin{array}{r} 5\frac{3}{8} = 4\frac{3}{8} + \frac{8}{8} = 4\frac{11}{8} \\ -2\frac{7}{8} \qquad\qquad = 2\frac{7}{8} \\ \hline 2\frac{4}{8} = \mathbf{2\frac{1}{2}} \end{array} \qquad \begin{array}{r} 7\frac{1}{6} = 6\frac{1}{6} + \frac{6}{6} = 6\frac{7}{6} \\ -3\frac{5}{6} \qquad\qquad = 3\frac{5}{6} \\ \hline 3\frac{2}{6} = \mathbf{3\frac{1}{3}} \end{array}$$

$\mathbf{5\frac{3}{5}}$ $\qquad$ $\mathbf{\frac{5}{9}}$ $\qquad$ $\mathbf{\frac{2}{5}}$

$$\begin{array}{r} 10\frac{2}{5} = 9\frac{2}{5} + \frac{5}{5} = 9\frac{7}{5} \\ -\ 4\frac{4}{5} \qquad\qquad = 4\frac{4}{5} \\ \hline \mathbf{5\frac{3}{5}} \end{array} \qquad \begin{array}{r} 6\frac{4}{9} = 5\frac{4}{9} + \frac{9}{9} = 5\frac{13}{9} \\ -5\frac{8}{9} \qquad\qquad = 5\frac{8}{9} \\ \hline \mathbf{\frac{5}{9}} \end{array} \qquad \begin{array}{r} 2\frac{3}{10} = 1\frac{3}{10} + \frac{10}{10} = 1\frac{13}{10} \\ -1\frac{9}{10} \qquad\qquad = 1\frac{9}{10} \\ \hline \frac{4}{10} = \mathbf{\frac{2}{5}} \end{array}$$

4. $\mathbf{1\frac{3}{7}}$ $\qquad$ $\mathbf{3\frac{1}{2}}$ $\qquad$ $\mathbf{6\frac{2}{3}}$

$$\begin{array}{r} 8\frac{2}{7} = 7\frac{2}{7} + \frac{7}{7} = 7\frac{9}{7} \\ -6\frac{6}{7} \qquad\qquad = 6\frac{6}{7} \\ \hline \mathbf{1\frac{3}{7}} \end{array} \qquad \begin{array}{r} 6\frac{5}{12} = 5\frac{5}{12} + \frac{12}{12} = 5\frac{17}{12} \\ -2\frac{11}{12} \qquad\qquad = 2\frac{11}{12} \\ \hline 3\frac{6}{12} = \mathbf{3\frac{1}{2}} \end{array} \qquad \begin{array}{r} 11\frac{1}{3} = 10\frac{1}{3} + \frac{3}{3} = 10\frac{4}{3} \\ -\ 4\frac{2}{3} \qquad\qquad = 4\frac{2}{3} \\ \hline \mathbf{6\frac{2}{3}} \end{array}$$

$$\mathbf{8\tfrac{1}{2}}$$

$$\begin{array}{r} 12\frac{1}{4} = 11\frac{1}{4} + \frac{4}{4} = 11\frac{5}{4} \\ -\ 3\frac{3}{4} \qquad\qquad = \ 3\frac{3}{4} \\ \hline 8\frac{2}{4} = \mathbf{8\frac{1}{2}} \end{array}$$

$$\mathbf{2\tfrac{1}{2}}$$

$$\begin{array}{r} 4\frac{7}{16} = 3\frac{7}{16} + \frac{16}{16} = 3\frac{23}{16} \\ -1\frac{15}{16} \qquad\qquad = 1\frac{15}{16} \\ \hline 2\frac{8}{16} = \mathbf{2\frac{1}{2}} \end{array}$$

EXERCISE 12

1.

$$\mathbf{4\tfrac{13}{20}}$$

$$\begin{array}{r} 8\frac{2}{5} = 8\frac{8}{20} = 7\frac{8}{20} + \frac{20}{20} = 7\frac{28}{20} \\ -3\frac{3}{4} = 3\frac{15}{20} \qquad\qquad = 3\frac{15}{20} \\ \hline \mathbf{4\frac{13}{20}} \end{array}$$

$$\mathbf{3\tfrac{5}{8}}$$

$$\begin{array}{r} 5\frac{1}{2} = 5\frac{4}{8} = 4\frac{4}{8} + \frac{8}{8} = 4\frac{12}{8} \\ -1\frac{7}{8} = 1\frac{7}{8} \qquad\qquad = 1\frac{7}{8} \\ \hline \mathbf{3\frac{5}{8}} \end{array}$$

$$\mathbf{3\tfrac{16}{21}}$$

$$\begin{array}{r} 6\frac{3}{7} = 6\frac{9}{21} = 5\frac{9}{21} + \frac{21}{21} = 5\frac{30}{21} \\ -2\frac{2}{3} = 2\frac{14}{21} \qquad\qquad = 2\frac{14}{21} \\ \hline \mathbf{3\frac{16}{21}} \end{array}$$

$$\mathbf{6\tfrac{7}{18}}$$

$$\begin{array}{r} 10\frac{2}{9} = 10\frac{4}{18} = 9\frac{4}{18} + \frac{18}{18} = 9\frac{22}{18} \\ -\ 3\frac{5}{6} = \ 3\frac{15}{18} \qquad\qquad = 3\frac{15}{18} \\ \hline \mathbf{6\frac{7}{18}} \end{array}$$

$$\mathbf{1\tfrac{7}{12}}$$

$$\begin{array}{r} 4\frac{1}{3} = 4\frac{4}{12} = 3\frac{4}{12} + \frac{12}{12} = 3\frac{16}{12} \\ -2\frac{3}{4} = 2\frac{9}{12} \qquad\qquad = 2\frac{9}{12} \\ \hline \mathbf{1\frac{7}{12}} \end{array}$$

2.

$$\mathbf{1\tfrac{23}{30}}$$

$$\begin{array}{r} 9\frac{1}{2} = 9\frac{15}{30} = 8\frac{15}{30} + \frac{30}{30} = 8\frac{45}{30} \\ -7\frac{11}{15} = 7\frac{22}{30} \qquad\qquad = 7\frac{22}{30} \\ \hline \mathbf{1\frac{23}{30}} \end{array}$$

$$\mathbf{5\tfrac{5}{6}}$$

$$\begin{array}{r} 11\frac{1}{2} = 11\frac{3}{6} = 10\frac{3}{6} + \frac{6}{6} = 10\frac{9}{6} \\ -\ 5\frac{2}{3} = \ 5\frac{4}{6} \qquad\qquad = \ 5\frac{4}{6} \\ \hline \mathbf{5\frac{5}{6}} \end{array}$$

$$\mathbf{2\tfrac{19}{30}}$$

$$\begin{array}{r} 7\frac{3}{10} = 7\frac{9}{30} = 6\frac{9}{30} + \frac{30}{30} = 6\frac{39}{30} \\ -4\frac{2}{3} = 4\frac{20}{30} \qquad\qquad = 4\frac{20}{30} \\ \hline \mathbf{2\frac{19}{30}} \end{array}$$

$$\mathbf{3\tfrac{7}{24}}$$

$$\begin{array}{r} 12\frac{1}{6} = 12\frac{4}{24} = 11\frac{4}{24} + \frac{24}{24} = 11\frac{28}{24} \\ -\ 8\frac{7}{8} = \ 8\frac{21}{24} \qquad\qquad = \ 8\frac{21}{24} \\ \hline \mathbf{3\frac{7}{24}} \end{array}$$

$\mathbf{8\frac{7}{9}}$

$$\begin{array}{rcl} 15\frac{4}{9} = 15\frac{4}{9} = 14\frac{4}{9} + \frac{9}{9} & = & 14\frac{13}{9} \\ -\ 6\frac{2}{3} = \ \ 6\frac{6}{9} \qquad\qquad\quad & = & 6\frac{6}{9} \\ \hline & & \mathbf{8\frac{7}{9}} \end{array}$$

EXERCISE 13

1. $\mathbf{\frac{12}{25}}$ $\quad \frac{4}{5} \times \frac{3}{5} = \mathbf{\frac{12}{25}}$

   $\mathbf{\frac{10}{21}}$ $\quad \frac{5}{7} \times \frac{2}{3} = \mathbf{\frac{10}{21}}$

   $\mathbf{\frac{21}{64}}$ $\quad \frac{3}{8} \times \frac{7}{8} = \mathbf{\frac{21}{64}}$

   $\mathbf{\frac{9}{20}}$ $\quad \frac{1}{2} \times \frac{9}{10} = \mathbf{\frac{9}{20}}$

   $\mathbf{\frac{4}{27}}$ $\quad \frac{2}{3} \times \frac{2}{9} = \mathbf{\frac{4}{27}}$

2. $\mathbf{\frac{18}{77}}$ $\quad \frac{3}{7} \times \frac{6}{11} = \mathbf{\frac{18}{77}}$

   $\mathbf{\frac{15}{32}}$ $\quad \frac{3}{4} \times \frac{5}{8} = \mathbf{\frac{15}{32}}$

   $\mathbf{\frac{8}{45}}$ $\quad \frac{4}{9} \times \frac{2}{5} = \mathbf{\frac{8}{45}}$

   $\mathbf{\frac{8}{27}}$ $\quad \frac{8}{9} \times \frac{1}{3} = \mathbf{\frac{8}{27}}$

   $\mathbf{\frac{20}{63}}$ $\quad \frac{5}{7} \times \frac{4}{9} = \mathbf{\frac{20}{63}}$

EXERCISE 14

1. $\mathbf{\frac{11}{24}}$ $\quad \frac{\overset{1}{\cancel{5}}}{8} \times \frac{11}{\underset{3}{\cancel{15}}} = \mathbf{\frac{11}{24}}$

   $\mathbf{\frac{5}{12}}$ $\quad \frac{\overset{1}{\cancel{4}}}{\underset{3}{\cancel{9}}} \times \frac{\overset{5}{\cancel{15}}}{\underset{4}{\cancel{16}}} = \mathbf{\frac{5}{12}}$

   $\mathbf{\frac{3}{5}}$ $\quad \frac{\overset{1}{\cancel{2}}}{\underset{1}{\cancel{3}}} \times \frac{\overset{3}{\cancel{9}}}{\underset{5}{\cancel{10}}} = \mathbf{\frac{3}{5}}$

   $\mathbf{\frac{2}{7}}$ $\quad \frac{\overset{1}{\cancel{3}}}{\underset{1}{\cancel{10}}} \times \frac{\overset{2}{\cancel{20}}}{\underset{7}{\cancel{21}}} = \mathbf{\frac{2}{7}}$

   $\mathbf{\frac{3}{10}}$ $\quad \frac{\overset{3}{\cancel{9}}}{\underset{2}{\cancel{16}}} \times \frac{\overset{1}{\cancel{8}}}{\underset{5}{\cancel{15}}} = \mathbf{\frac{3}{10}}$

2. $\mathbf{\frac{1}{16}}$ $\quad \frac{\overset{1}{\cancel{7}}}{\underset{8}{\cancel{24}}} \times \frac{\overset{1}{\cancel{3}}}{\underset{2}{\cancel{14}}} = \mathbf{\frac{1}{16}}$

   $\mathbf{\frac{9}{25}}$ $\quad \frac{9}{\underset{5}{\cancel{20}}} \times \frac{\overset{1}{\cancel{4}}}{5} = \mathbf{\frac{9}{25}}$

   $\mathbf{\frac{1}{16}}$ $\quad \frac{\overset{1}{\cancel{5}}}{\underset{4}{\cancel{12}}} \times \frac{\overset{1}{\cancel{3}}}{\underset{4}{\cancel{20}}} = \mathbf{\frac{1}{16}}$

   $\mathbf{\frac{8}{27}}$ $\quad \frac{\overset{1}{\cancel{7}}}{\underset{9}{\cancel{18}}} \times \frac{\overset{8}{\cancel{16}}}{\underset{3}{\cancel{21}}} = \mathbf{\frac{8}{27}}$

   $\mathbf{\frac{2}{5}}$ $\quad \frac{\overset{2}{\cancel{8}}}{\underset{5}{\cancel{15}}} \times \frac{\overset{1}{\cancel{3}}}{\underset{1}{\cancel{4}}} = \mathbf{\frac{2}{5}}$

3. $\mathbf{\frac{5}{14}}$ $\quad \frac{5}{\underset{\underset{1}{\cancel{3}}}{\cancel{9}}} \times \frac{\overset{\overset{1}{\cancel{2}}}{\cancel{6}}}{7} \times \frac{\overset{1}{\cancel{3}}}{\underset{2}{\cancel{4}}} = \mathbf{\frac{5}{14}}$

   $\mathbf{\frac{3}{25}}$ $\quad \frac{1}{\underset{1}{\cancel{2}}} \times \frac{\overset{3}{\cancel{9}}}{\underset{5}{\cancel{10}}} \times \frac{\overset{\overset{1}{\cancel{2}}}{\cancel{4}}}{\underset{5}{\cancel{15}}} = \mathbf{\frac{3}{25}}$

   $\mathbf{\frac{5}{18}}$ $\quad \frac{5}{\underset{2}{\cancel{6}}} \times \frac{\overset{1}{\cancel{4}}}{9} \times \frac{\overset{1}{\cancel{3}}}{\underset{1}{\cancel{4}}} = \mathbf{\frac{5}{18}}$

   $\mathbf{\frac{2}{15}}$ $\quad \frac{\overset{1}{\cancel{3}}}{\underset{\underset{1}{\cancel{2}}}{\cancel{16}}} \times \frac{\overset{1}{\cancel{8}}}{\underset{3}{\cancel{9}}} \times \frac{\overset{2}{\cancel{4}}}{5} = \mathbf{\frac{2}{15}}$

4. $\mathbf{\frac{2}{7}}$ $\quad \frac{\overset{1}{\cancel{3}}}{\underset{1}{\cancel{4}}} \times \frac{1}{\underset{1}{\cancel{2}}} \times \frac{\overset{\overset{2}{\cancel{4}}}{\cancel{16}}}{\underset{7}{\cancel{21}}} = \mathbf{\frac{2}{7}}$

   $\mathbf{\frac{1}{12}}$ $\quad \frac{\overset{1}{\cancel{8}}}{\underset{3}{\cancel{15}}} \times \frac{1}{4} \times \frac{\overset{1}{\cancel{5}}}{\underset{1}{\cancel{8}}} = \mathbf{\frac{1}{12}}$

   $\mathbf{\frac{1}{15}}$ $\quad \frac{\overset{1}{\cancel{3}}}{\underset{\underset{1}{\cancel{4}}}{\cancel{8}}} \times \frac{\overset{1}{\cancel{2}}}{\underset{3}{\cancel{9}}} \times \frac{\overset{1}{\cancel{4}}}{5} = \mathbf{\frac{1}{15}}$

   $\mathbf{\frac{1}{4}}$ $\quad \frac{\overset{1}{\cancel{2}}}{\underset{1}{\cancel{3}}} \times \frac{\overset{\overset{1}{\cancel{3}}}{\cancel{9}}}{\underset{4}{\cancel{20}}} \times \frac{\overset{1}{\cancel{5}}}{\underset{\underset{1}{\cancel{2}}}{\cancel{6}}} = \mathbf{\frac{1}{4}}$

EXERCISE 15

1.

4

$\frac{2}{3} \times 6 =$

$\frac{2}{\underset{1}{\cancel{3}}} \times \frac{\overset{2}{\cancel{6}}}{1} = \frac{4}{1} = \mathbf{4}$

$\mathbf{7\frac{1}{2}}$

$\frac{5}{6} \times 9 =$

$\frac{5}{\underset{2}{\cancel{6}}} \times \frac{\overset{3}{\cancel{9}}}{1} = \frac{15}{2} = \mathbf{7\frac{1}{2}}$

$\mathbf{1\frac{1}{2}}$

$\frac{3}{10} \times 5 =$

$\frac{3}{\underset{2}{\cancel{10}}} \times \frac{\overset{1}{\cancel{5}}}{1} = \frac{3}{2} = \mathbf{1\frac{1}{2}}$

$\mathbf{\frac{3}{4}}$

$\frac{1}{4} \times 3 =$

$\frac{1}{4} \times \frac{3}{1} = \mathbf{\frac{3}{4}}$

$\mathbf{6\frac{2}{3}}$

$\frac{5}{9} \times 12 =$

$\frac{5}{\underset{3}{\cancel{9}}} \times \frac{\overset{4}{\cancel{12}}}{1} = \frac{20}{3} = \mathbf{6\frac{2}{3}}$

2.

$\mathbf{3\frac{1}{3}}$

$4 \times \frac{5}{6} =$

$\frac{\overset{2}{\cancel{4}}}{1} \times \frac{5}{\underset{3}{\cancel{6}}} = \frac{10}{3} = \mathbf{3\frac{1}{3}}$

$\mathbf{1\frac{1}{5}}$

$2 \times \frac{3}{5} =$

$\frac{2}{1} \times \frac{3}{5} = \frac{6}{5} = \mathbf{1\frac{1}{5}}$

$\mathbf{4\frac{2}{3}}$

$8 \times \frac{7}{12} =$

$\frac{\overset{2}{\cancel{8}}}{1} \times \frac{7}{\underset{3}{\cancel{12}}} = \frac{14}{3} = \mathbf{4\frac{2}{3}}$

$\mathbf{7\frac{1}{3}}$

$11 \times \frac{2}{3} =$

$\frac{11}{1} \times \frac{2}{3} = \frac{22}{3} = \mathbf{7\frac{1}{3}}$

$\mathbf{4\frac{1}{2}}$

$7 \times \frac{9}{14} =$

$\frac{\overset{1}{\cancel{7}}}{1} \times \frac{9}{\underset{2}{\cancel{14}}} = \frac{9}{2} = \mathbf{4\frac{1}{2}}$

3.

$\mathbf{1\frac{2}{3}}$

$2\frac{2}{3} \times \frac{5}{8} =$

$\frac{\overset{1}{\cancel{8}}}{3} \times \frac{5}{\underset{1}{\cancel{8}}} = \frac{5}{3} = \mathbf{1\frac{2}{3}}$

$\mathbf{1\frac{1}{6}}$

$1\frac{5}{9} \times \frac{3}{4} =$

$\frac{\overset{7}{\cancel{14}}}{\underset{3}{\cancel{9}}} \times \frac{\overset{1}{\cancel{3}}}{\underset{2}{\cancel{4}}} = \frac{7}{6} = \mathbf{1\frac{1}{6}}$

2

$2\frac{6}{7} \times \frac{7}{10} =$

$\frac{\overset{2}{\cancel{20}}}{\underset{1}{\cancel{7}}} \times \frac{\overset{1}{\cancel{7}}}{\underset{1}{\cancel{10}}} = \frac{2}{1} = \mathbf{2}$

$\mathbf{\frac{5}{9}}$

$\frac{2}{9} \times 2\frac{1}{2} =$

$\frac{\overset{1}{\cancel{2}}}{9} \times \frac{5}{\underset{1}{\cancel{2}}} = \mathbf{\frac{5}{9}}$

$\mathbf{1\frac{1}{5}}$

$\frac{8}{15} \times 2\frac{1}{4} =$

$\frac{\overset{2}{\cancel{8}}}{\underset{5}{\cancel{15}}} \times \frac{\overset{3}{\cancel{9}}}{\underset{1}{\cancel{4}}} = \frac{6}{5} = \mathbf{1\frac{1}{5}}$

4.

$\mathbf{3\frac{3}{4}}$

$$2\frac{1}{4} \times 1\frac{2}{3} =$$

$$\frac{\overset{3}{\cancel{9}}}{4} \times \frac{5}{\underset{1}{\cancel{3}}} = \frac{15}{4} = \mathbf{3\frac{3}{4}}$$

$\mathbf{3}$

$$1\frac{3}{4} \times 1\frac{5}{7} =$$

$$\frac{\overset{1}{\cancel{7}}}{\underset{1}{\cancel{4}}} \times \frac{\overset{3}{\cancel{12}}}{\underset{1}{\cancel{7}}} = \frac{3}{1} = \mathbf{3}$$

$\mathbf{4}$

$$3\frac{3}{5} \times 1\frac{1}{9} =$$

$$\frac{\overset{2}{\cancel{18}}}{\underset{1}{\cancel{5}}} \times \frac{\overset{2}{\cancel{10}}}{\underset{1}{\cancel{9}}} = \frac{4}{1} = \mathbf{4}$$

$\mathbf{8\frac{3}{4}}$

$$1\frac{5}{16} \times 6\frac{2}{3} =$$

$$\frac{\overset{7}{\cancel{21}}}{\underset{4}{\cancel{16}}} \times \frac{\overset{5}{\cancel{20}}}{\underset{1}{\cancel{3}}} = \frac{35}{4} = \mathbf{8\frac{3}{4}}$$

5.

$\mathbf{1}$

$$2\frac{2}{5} \times \frac{3}{8} \times 1\frac{1}{9} =$$

$$\frac{\overset{\overset{1}{\cancel{3}}}{\cancel{12}}}{\underset{1}{\cancel{5}}} \times \frac{\overset{1}{\cancel{3}}}{\underset{\underset{1}{\cancel{2}}}{\cancel{8}}} \times \frac{\overset{\overset{1}{\cancel{2}}}{\cancel{10}}}{\underset{\underset{1}{\cancel{3}}}{\cancel{9}}} = \frac{1}{1} = \mathbf{1}$$

$\mathbf{8}$

$$2\frac{2}{3} \times \frac{5}{6} \times 3\frac{3}{5} =$$

$$\frac{8}{\underset{1}{\cancel{3}}} \times \frac{\overset{1}{\cancel{5}}}{\underset{1}{\cancel{6}}} \times \frac{\overset{\overset{1}{\cancel{6}}}{\cancel{18}}}{\underset{1}{\cancel{5}}} = \frac{8}{1} = \mathbf{8}$$

$\mathbf{4\frac{1}{2}}$

$$1\frac{1}{6} \times 1\frac{1}{2} \times 2\frac{4}{7} =$$

$$\frac{\overset{1}{\cancel{7}}}{\underset{2}{\cancel{6}}} \times \frac{\overset{1}{\cancel{3}}}{\underset{1}{\cancel{2}}} \times \frac{\overset{9}{\cancel{18}}}{\underset{1}{\cancel{7}}} = \frac{9}{2} = \mathbf{4\frac{1}{2}}$$

EXERCISE 16

1.

$\mathbf{\frac{2}{3}}$

$$\frac{3}{5} \div \frac{9}{10} =$$

$$\frac{\overset{1}{\cancel{3}}}{\underset{1}{\cancel{5}}} \times \frac{\overset{2}{\cancel{10}}}{\underset{3}{\cancel{9}}} = \mathbf{\frac{2}{3}}$$

$\mathbf{\frac{5}{6}}$

$$\frac{4}{9} \div \frac{8}{15} =$$

$$\frac{\overset{1}{\cancel{4}}}{\underset{3}{\cancel{9}}} \times \frac{\overset{5}{\cancel{15}}}{\underset{2}{\cancel{8}}} = \mathbf{\frac{5}{6}}$$

$\mathbf{1\frac{1}{2}}$

$$\frac{2}{3} \div \frac{4}{9} =$$

$$\frac{\overset{1}{\cancel{2}}}{\underset{1}{\cancel{3}}} \times \frac{\overset{3}{\cancel{9}}}{\underset{2}{\cancel{4}}} = \frac{3}{2} = \mathbf{1\frac{1}{2}}$$

$\mathbf{1\frac{1}{9}}$

$$\frac{5}{12} \div \frac{3}{8} =$$

$$\frac{5}{\underset{3}{\cancel{12}}} \times \frac{\overset{2}{\cancel{8}}}{3} = \frac{10}{9} = \mathbf{1\frac{1}{9}}$$

$\mathbf{2\frac{1}{4}}$

$$\frac{9}{10} \div \frac{2}{5} =$$

$$\frac{9}{\underset{2}{\cancel{10}}} \times \frac{\overset{1}{\cancel{5}}}{2} = \frac{9}{4} = \mathbf{2\frac{1}{4}}$$

2. $\mathbf{1\frac{1}{2}}$

$\frac{7}{8} \div \frac{7}{12} =$

$\frac{\overset{1}{\cancel{7}}}{\underset{2}{\cancel{8}}} \times \frac{\overset{3}{\cancel{12}}}{\underset{1}{\cancel{7}}} = \frac{3}{2} = \mathbf{1\frac{1}{2}}$

$\mathbf{3\frac{1}{3}}$

$\frac{8}{15} \div \frac{4}{25} =$

$\frac{\overset{2}{\cancel{8}}}{\underset{3}{\cancel{15}}} \times \frac{\overset{5}{\cancel{25}}}{\underset{1}{\cancel{4}}} = \frac{10}{3} = \mathbf{3\frac{1}{3}}$

$\mathbf{1\frac{3}{8}}$

$\frac{3}{4} \div \frac{6}{11} =$

$\frac{\overset{1}{\cancel{3}}}{4} \times \frac{11}{\underset{2}{\cancel{6}}} = \frac{11}{8} = \mathbf{1\frac{3}{8}}$

$\mathbf{\frac{5}{14}}$

$\frac{2}{7} \div \frac{4}{5} =$

$\frac{\overset{1}{\cancel{2}}}{7} \times \frac{5}{\underset{2}{\cancel{4}}} = \mathbf{\frac{5}{14}}$

$\mathbf{\frac{13}{15}}$

$\frac{4}{5} \div \frac{12}{13} =$

$\frac{\overset{1}{\cancel{4}}}{5} \times \frac{13}{\underset{3}{\cancel{12}}} = \mathbf{\frac{13}{15}}$

EXERCISE 17

1. $\mathbf{10}$

$8 \div \frac{4}{5} =$

$\frac{\overset{2}{\cancel{8}}}{1} \times \frac{5}{\underset{1}{\cancel{4}}} = \frac{10}{1} = \mathbf{10}$

$\mathbf{3\frac{1}{2}}$

$3 \div \frac{6}{7} =$

$\frac{\overset{1}{\cancel{3}}}{1} \times \frac{7}{\underset{2}{\cancel{6}}} = \frac{7}{2} = \mathbf{3\frac{1}{2}}$

$\mathbf{15}$

$6 \div \frac{2}{5} =$

$\frac{\overset{3}{\cancel{6}}}{1} \times \frac{5}{\underset{1}{\cancel{2}}} = \frac{15}{1} = \mathbf{15}$

$\mathbf{16}$

$10 \div \frac{5}{8} =$

$\frac{\overset{2}{\cancel{10}}}{1} \times \frac{8}{\underset{1}{\cancel{5}}} = \frac{16}{1} = \mathbf{16}$

$\mathbf{2\frac{2}{3}}$

$2 \div \frac{3}{4} =$

$\frac{2}{1} \times \frac{4}{3} = \frac{8}{3} = \mathbf{2\frac{2}{3}}$

2. $\mathbf{5\frac{1}{3}}$

$5 \div \frac{15}{16} =$

$\frac{\overset{1}{\cancel{5}}}{1} \times \frac{16}{\underset{3}{\cancel{15}}} = \frac{16}{3} = \mathbf{5\frac{1}{3}}$

$\mathbf{10\frac{1}{2}}$

$9 \div \frac{6}{7} =$

$\frac{\overset{3}{\cancel{9}}}{1} \times \frac{7}{\underset{2}{\cancel{6}}} = \frac{21}{2} = \mathbf{10\frac{1}{2}}$

$\mathbf{10}$

$7 \div \frac{7}{10} =$

$\frac{\overset{1}{\cancel{7}}}{1} \times \frac{10}{\underset{1}{\cancel{7}}} = \frac{10}{1} = \mathbf{10}$

$\mathbf{5\frac{3}{7}}$

$4 \div \frac{14}{19} =$

$\frac{\overset{2}{\cancel{4}}}{1} \times \frac{19}{\underset{7}{\cancel{14}}} = \frac{38}{7} = \mathbf{5\frac{3}{7}}$

$\mathbf{6\frac{1}{2}}$

$3 \div \frac{6}{13} =$

$\frac{\overset{1}{\cancel{3}}}{1} \times \frac{13}{\underset{2}{\cancel{6}}} = \frac{13}{2} = \mathbf{6\frac{1}{2}}$

3. $\mathbf{2\frac{1}{2}}$

$1\frac{7}{8} \div \frac{3}{4} =$

$\frac{\overset{5}{\cancel{15}}}{\underset{2}{\cancel{8}}} \times \frac{\overset{1}{\cancel{4}}}{\underset{1}{\cancel{3}}} = \frac{5}{2} = \mathbf{2\frac{1}{2}}$

$\mathbf{2\frac{2}{3}}$

$2\frac{2}{9} \div \frac{5}{6} =$

$\frac{\overset{4}{\cancel{20}}}{\underset{3}{\cancel{9}}} \times \frac{\overset{2}{\cancel{6}}}{\underset{1}{\cancel{5}}} = \frac{8}{3} = \mathbf{2\frac{2}{3}}$

$\mathbf{5\frac{1}{3}}$

$2\frac{2}{3} \div \frac{1}{2} =$

$\frac{8}{3} \times \frac{2}{1} = \frac{16}{3} = \mathbf{5\frac{1}{3}}$

$\mathbf{3}$

$1\frac{5}{7} \div \frac{4}{7} =$

$\frac{\overset{3}{\cancel{12}}}{\underset{1}{\cancel{7}}} \times \frac{\overset{1}{\cancel{7}}}{\underset{1}{\cancel{4}}} = \frac{3}{1} = \mathbf{3}$

$\mathbf{10}$

$6\frac{1}{4} \div \frac{5}{8} =$

$\frac{\overset{5}{\cancel{25}}}{\underset{1}{\cancel{4}}} \times \frac{\overset{2}{\cancel{8}}}{\underset{1}{\cancel{5}}} = \frac{10}{1} = \mathbf{10}$

4. $\mathbf{4\frac{1}{2}}$

$2\frac{5}{8} \div \frac{7}{12} =$

$\frac{\overset{3}{\cancel{21}}}{\underset{2}{\cancel{8}}} \times \frac{\overset{3}{\cancel{12}}}{\underset{1}{\cancel{7}}} = \frac{9}{2} = \mathbf{4\frac{1}{2}}$

$\mathbf{8\frac{3}{4}}$

$1\frac{5}{16} \div \frac{3}{20} =$

$\frac{\overset{7}{\cancel{21}}}{\underset{4}{\cancel{16}}} \times \frac{\overset{5}{\cancel{20}}}{\underset{1}{\cancel{3}}} = \frac{35}{4} = \mathbf{8\frac{3}{4}}$

$\mathbf{2\frac{2}{3}}$

$1\frac{7}{9} \div \frac{2}{3} =$

$\frac{\overset{8}{\cancel{16}}}{\underset{3}{\cancel{9}}} \times \frac{\overset{1}{\cancel{3}}}{\underset{1}{\cancel{2}}} = \frac{8}{3} = \mathbf{2\frac{2}{3}}$

$\mathbf{13\frac{1}{2}}$

$3\frac{3}{8} \div \frac{1}{4} =$

$\frac{27}{\underset{2}{\cancel{8}}} \times \frac{\overset{1}{\cancel{4}}}{1} = \frac{27}{2} = \mathbf{13\frac{1}{2}}$

$\mathbf{5\frac{1}{3}}$

$2\frac{2}{15} \div \frac{2}{5} =$

$\frac{\overset{16}{\cancel{32}}}{\underset{3}{\cancel{15}}} \times \frac{\overset{1}{\cancel{5}}}{\underset{1}{\cancel{2}}} = \frac{16}{3} = \mathbf{5\frac{1}{3}}$

EXERCISE 18

1. $\mathbf{\frac{1}{12}}$

$\frac{2}{3} \div 8 =$

$\frac{2}{3} \div \frac{8}{1} =$

$\frac{\overset{1}{\cancel{2}}}{3} \times \frac{1}{\underset{4}{\cancel{8}}} = \mathbf{\frac{1}{12}}$

$\mathbf{\frac{3}{14}}$

$\frac{6}{7} \div 4 =$

$\frac{6}{7} \div \frac{4}{1} =$

$\frac{\overset{3}{\cancel{6}}}{7} \times \frac{1}{\underset{2}{\cancel{4}}} = \mathbf{\frac{3}{14}}$

$\mathbf{\frac{1}{24}}$

$\frac{3}{8} \div 9 =$

$\frac{3}{8} \div \frac{9}{1} =$

$\frac{\overset{1}{\cancel{3}}}{8} \times \frac{1}{\underset{3}{\cancel{9}}} = \mathbf{\frac{1}{24}}$

$\mathbf{\frac{1}{15}}$

$\frac{4}{5} \div 12 =$

$\frac{4}{5} \div \frac{12}{1} =$

$\frac{\overset{1}{\cancel{4}}}{5} \times \frac{1}{\underset{3}{\cancel{12}}} = \mathbf{\frac{1}{15}}$

$\mathbf{\frac{3}{10}}$

$\frac{9}{10} \div 3 =$

$\frac{9}{10} \div \frac{3}{1} =$

$\frac{\overset{3}{\cancel{9}}}{10} \times \frac{1}{\underset{1}{\cancel{3}}} = \mathbf{\frac{3}{10}}$

2. $\mathbf{\frac{3}{5}}$

$1\frac{1}{5} \div 2 =$

$\frac{6}{5} \div \frac{2}{1} =$

$\frac{\overset{3}{\cancel{6}}}{5} \times \frac{1}{\underset{1}{\cancel{2}}} = \mathbf{\frac{3}{5}}$

$\mathbf{\frac{2}{3}}$

$3\frac{1}{3} \div 5 =$

$\frac{10}{3} \div \frac{5}{1} =$

$\frac{\overset{2}{\cancel{10}}}{3} \times \frac{1}{\underset{1}{\cancel{5}}} = \mathbf{\frac{2}{3}}$

$\mathbf{\frac{1}{2}}$

$3\frac{1}{2} \div 7 =$

$\frac{7}{2} \div \frac{7}{1} =$

$\frac{\overset{1}{\cancel{7}}}{2} \times \frac{1}{\underset{1}{\cancel{7}}} = \mathbf{\frac{1}{2}}$

$\mathbf{\frac{2}{5}}$

$2\frac{2}{5} \div 6 =$

$\frac{12}{5} \div \frac{6}{1} =$

$\frac{\overset{2}{\cancel{12}}}{5} \times \frac{1}{\underset{1}{\cancel{6}}} = \mathbf{\frac{2}{5}}$

$\mathbf{\frac{1}{9}}$

$1\frac{2}{3} \div 15 =$

$\frac{5}{3} \div \frac{15}{1} =$

$\frac{\overset{1}{\cancel{5}}}{3} \times \frac{1}{\underset{3}{\cancel{15}}} = \mathbf{\frac{1}{9}}$

EXERCISE 19

1.

$\mathbf{\frac{1}{3}}$

$$\frac{5}{9} \div 1\frac{2}{3} =$$
$$\frac{5}{9} \div \frac{5}{3} =$$
$$\frac{\overset{1}{\cancel{5}}}{\underset{3}{\cancel{9}}} \times \frac{\overset{1}{\cancel{3}}}{\underset{1}{\cancel{5}}} = \mathbf{\frac{1}{3}}$$

$\mathbf{\frac{7}{36}}$

$$\frac{7}{8} \div 4\frac{1}{2} =$$
$$\frac{7}{8} \div \frac{9}{2} =$$
$$\frac{7}{\underset{4}{\cancel{8}}} \times \frac{\overset{1}{\cancel{2}}}{9} = \mathbf{\frac{7}{36}}$$

$\mathbf{\frac{1}{16}}$

$$\frac{3}{20} \div 2\frac{2}{5} =$$
$$\frac{3}{20} \div \frac{12}{5} =$$
$$\frac{\overset{1}{\cancel{3}}}{\underset{4}{\cancel{20}}} \times \frac{\overset{1}{\cancel{5}}}{\underset{4}{\cancel{12}}} = \mathbf{\frac{1}{16}}$$

$\mathbf{\frac{1}{4}}$

$$\frac{4}{7} \div 2\frac{2}{7} =$$
$$\frac{4}{7} \div \frac{16}{7} =$$
$$\frac{\overset{1}{\cancel{4}}}{\underset{1}{\cancel{7}}} \times \frac{\overset{1}{\cancel{7}}}{\underset{4}{\cancel{16}}} = \mathbf{\frac{1}{4}}$$

$\mathbf{\frac{1}{4}}$

$$\frac{5}{6} \div 3\frac{1}{3} =$$
$$\frac{5}{6} \div \frac{10}{3} =$$
$$\frac{\overset{1}{\cancel{5}}}{\underset{2}{\cancel{6}}} \times \frac{\overset{1}{\cancel{3}}}{\underset{2}{\cancel{10}}} = \mathbf{\frac{1}{4}}$$

2.

$\mathbf{4}$

$$14 \div 3\frac{1}{2} =$$
$$\frac{14}{1} \div \frac{7}{2} =$$
$$\frac{\overset{2}{\cancel{14}}}{1} \times \frac{2}{\underset{1}{\cancel{7}}} = \frac{4}{1} = \mathbf{4}$$

$\mathbf{2\frac{2}{11}}$

$$4 \div 1\frac{5}{6} =$$
$$\frac{4}{1} \div \frac{11}{6} =$$
$$\frac{4}{1} \times \frac{6}{11} = \frac{24}{11} = \mathbf{2\frac{2}{11}}$$

$\mathbf{5\frac{1}{3}}$

$$6 \div 1\frac{1}{8} =$$
$$\frac{6}{1} \div \frac{9}{8} =$$
$$\frac{\overset{2}{\cancel{6}}}{1} \times \frac{8}{\underset{3}{\cancel{9}}} = \frac{16}{3} = \mathbf{5\frac{1}{3}}$$

$\mathbf{2\frac{2}{3}}$

$$10 \div 3\frac{3}{4} =$$
$$\frac{10}{1} \div \frac{15}{4} =$$
$$\frac{\overset{2}{\cancel{10}}}{1} \times \frac{4}{\underset{3}{\cancel{15}}} = \frac{8}{3} = \mathbf{2\frac{2}{3}}$$

$\mathbf{2\frac{1}{2}}$

$$9 \div 3\frac{3}{5} =$$
$$\frac{9}{1} \div \frac{18}{5} =$$
$$\frac{\overset{1}{\cancel{9}}}{1} \times \frac{5}{\underset{2}{\cancel{18}}} = \frac{5}{2} = \mathbf{2\frac{1}{2}}$$

3.

$\mathbf{2\frac{3}{16}}$

$$7\frac{1}{2} \div 3\frac{3}{7} =$$
$$\frac{15}{2} \div \frac{24}{7} =$$
$$\frac{\overset{5}{\cancel{15}}}{2} \times \frac{7}{\underset{8}{\cancel{24}}} = \frac{35}{16} = \mathbf{2\frac{3}{16}}$$

$\mathbf{\frac{14}{45}}$

$$2\frac{1}{10} \div 6\frac{3}{4} =$$
$$\frac{21}{10} \div \frac{27}{4} =$$
$$\frac{\overset{7}{\cancel{21}}}{\underset{5}{\cancel{10}}} \times \frac{\overset{2}{\cancel{4}}}{\underset{9}{\cancel{27}}} = \mathbf{\frac{14}{45}}$$

$\mathbf{\frac{5}{6}}$

$$3\frac{8}{9} \div 4\frac{2}{3} =$$
$$\frac{35}{9} \div \frac{14}{3} =$$
$$\frac{\overset{5}{\cancel{35}}}{\underset{3}{\cancel{9}}} \times \frac{\overset{1}{\cancel{3}}}{\underset{2}{\cancel{14}}} = \mathbf{\frac{5}{6}}$$

$\mathbf{\frac{3}{5}}$

$$1\frac{1}{4} \div 2\frac{1}{12} =$$
$$\frac{5}{4} \div \frac{25}{12} =$$
$$\frac{\overset{1}{\cancel{5}}}{\underset{1}{\cancel{4}}} \times \frac{\overset{3}{\cancel{12}}}{\underset{5}{\cancel{25}}} = \mathbf{\frac{3}{5}}$$

$\mathbf{\frac{3}{7}}$

$$1\frac{7}{8} \div 4\frac{3}{8} =$$
$$\frac{15}{8} \div \frac{35}{8} =$$
$$\frac{\overset{3}{\cancel{15}}}{\underset{1}{\cancel{8}}} \times \frac{\overset{1}{\cancel{8}}}{\underset{7}{\cancel{35}}} = \mathbf{\frac{3}{7}}$$

EXERCISE 20 (WORD PROBLEMS)

1. (2) **$6\frac{7}{10}$ mi.**

$$\begin{array}{r} 15 = 14\frac{10}{10} \\ -\ 8\frac{3}{10} = 8\frac{3}{10} \\ \hline \mathbf{6\frac{7}{10}} \end{array}$$

2. (5) **$3,375**

$$\frac{1}{\cancel{4}_1} \times \frac{\cancel{13{,}500}^{3{,}375}}{1} = \mathbf{\$3{,}375}$$

3. (2) **15 mi.**

$$159 \div 10\frac{3}{5} =$$
$$\frac{159}{1} \div \frac{53}{5} =$$
$$\frac{\cancel{159}^{3}}{1} \times \frac{5}{\cancel{53}_1} = \frac{15}{1} = \mathbf{15}$$

4. (4) **$\frac{13}{16}$**

$$\begin{array}{r} \frac{3}{8} = \frac{6}{16} \\ \frac{1}{4} = \frac{4}{16} \\ +\frac{3}{16} = \frac{3}{16} \\ \hline \mathbf{\frac{13}{16}} \end{array}$$

5. (2) **$25\frac{1}{4}$ hrs.**

$$\begin{array}{r} 35 = 34\frac{4}{4} \\ -\ 9\frac{3}{4} = 9\frac{3}{4} \\ \hline \mathbf{25\frac{1}{4}} \end{array}$$

6. (3) **$1\frac{1}{3}$ cups**

$$2 \times \frac{2}{3} =$$
$$\frac{2}{1} \times \frac{2}{3} = \frac{4}{3} = \mathbf{1\frac{1}{3}}$$

7. (5) **96**

$$72 \div \frac{3}{4} =$$
$$\frac{\cancel{72}^{24}}{1} \times \frac{4}{\cancel{3}_1} = \mathbf{96}$$

8. (3) **40 mph**

$$140 \div 3\frac{1}{2} =$$
$$\frac{140}{1} \div \frac{7}{2} =$$
$$\frac{\cancel{140}^{20}}{1} \times \frac{2}{\cancel{7}_1} = \mathbf{40}$$

9. (3) **$10\frac{5}{16}$ lbs.**

$$\begin{array}{r} 2\frac{1}{2} = 2\frac{8}{16} \\ 3\frac{7}{16} = 3\frac{7}{16} \\ +4\frac{3}{8} = 4\frac{6}{16} \\ \hline 9\frac{21}{16} = \mathbf{10\frac{5}{16}} \end{array}$$

10. (2) **$46**

$$\frac{1}{\cancel{3}_1} \times \frac{\cancel{69}^{23}}{1} = \$23$$

$$\begin{array}{r} \$69 \\ -\ 23 \\ \hline \mathbf{\$46} \end{array}$$

11. (1) **$24\frac{1}{4}$**

$$\begin{array}{r} 24\frac{3}{8} = 23\frac{3}{8} + \frac{8}{8} = 23\frac{11}{8} \\ -\ \frac{1}{2} = \frac{4}{8} \\ \hline 23\frac{7}{8} \end{array}$$

$$\begin{array}{r} 23\frac{7}{8} \\ +\ \frac{5}{8} \\ \hline 23\frac{12}{8} = 24\frac{4}{8} = 24\frac{1}{2} \end{array}$$

$$\begin{array}{r} 24\frac{1}{2} = 24\frac{2}{4} \\ -\ \frac{1}{4} = \frac{1}{4} \\ \hline \mathbf{24\frac{1}{4}} \end{array}$$

12. (4) **$4\frac{7}{16}$ lbs.**

$$17\frac{3}{4} \div 4 =$$
$$\frac{71}{4} \times \frac{1}{4} = \frac{71}{16} = \mathbf{4\frac{7}{16}}$$

13. (3) **$57,000**

$$\frac{\cancel{3}^{1}}{\cancel{5}_1} \times \frac{1}{\cancel{3}_1} \times \frac{\cancel{285{,}000}^{57{,}000}}{1} = \mathbf{\$57{,}000}$$

14. (1) **$2\frac{3}{4}$ yds.**

$$\begin{array}{rcccl} 6\frac{1}{2} = & 6\frac{2}{4} = & 5\frac{2}{4} + \frac{4}{4} = & 5\frac{6}{4} \\ -3\frac{3}{4} = & 3\frac{3}{4} = & & 3\frac{3}{4} \\ \hline & & & \mathbf{2\frac{3}{4}} \end{array}$$

15. (4) **$10\frac{11}{12}$ hrs.**

$$\begin{array}{rl} 2\frac{1}{2} = & 2\frac{6}{12} \\ 1\frac{3}{4} = & 1\frac{9}{12} \\ +6\frac{2}{3} = & 6\frac{8}{12} \\ \hline & 9\frac{23}{12} = \mathbf{10\frac{11}{12}} \end{array}$$

16. (5) **18**

$$9 \div \frac{1}{2} =$$
$$\frac{9}{1} \times \frac{2}{1} = \mathbf{18}$$

17. (3) **$4\frac{1}{4}$ in.**

$$2 \times 10\frac{1}{4} =$$
$$\frac{\overset{1}{\cancel{2}}}{1} \times \frac{41}{\underset{2}{\cancel{4}}} = \frac{41}{2} = 20\frac{1}{2}$$

$$3 \times 15\frac{3}{4} =$$
$$\frac{3}{1} \times \frac{63}{4} = \frac{189}{4} = 47\frac{1}{4}$$

$$\begin{array}{rl} 20\frac{1}{2} = & 20\frac{2}{4} \\ +47\frac{1}{4} = & 47\frac{1}{4} \\ \hline & 67\frac{3}{4} \end{array} \qquad \begin{array}{rl} 72 = & 71\frac{4}{4} \\ -67\frac{3}{4} = & 67\frac{3}{4} \\ \hline & \mathbf{4\frac{1}{4}} \end{array}$$

18. (1) **$1\frac{5}{6}$ million**

$$\begin{array}{rcccl} 4\frac{1}{3} = & 4\frac{2}{6} = & 3\frac{2}{6} + \frac{6}{6} = & 3\frac{8}{6} \\ -2\frac{1}{2} = & 2\frac{3}{6} = & & 2\frac{3}{6} \\ \hline & & & \mathbf{1\frac{5}{6}} \end{array}$$

19. (5) **$594**

$$2\frac{3}{4} \times 216 =$$
$$\frac{11}{\underset{1}{\cancel{4}}} \times \frac{\overset{54}{\cancel{216}}}{1} = \mathbf{\$594}$$

20. (2) **$14\frac{9}{10}$ mi.**

$$\begin{array}{r} 2\frac{3}{10} \\ 5\frac{9}{10} \\ +6\frac{7}{10} \\ \hline 13\frac{19}{10} = \mathbf{14\frac{9}{10}} \end{array}$$

21. (5) **80**

5 ft. = 60 in.

$$60 \div \frac{3}{4} = \frac{\overset{20}{\cancel{60}}}{1} \times \frac{4}{\underset{1}{\cancel{3}}} = \mathbf{80}$$

22. (3) **$21**

$$\frac{1}{4} \times \$28 =$$
$$\frac{1}{\underset{1}{\cancel{4}}} \times \frac{\overset{7}{\cancel{28}}}{1} = \$7$$

$$\begin{array}{r} \$28 \\ -\ 7 \\ \hline \mathbf{\$21} \end{array}$$

23. **(3)** $\mathbf{182\frac{1}{2}}$ **lbs.**

$$\begin{array}{rcr} 216 & = & 215\frac{2}{2} \\ -\ 33\frac{1}{2} & = & 33\frac{1}{2} \\ \hline & & \mathbf{182\frac{1}{2}} \end{array}$$

24. **(2)** **$63**

$7 \times \$6 = \$42$

$2\frac{1}{3} \times \$9 =$

$\frac{7}{\cancel{3}_1} \times \frac{\cancel{9}^3}{1} = \frac{21}{1} = \$21$

$$\begin{array}{r} \$42 \\ +21 \\ \hline \mathbf{\$63} \end{array}$$

25. **(4)** $\mathbf{74\frac{1}{4}}$ **in.**

$$\begin{array}{rcr} 71\frac{3}{4} & = & 71\frac{3}{4} \\ +\ 2\frac{1}{2} & = & 2\frac{2}{4} \\ \hline & & 73\frac{5}{4} = \mathbf{74\frac{1}{4}} \end{array}$$

26. **(2)** **$235**

$\frac{1}{6} \times \$282 =$

$\frac{1}{\cancel{6}_1} \times \frac{\cancel{282}^{47}}{1} = \$47$

$$\begin{array}{r} \$282 \\ -47 \\ \hline \mathbf{\$235} \end{array}$$

27. **(2)** **17,900**

$\frac{5}{8} \times 28{,}640 =$

$\frac{5}{\cancel{8}_1} \times \frac{\cancel{28{,}640}^{3{,}580}}{1} = \mathbf{17{,}900}$

28. **(1)** **24**

$36 \div 1\frac{1}{2} =$

$\frac{36}{1} \div \frac{3}{2} =$

$\frac{\cancel{36}^{12}}{1} \times \frac{2}{\cancel{3}_1} = \mathbf{24}$

29. **(2)** $\mathbf{\$\frac{5}{8}}$ **million**

$$\begin{array}{rcccr} 2\frac{1}{2} = 2\frac{4}{8} & = & 1\frac{4}{8} + \frac{8}{8} & = & 1\frac{12}{8} \\ -1\frac{7}{8} = 1\frac{7}{8} & = & & & 1\frac{7}{8} \\ \hline & & & & \mathbf{\frac{5}{8}} \end{array}$$

30. **(4)** **17**

$76\frac{1}{2} \div 4\frac{1}{2} =$

$\frac{153}{2} \div \frac{9}{2} =$

$\frac{\cancel{153}^{17}}{\cancel{2}_1} \times \frac{\cancel{2}^1}{\cancel{9}_1} = \mathbf{17}$

FRACTIONS REVIEW

1. $\mathbf{\frac{17}{36}}$

(Writing Fractions)

2. $\mathbf{\frac{4}{11}}$

$\frac{16 \div 4}{44 \div 4} = \mathbf{\frac{4}{11}}$

(Reducing Fractions)

3. $\mathbf{\frac{27}{48}}$

$\frac{9 \times 3}{16 \times 3} = \mathbf{\frac{27}{48}}$

(Raising Fractions to Higher Terms)

4. $\mathbf{6\frac{2}{3}}$

$\frac{60}{9} = 9\overline{)\,60}$; $60 - 54 = 6$; $6\frac{6}{9} = \mathbf{6\frac{2}{3}}$

(Changing Improper Fractions to Whole or Mixed Numbers)

5. $\mathbf{\frac{77}{9}}$

$8\frac{5}{9} = \mathbf{\frac{77}{9}}$

(Changing Mixed Numbers to Improper Fractions)

6. $\mathbf{7\frac{1}{5}}$

$$\begin{array}{r} 2\frac{7}{15} \\ +4\frac{11}{15} \\ \hline 6\frac{18}{15} = 7\frac{3}{15} = \mathbf{7\frac{1}{5}} \end{array}$$

(Adding Fractions with the Same Denominators)

7. $\mathbf{10\frac{19}{24}}$

$$\begin{array}{r} 3\frac{5}{8} = 3\frac{15}{24} \\ 2\frac{1}{3} = 2\frac{8}{24} \\ +4\frac{5}{6} = 4\frac{20}{24} \\ \hline 9\frac{43}{24} = \mathbf{10\frac{19}{24}} \end{array}$$

(Adding Fractions with Different Denominators)

8. $\mathbf{5\frac{3}{10}}$

$$\begin{array}{r} 9\frac{13}{20} \\ -4\frac{7}{20} \\ \hline 5\frac{6}{20} = \mathbf{5\frac{3}{10}} \end{array}$$

(Subtracting Fractions with the Same Denominators)

9. $\mathbf{2\frac{5}{16}}$

$$\begin{array}{r} 5\frac{7}{8} = 5\frac{14}{16} \\ -3\frac{9}{16} = 3\frac{9}{16} \\ \hline \mathbf{2\frac{5}{16}} \end{array}$$

(Subtracting Fractions with Different Denominators)

10. $\mathbf{5\frac{17}{24}}$

$$\begin{array}{r} 8\frac{3}{8} = 8\frac{9}{24} = 7\frac{9}{24} + \frac{24}{24} = 7\frac{33}{24} \\ -2\frac{2}{3} = 2\frac{16}{24} = \qquad\qquad 2\frac{16}{24} \\ \hline \mathbf{5\frac{17}{24}} \end{array}$$

(Borrowing)

11. $\mathbf{\frac{15}{56}}$

$\frac{5}{8} \times \frac{3}{7} = \mathbf{\frac{15}{56}}$

(Multiplying Fractions by Fractions)

12. $\mathbf{\frac{2}{15}}$

$\frac{\overset{1}{\cancel{7}}}{\underset{3}{\cancel{12}}} \times \frac{\overset{2}{\cancel{8}}}{\underset{5}{\cancel{35}}} = \mathbf{\frac{2}{15}}$

(Canceling)

13. $\mathbf{9}$

$\frac{3}{\underset{1}{\cancel{8}}} \times \frac{\overset{3}{\cancel{24}}}{1} = \frac{9}{1} = \mathbf{9}$

(Multiplying Fractions and Whole Numbers)

14. $\mathbf{21}$

$3\frac{3}{4} \times 5\frac{3}{5} =$

$\frac{\overset{3}{\cancel{15}}}{\underset{1}{\cancel{4}}} \times \frac{\overset{7}{\cancel{28}}}{\underset{1}{\cancel{5}}} = \frac{21}{1} = \mathbf{21}$

(Multiplying Mixed Numbers)

15. $\mathbf{\frac{14}{15}}$

$\frac{7}{10} \div \frac{3}{4} =$

$\frac{7}{\cancel{10}_{5}} \times \frac{\cancel{4}^{2}}{3} = \mathbf{\frac{14}{15}}$

(Dividing by Fractions)

16. $\mathbf{14\frac{2}{5}}$

$12 \div \frac{5}{6} =$

$\frac{12}{1} \times \frac{6}{5} = \frac{72}{5} = \mathbf{14\frac{2}{5}}$

(Dividing Whole Numbers by Fractions)

17. $\mathbf{6\frac{2}{3}}$

$4\frac{1}{6} \div \frac{5}{8} =$

$\frac{25}{6} \div \frac{5}{8} =$

$\frac{\cancel{25}^{5}}{\cancel{6}_{3}} \times \frac{\cancel{8}^{4}}{\cancel{5}_{1}} = \frac{20}{3} = \mathbf{6\frac{2}{3}}$

(Dividing Mixed Numbers by Fractions)

18. $\mathbf{1\frac{1}{5}}$

$8\frac{2}{5} \div 7 =$

$\frac{42}{5} \div \frac{7}{1} =$

$\frac{\cancel{42}^{6}}{5} \times \frac{1}{\cancel{7}_{1}} = \frac{6}{5} = \mathbf{1\frac{1}{5}}$

(Dividing Fractions and Mixed Numbers by Whole Numbers)

19. $\mathbf{\frac{2}{3}}$

$4\frac{4}{9} \div 6\frac{2}{3} =$

$\frac{40}{9} \div \frac{20}{3} =$

$\frac{\cancel{40}^{2}}{\cancel{9}_{3}} \times \frac{\cancel{3}^{1}}{\cancel{20}_{1}} = \mathbf{\frac{2}{3}}$

(Dividing by Mixed Numbers)

20. $\mathbf{6\frac{1}{16}}$ **lbs.**

$$\begin{array}{rcl} 2\frac{1}{2} & = & 2\frac{8}{16} \\ 1\frac{5}{16} & = & 1\frac{5}{16} \\ +2\frac{1}{4} & = & 2\frac{4}{16} \\ \hline & & 5\frac{17}{16} = \mathbf{6\frac{1}{16}} \end{array}$$

(Fraction Word Problems)

21. $\mathbf{62\frac{5}{8}}$

$$\begin{array}{rcl} 62\frac{1}{2} & = & 62\frac{2}{4} \\ -\ \frac{1}{4} & = & \frac{1}{4} \\ \hline & & 62\frac{1}{4} = 62\frac{2}{8} \\ & & +\ \frac{3}{8} = \frac{3}{8} \\ \hline & & \mathbf{62\frac{5}{8}} \end{array}$$

(Fraction Word Problems)

22. **$1,050,000**

$\frac{5}{\cancel{16}_{1}} \times \frac{\cancel{\$3,360,000}^{210,000}}{1} = \mathbf{\$1,050,000}$

(Fraction Word Problems)

Decimals

A decimal is a kind of fraction. When you use money, you are using the decimal system. $.20 is a decimal. It is 20 of the 100 pennies in a dollar. $.20 written in the form of a common fraction is $\frac{20}{100}$. The denominator (100) tells how many parts are in one dollar. The numerator (20) tells how many parts you have.

Decimals are different from common fractions in two ways. One difference is that the denominators of decimals are not written. The other difference is that only certain numbers—10, 100, 1,000, etc.—can be decimal denominators. Decimal denominators get their names from the number of **places** at the right of the decimal point. A place is the position of a digit. The decimal point itself does not take up a decimal place. .20 is a decimal with two places. .40935 is a decimal with five places.

The list below gives the names of the first six decimal places and the number of places they use. Be sure you know the six decimal names before you go on.

Decimal name	**Number of places**
tenths	one place
hundredths	two places
thousandths	three places
ten-thousandths	four places
hundred-thousandths	five places
millionths	six places

The next list gives an example of each of the first six decimals and the common fractions each decimal is equal to.

Decimal	**Value**	**Common fraction**
.3	three tenths	$\frac{3}{10}$
.09	nine hundredths	$\frac{9}{100}$
.017	seventeen thousandths	$\frac{17}{1{,}000}$
.0004	four ten-thousandths	$\frac{4}{10{,}000}$
.00018	eighteen hundred-thousandths	$\frac{18}{100{,}000}$
.000503	five hundred three millionths	$\frac{503}{1{,}000{,}000}$

Mixed decimals are numbers with digits on both sides of the decimal point. $1.25 is a mixed decimal. It means one whole dollar and $\frac{25}{100}$ of a dollar. Mixed decimals have whole numbers at the left of the decimal point.

Zeros often cause trouble in decimals. Any zero at the right of a decimal point and, at the same time, at the left of another digit is important. For example, all the zeros in .00408 are important. These zeros keep 4 in the thousandths place and 8 in the hundred-thousandths place. In the decimal .02600, the two zeros at the right are not needed. They have nothing to do with the value of 2 and 6. You could write this decimal as .026.

EXERCISE 1

DIRECTIONS: Circle the mixed decimals in this list.

1. .496 380.2 6.8 .429 65.3

.36 19.44

Cross out the zeros that are not needed in each decimal or mixed decimal.

2. 30.0490 .2070 013.50 08.06 3.0

06.1090 .002040

Check your answers on page 106.

Reading Decimals

A decimal gets its name from the number of places at the **right** of the decimal point. To read a decimal, count the places at the right of the point.

EXAMPLE: Read the decimal .081.

.081

Step 1: Notice that every digit is at the right of the decimal point. There is no whole number with this decimal.
Step 2: Count the decimal places. The decimal has three places. Three decimal places are thousandths.
Step 3: Read .081 as **eighty-one thousandths.**

With mixed decimals, separate the whole number and the decimal fraction with the word **and.**

EXAMPLE: Read 9.02.

9.02

Step 1: Notice that there are digits on both sides of the decimal point. This number is a mixed decimal.
Step 2: Read the whole number as **nine.**
Step 3: Count the decimal places. The decimal part has two places. Two decimal places are hundredths.
Step 4: Read 9.02 as **nine and two hundredths.**

EXERCISE 2

DIRECTIONS: Write these decimals or mixed decimals in words.

1. .9 .03 .015 .27

2. .0039 .008 .00361 .000052

3. 12.6 2.09 120.05 7.0538

4. 1.00007 6.000123 3.02806 19.0571

Check your answers on page 106.

Writing Decimals

When you write decimals, decide how many places you need. Use zeros in places that are not filled.

EXAMPLE: Write six thousandths as a decimal.

.006

Step 1: Decide how many places you need. Thousandths need three places.
Step 2: The digit 6 needs only one place. Use zeros in the first two decimal places.

When you write mixed numbers, put a decimal point in place of the word **and.**

EXAMPLE: Write thirty-four and three hundredths as a mixed decimal.

34.03

Step 1: Write the whole number 34.
Step 2: Decide how many decimal places you need. Hundredths need two places.

Step 3: The digit 3 needs only one place. Use a zero in the first decimal place.
Step 4: Separate the whole number and the decimal fraction with a decimal point.

EXERCISE 3

DIRECTIONS: Write each number as a decimal or mixed decimal.

1. nine tenths
2. twelve and six hundredths
3. twenty-two thousandths
4. thirteen and eight thousandths
5. three hundred seventeen ten-thousandths
6. forty-one and two ten-thousandths
7. one hundred ninety-three hundred-thousandths
8. five and seven millionths
9. two thousand four hundred fifty-two millionths
10. seventy and eighty-five hundred-thousandths

Check your answers on page 106.

Comparing Decimals

To compare decimals, first change the decimals so that they have the same number of places. Decimals with the same number of places have a common denominator. You can put zeros to the right of a decimal without changing its value. .6 and .60 have the same value. The 6 is in the tenths place in each decimal.

EXAMPLE: Which decimal is bigger, .04 or .2?

.2 = **.20**	**.20** .04
Step 1	Step 2

Step 1: Put a zero at the right of .2 to change it to .20. Both decimals are now in hundredths.
Step 2: Decide which decimal is bigger. 20 hundredths is bigger than 4 hundredths. .2 is the bigger of the two decimals.

EXAMPLE: Which decimal is biggest, .38, .3, or .308?

.38 = **.380**	.3 = **.300**	**.380** .300 .308
Step 1	Step 2	Step 3

Step 1: Put a zero at the right of .38 to change it to thousandths.
Step 2: Put two zeros at the right of .3 to change it to thousandths.
Step 3: Decide which decimal is biggest. 380 thousandths is bigger than either 300 thousandths or 308 thousandths. .38 is the biggest of the three decimals.

EXERCISE 4

DIRECTIONS: Find the bigger decimal in each pair.

1.	.3 or .32	.29 or .3	.075 or .2	.09 or .074
2.	.007 or .07	.08 or .093	.64 or .626	.424 or .41
3.	.897 or .88	.2 or .0347	.72 or .278	.08 or .1032

Find the biggest decimal in each group.

4.	.7, .07, or .57	.108, .12, or .1	.0056, .605, or .56
5.	.29, .3, or .302	.48, .084, or .408	.03, .033, or .3303
6.	.05, .015, or .51	.06, .406, or .6	.93, .9309, or .9

Check your answers on page 107.

Changing Decimals to Fractions

To change a decimal to a common fraction, write the digits in the decimal as the numerator. Write the name of the decimal (tenths, hundredths, etc.) as the denominator. Then reduce the fraction.

EXAMPLE: Change .06 to a fraction.

$\underline{\mathbf{06}}$	$\frac{06}{\mathbf{100}}$	$\frac{06}{100} = \frac{\mathbf{3}}{\mathbf{50}}$
Step 1	Step 2	Step 3

Step 1: Write 06 as the numerator.
Step 2: .06 has two decimal places. Two decimal places are hundredths. Write 100 as the denominator.
Step 3: Reduce $\frac{6}{100}$ to lowest terms. (Notice that the zero in front of the 6 is left out.)

EXAMPLE: Change 4.2 to a mixed number.

$4^{\underline{2}}$	$4\frac{2}{10}$	$4\frac{2}{10} = 4\frac{1}{5}$
Step 1	Step 2	Step 3

Step 1: Write 4 as the whole number and 2 as the numerator.
Step 2: 4.2 has one decimal place. One decimal place is tenths. Write 10 as the denominator.
Step 3: Reduce the fraction to lowest terms.

EXERCISE 5

DIRECTIONS: Change each decimal or mixed decimal to a fraction or mixed number. Reduce each fraction.

1.	.4 =	.02 =	.375 =	6.8 =	9.16 =
2.	.125 =	.008 =	.00085 =	2.004 =	5.96 =
3.	.64 =	.015 =	.0024 =	10.625 =	12.084 =

Check your answers on page 107.

Changing Fractions to Decimals

A fraction is an instruction to divide. The line separating the numerator from the denominator means "divided by." For example, $\frac{9}{20}$ means 9 divided by 20. To change a fraction to a decimal, divide the numerator by the denominator. Put a decimal point and zeros to the right of the numerator.

EXAMPLE: Change $\frac{9}{20}$ to a decimal.

Step 1	Step 2	Step 3
$20\overline{)9}$	$20\overline{)9.00}$	$\begin{array}{r} .45 \\ 20\overline{)\ 9.00} \\ -8\,0 \\ 1\,00 \\ -1\,00 \end{array}$

Step 1: Set up the problem. Divide 9 by 20.
Step 2: Put a decimal point and two zeros to the right of 9.
Step 3: Divide. Be sure to bring the decimal point up into the answer. It is important to line the numbers up correctly.

Sometimes the division will come out even with just one zero at the right of the point. Sometimes the division will not come out even with many zeros.

EXAMPLE: Change $\frac{1}{6}$ to a decimal.

$6\overline{)1}$	$6\overline{)1.00}$	$.16\frac{4}{6}$ $6\overline{)\ 1.00}$ $-\ \ 6$ 40 -36 4	$.16\frac{4}{6} = .16\frac{2}{3}$
Step 1	Step 2	Step 3	Step 4

Step 1: Set up the problem. Divide 1 by 6.
Step 2: Put a decimal point and two zeros to the right of 1.
Step 3: Divide. Bring the decimal point up into the answer. (Notice that, no matter how many times you divide, the answer will not come out even.)
Step 4: Reduce the fraction part of the decimal to lowest terms.

EXERCISE 6

DIRECTIONS: Change each fraction to a decimal.

1. $\frac{3}{4} =$ $\quad \frac{5}{6} =$ $\quad \frac{1}{7} =$ $\quad \frac{2}{9} =$ $\quad \frac{3}{10} =$

2. $\frac{5}{8} =$ $\quad \frac{1}{2} =$ $\quad \frac{2}{5} =$ $\quad \frac{7}{20} =$ $\quad \frac{11}{100} =$

3. $\frac{3}{50} =$ $\quad \frac{4}{25} =$ $\quad \frac{1}{4} =$ $\quad \frac{2}{3} =$ $\quad \frac{1}{12} =$

Check your answers on page 107.

Rounding Off Decimals

The answers to many decimal problems have more places than they need. **Rounding off** is a way of making decimals easier to read.

To round off a decimal, mark the digit in the place you want to round off to. If the digit at the right of this place is more than 4, add 1 to the digit you marked. If the digit at the right is less than 5, leave the digit you marked as it is. Then drop the digits at the right of the digit you marked.

EXAMPLE: Round off 5.284 to the nearest tenth.

$5.\underline{2}84$	$5.\underline{2}\mathbf{8}4$	$5.\underline{3}$
Step 1	Step 2	Step 3

Step 1: Find the digit in the tenths place, 2.
Step 2: Now, look at the digit to the right of 2 (8). Since 8 is more than 4, add 1 to the 2: 1 + 2 = 3.
Step 3: Put a 3 in the tenths place, and drop the digits to the right of 3.

EXAMPLE: Round off 1.992 to the nearest tenth.

$1.\underline{9}92$	$1.\underline{9}\mathbf{9}2$	2.0 = **2**
Step 1	Step 2	Step 3

Step 1: Find the digit in the tenths place, 9.
Step 2: Now, look at the digit to the right of 9 (9). Since 9 is more than 4, add 1 to the 9 in the tenths place: 9 + 1 = 10. Carry the 1 into the units place.
Step 3: Since the zero in the tenths place is not needed, drop it.

EXERCISE 7

DIRECTIONS: Round off each decimal to the nearest tenth.

1. .263 .42 5.057 3.8076 6.981 4.74

Round off each decimal to the nearest hundredth.

2. .358 .0347 1.222 4.9152

12.188 2.1948

Round off each decimal to the nearest thousandth.

3. .0362 .1295 4.8327 16.0077

2.5832 1.06382

Check your answers on page 108.

ADDING DECIMALS

To add decimals, line up the numbers with the decimal points under each other.

EXAMPLE: .28 + .9 + .327 =

Step 1	Step 2
.28 **.9** **+.327**	.28 .9 +.327 **1.507**

Step 1: Set up the problem. Line up the numbers with the decimal points in one column.
Step 2: Add each column. In the thousandths column, the only digit is 7. In the hundredths column, the digits are 8 and 2. In the tenths column, the digits are 2, 9, and 3. Make sure to bring the decimal point down in its column.

In the last example, the total of the tenths column was 15. Only one digit can fit in each column. Write the 5 in the tenths column and carry the 1 over to the units column.

When you add whole numbers with decimals or mixed decimals, put a point at the right of the whole number. Then line up the numbers with the decimal points under each other.

EXAMPLE: Add 26, 3.85, and .907.

Step 1	Step 2	Step 3
26 = **26.0**	**26.** **3.85** **+ .907**	26. 3.85 + .907 **30.757**

Step 1: Put a decimal point at the right of 26.
Step 2: Set up the problem. Line up the numbers with the decimal points in one column.
Step 3: Add each column.

EXERCISE 8

DIRECTIONS: Add each problem.

1. .48 + .6 + .709 =   .27 + .726 + .4 =   .6 + .8 + .3 =

2. .0764 + .3 + .125 =   .009 + .28 + .4736 =

3. 4.7 + 26 + 1.19 =   .804 + 2.5 + 7 =   .229 + 6 + .83 =

4. 1.0025 + 9 + .36 =   8 + .0685 + 2.6 =

Check your answers on page 108.

SUBTRACTING DECIMALS

To subtract decimals, line up the numbers with the decimal points under each other. Put a point at the right of whole numbers. Put zeros at the right until each decimal has the same number of places. You will need the zeros for borrowing.

EXAMPLE: Subtract .32 from 8.

Step 1	Step 2	Step 3	Step 4
	8.	8.**00**	8.00
8 = **8.**	**− .32**	− .32	− .32
			7.68

Step 1: Put a decimal point at the right of 8.
Step 2: Set up the problem. Line up the decimal points in one column.
Step 3: Put two zeros at the right of 8 to give each decimal the same number of places.
Step 4: Borrow and subtract.

EXAMPLE: Subtract .029 from .06.

Step 1	Step 2	Step 3
.06	.060	.060
−.029	−.029	−.029
		.031

Step 1: Set up the problem. Line up the decimal points in one column.

Step 2: Put a zero at the right of .06 to give each decimal the same number of places.
Step 3: Borrow and subtract.

EXERCISE 9

DIRECTIONS: Subtract each problem.

1. 9 − .357 =   12 − .26 =   .4 − .138 =   .36 − .029 =

2. 8 − .049 =   .02 − .005 =   50 − .6 =   8.93 − 4 =

3. .036 − .008 =   1 − .2305 =   .7 − .281 =   9 − .508 =

Check your answers on page 109.

MULTIPLYING DECIMALS

When you multiply decimals, you do not have to line up decimal points. Set up the numbers for easy multiplication. Then count the number of decimal places in each number. Put the total number of places from the two numbers in the answer.

EXAMPLE: 2.36 × 4 =

2.36 **× 4**	2.36 × 4 **944**	2.**36** 4	2.36 × 4 **9.44**
Step 1	Step 2	Step 3	Step 4

Step 1: Set up the problem. Put the number with the fewer digits (4) below.
Step 2: Multiply.
Step 3: Count the decimal places in each number. 2.36 has two decimal places. 4 has no decimal places.
Step 4: Put the total number of decimal places (2) in the answer. When you put the decimal in, count the places from right to left.

Sometimes you will need to put extra zeros in your answer.

EXAMPLE: Multiply .08 by .6.

Step 1	Step 2	Step 3	Step 4
.08 **× .6**	.08 × .6 48	**.08** **.6**	.08 × .6 **.048**

Step 1: Set up the problem. Put the number with the fewer digits (.6) below.
Step 2: Multiply.
Step 3: Count the decimal places in each number. .08 has two decimal places. .6 has one decimal place.
Step 4: Put the total number of decimal places (3) in the answer. Count from right to left. Put a zero at the left of 48 to make three places.

EXERCISE 10

DIRECTIONS: Multiply each problem.

1. $8 \times 2.6 =$ $\qquad$ $.47 \times 6 =$ $\qquad$ $14 \times .05 =$ $\qquad$ $.09 \times .6 =$

2. $.35 \times 24 =$ $\qquad$ $.792 \times .4 =$ $\qquad$ $.93 \times 8.8 =$ $\qquad$ $.37 \times .009 =$

3. $50 \times 4.17 =$ $\qquad$ $.059 \times .38 =$ $\qquad$ $.0814 \times .6 =$ $\qquad$ $423 \times .04 =$

Check your answers on page 109.

DIVIDING DECIMALS

Dividing Decimals by Whole Numbers

You started dividing decimals when you changed fractions into decimals. There, you divided a whole number by a bigger whole number. When you divide a decimal by a whole number, line up your problem carefully. Then bring the decimal point up into the answer above its position in the problem.

EXAMPLE: Divide 2.24 by 4.

Step 1	Step 2	Step 3
4)2.24	**56** 4) 2.24 −2 0 24 − 24	**.56** 4) 2.24 −2 0 24 − 24

Step 1: Set up the problem.
Step 2: Divide.
Step 3: Bring the decimal point up into the answer above its position in the problem.

Sometimes you will need to put zeros in your answer.

EXAMPLE: .365 ÷ 5 =

Step 1	Step 2	Step 3
5).365	**0** 5).365	**.073** 5) .365 − 35 15 − 15

Step 1: Set up the problem.
Step 2: Divide. 5 does not go into 3. Put a zero above the 3 to show that 5 does not divide into it.
Step 3: Divide the rest of the problem. Bring the decimal point up into the answer above its position in the problem.

EXERCISE 11

DIRECTIONS: Divide each problem.

1. 39.2 ÷ 8 = 1.74 ÷ 3 = .765 ÷ 9 = 55.2 ÷ 6 =

2. 33.6 ÷ 12 = 8.88 ÷ 24 = 348.4 ÷ 52 =

3. .322 ÷ 7 = 29.92 ÷ 16 = 15.48 ÷ 36 =

Check your answers on page 109.

Dividing Decimals by Decimals

Dividing by decimals is a complicated operation. To divide by a decimal, first change the decimal to a whole number. Move the decimal point to the right end. Next, move the decimal point in the number that is being divided the same number of places.

These steps are easier to understand with whole numbers. Look at the problem 8 ÷ 4 = 2. The answer is the same if the decimal point is moved one place to the right in both 8 and 4: 80 ÷ 40 = 2. The problems are different, but the answers are the same.

EXAMPLE: Divide 3.76 by .8.

Step 1	Steps 2-3	Step 4
.8)3.76	8.)37.6	4.7 8) 37.6 −32 56 56

Step 1: Set up the problem.

Step 2: Move the decimal point in the number you are dividing by (.8) to make it a whole number. Count the number of places you moved the decimal point (1 place).

Step 3: Now, move the decimal point in the number that you are dividing into. Move the decimal point the same number of places (1).

Step 4: Divide and bring the decimal point up into the answer above its new position.

It is a good idea to check decimal division problems. Multiply the answer by the decimal you divided by. The product should equal the number that was divided.

$$\begin{array}{r} 4.7 \\ \times\ .8 \\ \hline 3.76 \end{array}$$

EXERCISE 12

DIRECTIONS: Divide each problem.

1. .54 ÷ .9 = .096 ÷ .4 =

2. 1.94 ÷ .2 = .567 ÷ .7 =

3. .516 ÷ .06 = 2.96 ÷ .08 =

4. 2.507 ÷ 2.3 = .0432 ÷ .54 =

5. .384 ÷ .024 = 10.591 ÷ 6.23 =

6. 33.12 ÷ 41.4 = 3.2067 ÷ .509 =

Check your answers on page 110.

Dividing Whole Numbers by Decimals

To divide a whole number by a decimal, first put a decimal point at the right of the whole number. Then move the points in both numbers. You will have to put zeros in the number being divided.

EXAMPLE: 36 ÷ .009 =

Step 1	Steps 2-3	Step 4	Step 5
.009)36.	009.)36 000.	**4,000** 9) 36,000 −36 0 000	4,000 ×.009 36.000 = **36**

Step 1: Set up the problem. Put a decimal point at the right of 36.

Step 2: Move the decimal point in the number you are dividing by (.009) to make it a whole number.

Step 3: Now, move the decimal point in the number that you are dividing into. Move the decimal point the same number of places (3).

Step 4: Divide and bring the decimal point up into the answer above its new position. In this problem, the answer is a whole number, so the decimal point is dropped.

Step 5: Check by multiplying: 4,000 × .009 = 36.

EXERCISE 13

DIRECTIONS: Divide each problem.

1. 84 ÷ 2.4 = 114 ÷ .38 = 36 ÷ .09 = 128 ÷ 1.6 =

2. $1{,}656 \div 7.2 =$ $\quad 14 \div .007 =$ $\quad 862 \div .4 =$ $\quad 187 \div 8.5 =$

3. $6 \div .015 =$ $\quad 192 \div .6 =$ $\quad 624 \div .48 =$ $\quad 198 \div 6.6 =$

Check your answers on page 111.

Decimal Word Problems

Before you begin the decimal word problems, read the section about fraction word problems again (page 59). Decimal word problems are very much like fraction word problems. The key words and situations for fractions are the same for decimals.

In division of decimals problems, remember that the number being divided must come inside the sign.

EXAMPLE: Ann paid $3.91 for 2.3 pounds of ground beef. Find the price of one pound of ground beef.

Here the money is being divided into smaller amounts. Put the money inside the $\overline{)\quad}$ sign.

```
         $  1.70
2.3.)$3.9.10
        2 3
        1 6 1
        1 6 1
           00
```

EXERCISE 14 (WORD PROBLEMS)

DIRECTIONS: Choose the correct answer to each problem.

1. Carl weighs 165 pounds. One pound equals 0.45 kilograms. Find Carl's weight in kilograms.

   (1) 7.43 kg. (2) 36.7 kg. (3) 74.25 kg.
   (4) 120 kg. (5) 366.7 kg.

2. Lois bought 2.8 pounds of ground beef for $4.76. How much did one pound of ground beef cost?

   (1) $1.33 (2) $1.68 (3) $1.70
   (4) $1.88 (5) $1.96

3. In August 1980, 8 million Americans were officially unemployed. In September 1980, 7.8 million were unemployed. How many fewer Americans were unemployed in September than in August?

   (1) 2,000,000 (2) 1,800,000 (3) 1,200,000
   (4) 200,000 (5) 20,000

4. A batting average is the number of hits a player gets divided by the number of times he goes to bat. Last season, Chuck got 23 hits in 80 times at bat. To the nearest thousandth, what was Chuck's batting average?

   (1) .103 (2) .184 (3) .248 (4) .288 (5) .318

5. Before their vacation, the Smiths' car had a mileage reading of exactly 18,450 miles. They drove a total of 1,963.6 miles on their trip. What was the mileage reading after the trip?

   (1) 16,486.4 (2) 20,413.6 (3) 28,086
   (4) 31,313.6 (5) 37,573.6

6. Frank makes $6.80 an hour. How much does he make in a week when he works 37.5 hours?

   (1) $44.35 (2) $228.00 (3) $255.00
   (4) $272.00 (5) $525.00

7. Pete wants to cut a wire 9.75 yards long into 13 equal pieces. How long will each piece be?

   (1) .65 yd. (2) .75 yd. (3) .85 yd.
   (4) 1.28 yds. (5) 9.62 yds.

8. Rachel's normal temperature is 98.6°. When she was sick, her temperature went up 4.5°. What was her temperature when she was sick?

   (1) 104.2° (2) 104.1° (3) 103.1°
   (4) 102.1° (5) 94.1°

9. Faye bought .75 pound of cheese at $2.40 a pound and 2.3 pounds of bologna at $1.90 a pound. How much change did she get from $10?

   (1) $6.17 (2) $4.40 (3) $4.17
   (4) $3.83 (5) $3.23

10. The express bus takes 12.7 minutes to get from Main Street to Maple Street. The local bus takes 16.2 minutes to go the same distance. The express bus is how much faster than the local bus?

    (1) 2.5 min. (2) 3.5 min. (3) 4.0 min.
    (4) 4.5 min. (5) 28.9 min.

11. Rick bought 8.25 yards of lumber. The lumber cost $3.36 a yard. Find the total cost of the lumber.

    (1) $26.88 (2) $27.13 (3) $27.72
    (4) $27.88 (5) $28.13

12. In the U.S. in 1979, there were 41.1 million people living in the West, 71.6 million in the South, 58.4 million in the North Central states, and 49 million in the East. Find the total number of people living in these four areas.

    (1) 176 million (2) 219.1 million (3) 220.1 million
    (4) 221.0 million (5) 239.1 million

13. Don used 22 gallons of gas on a 400-mile trip. To the nearest tenth, what was the average number of miles Don drove on one gallon of gas?

    (1) 18.0 mi. (2) 18.2 mi. (3) 18.3 mi.
    (4) 19.0 mi. (5) 19.1 mi.

14. Pete bought 4.5 yards of lumber at $1.68 a yard. How much change did he get from $8?

    (1) $1.56 (2) $1.44 (3) $.56 (4) $.44 (5) $.20

15. Colin bought a used car that had a mileage reading of 38,460.8 miles. After he drove the car for three months, the mileage read 46,213.4 miles. How many miles did Colin drive the first three months he owned the car?

    (1) 2,584.2 mi. (2) 7,752.6 mi. (3) 7,842.2 mi.
    (4) 8,762.6 mi. (5) 17,842.2 mi.

16. Cecilia is 68 inches tall. One inch equals 2.54 centimeters. To the nearest tenth of a centimeter, how tall is Cecilia in centimeters?

    (1) 173.8 cm. (2) 172.7 cm. (3) 70.5 cm.
    (4) 26.8 cm. (5) 17.3 cm.

17. Sam shipped four packages. One weighed 2.35 kilograms; the second weighed 1.9 kilograms; the third weighed 1.4 kilograms; and the fourth weighed 2.75 kilograms. Find the average weight of the packages.

    (1) 1.1 kg. (2) 2.1 kg. (3) 2.2 kg.
    (4) 2.3 kg. (5) 8.4 kg.

18. In 1970, the number of people living in the Los Angeles-Long Beach area was 8.4 million. In 1980, the number of people living in the area was 2.3 million more than in 1970. Experts believe

that there will be another 2.3 million more people living in the area by 1990. According to the experts, how many people will be living in the Los Angeles-Long Beach area in 1990?

(1) 13,000,000 (2) 12,700,000 (3) 10,700,000
(4) 9,700,000 (5) 9,000,000

19. Frank, Phil, and Bill started a business. At the end of a year, the business made a profit of $18,924.75. The partners agreed to share the profit evenly among themselves. How much did each partner get?

(1) $4,731.19 (2) $6,308.25 (3) $6,382.25
(4) $9,462.38 (5) $12,616.50

20. From a pipe two meters long, Douglas cut off a piece 1.23 meters long. How long was the piece that was left?

(1) 1.77 m. (2) 1.23 m. (3) .87 m.
(4) .77 m. (5) .23 m.

21. Nora's car gets 13.5 miles per gallon of gas. How far can Nora drive on 15.4 gallons of gas?

(1) 20.8 mi. (2) 28.9 mi. (3) 197.9 mi.
(4) 207.9 mi. (5) 218.9 mi.

22. A batting average times a number of times at bat tells how many hits a player has gotten. Jack's batting average last season was .282. He was at bat 84 times. About how many hits did he get?

(1) 22 (2) 23 (3) 24 (4) 25 (5) 28

23. Phoebe makes $3.80 an hour. Every week, she makes $133 before taxes. How many hours does Phoebe work in a week?

(1) 40 hrs. (2) 37.5 hrs. (3) 35 hrs.
(4) 33.5 hrs. (5) 32 hrs.

24. Jason drove 184 miles in 4.5 hours. To the nearest tenth, what was his average speed in miles per hour?

(1) 48.9 mph (2) 46.0 mph (3) 41.2 mph
(4) 40.9 mph (5) 36.8 mph

25. To build a new city hall and community center, the town of Monsey got a $.75 million grant from the state, $.45 million in gifts, and a loan of $1.2 million. How much money did the town get for the project?

(1) $2.4 million (2) $2.2 million (3) $1.4 million
(4) $1.3 million (5) $.8 million

26. In 1960, about 7.1 million Americans worked on farms. By 1978, the number dropped to 3.9 million. How many fewer Americans worked on farms in 1978 than in 1960?

    (1) 4.8 million (2) 4.2 million (3) 3.8 million
    (4) 3.2 million (5) 2.8 million

27. It costs $.025 to run a TV for an hour. The Johnsons watch TV for an average of 130 hours a month. How much does it cost the Johnsons each month to watch TV?

    (1) $1.55 (2) $3.00 (3) $3.25
    (4) $8.13 (5) $32.50

28. A gallon is equal to about 3.8 liters. 6 gallons are about how many liters?

    (1) 19 l. (2) 20 l. (3) 21 l. (4) 23 l. (5) 25 l.

29. Sue bought 3.5 pounds of plums for $2.17. What was the price of a pound of plums?

    (1) 72¢ (2) 62¢ (3) 58¢ (4) 47¢ (5) 43¢

30. In 1969, the average American ate 46.7 pounds of poultry. In 1979, the average American ate 62 pounds of poultry. Find the increase in the amount of poultry an average American ate from 1969 to 1979.

    (1) 15.3 lbs. (2) 16.1 lbs. (3) 16.2 lbs.
    (4) 16.3 lbs. (5) 18.7 lbs.

Check your answers on page 111.

Decimal Review

These problems will give you a chance to practice the skills you learned in the decimal section of this book. With each answer is the name of the section in which the skills needed for the problem are explained. For every problem you miss, review the section in which the skills for that problem are explained. Then try the problem again.

DIRECTIONS: Solve each problem.

1. Write two hundred fifty-six hundred thousandths as a decimal.
2. Which decimal is biggest, .46, .503, or .5?
3. Change .0125 to a fraction and reduce.
4. Change $\frac{4}{15}$ to a decimal.
5. Round off 14.2954 to the nearest hundredth.
6. Add .285 + 9.66 + 14.
7. Subtract 8.407 from 13.3.
8. Multiply 6.08 by .47.
9. $8.51 \div 23 =$ 10. $1.176 \div .49 =$ 11. $54 \div .018 =$
12. The town of Southfield needs $4 million for a new municipal center. So far, it has raised $2.35 million. How much more money does it need?
13. John bought 4.8 meters of lumber at $6.50 a meter. Find the total cost of the lumber.
14. Dorothy cut a piece of material 4.5 yards long into 6 equal pieces. How long was each piece?

Check your answers on page 114.

ANSWERS & SOLUTIONS

EXERCISE 1

1. **380.2   6.8   65.3   19.44**

2. **30.049Ø   .207Ø   Ø13.5Ø   Ø8.06   3.Ø   Ø6.109Ø   .00204Ø**

EXERCISE 2

1. .9 = **nine tenths**
   .03 = **three hundredths**
   .015 = **fifteen thousandths**
   .27 = **twenty-seven hundredths**

2. .0039 = **thirty-nine ten-thousandths**
   .008 = **eight thousandths**
   .00361 = **three hundred sixty-one hundred-thousandths**
   .000052 = **fifty-two millionths**

3. 12.6 = **twelve and six tenths**
   2.09 = **two and nine hundredths**
   120.05 = **one hundred twenty and five hundredths**
   7.0538 = **seven and five hundred thirty-eight ten-thousandths**

4. 1.00007 = **one and seven hundred-thousandths**
   6.000123 = **six and one hundred twenty-three millionths**
   3.02806 = **three and two thousand eight hundred six hundred-thousandths**
   19.0571 = **nineteen and five hundred seventy-one ten-thousandths**

EXERCISE 3

1. nine tenths = **.9**

2. twelve and six hundredths = **12.06**

3. twenty-two thousandths = **.022**

4. thirteen and eight thousandths = **13.008**

5. three hundred seventeen ten-thousandths = **.0317**

6. forty-one and two ten-thousandths = **41.0002**

7. one hundred ninety-three hundred-thousandths = **.00193**

8. five and seven millionths = **5.000007**

9. two thousand four hundred fifty-two millionths = **.002452**

10. seventy and eighty-five hundred-thousandths = **70.00085**

EXERCISE 4

1. **.32** **.3** **.2** **.09** 2. **.07** **.093** **.64** **.424**

3. **.897** **.2** **.72** **.1032** 4. **.7** **.12** **.605**

5. **.302** **.48** **.3303** 6. **.51** **.6** **.9309**

EXERCISE 5

1. $\mathbf{\frac{2}{5}}$ $\mathbf{\frac{1}{50}}$ $\mathbf{\frac{3}{8}}$

$.4 = \frac{4}{10} = \mathbf{\frac{2}{5}}$ $.02 = \frac{2}{100} = \mathbf{\frac{1}{50}}$ $.375 = \frac{375}{1{,}000} = \mathbf{\frac{3}{8}}$

$\mathbf{6\frac{4}{5}}$ $\mathbf{9\frac{4}{25}}$

$6.8 = 6\frac{8}{10} = \mathbf{6\frac{4}{5}}$ $9.16 = 9\frac{16}{100} = \mathbf{9\frac{4}{25}}$

2. $\mathbf{\frac{1}{8}}$ $\mathbf{\frac{1}{125}}$ $\mathbf{\frac{17}{20{,}000}}$

$.125 = \frac{125}{1{,}000} = \mathbf{\frac{1}{8}}$ $.008 = \frac{8}{1{,}000} = \mathbf{\frac{1}{125}}$ $.00085 = \frac{85}{100{,}000} = \mathbf{\frac{17}{20{,}000}}$

$\mathbf{2\frac{1}{250}}$ $\mathbf{5\frac{24}{25}}$

$2.004 = 2\frac{4}{1{,}000} = \mathbf{2\frac{1}{250}}$ $5.96 = 5\frac{96}{100} = \mathbf{5\frac{24}{25}}$

3. $\mathbf{\frac{16}{25}}$ $\mathbf{\frac{3}{200}}$ $\mathbf{\frac{3}{1{,}250}}$

$.64 = \frac{64}{100} = \mathbf{\frac{16}{25}}$ $.015 = \frac{15}{1{,}000} = \mathbf{\frac{3}{200}}$ $.0024 = \frac{24}{10{,}000} = \mathbf{\frac{3}{1{,}250}}$

$\mathbf{10\frac{5}{8}}$ $\mathbf{12\frac{21}{250}}$

$10.625 = 10\frac{625}{1{,}000} = \mathbf{10\frac{5}{8}}$ $12.084 = 12\frac{84}{1{,}000} = \mathbf{12\frac{21}{250}}$

EXERCISE 6

1. **.75** $\mathbf{.83\frac{1}{3}}$ $\mathbf{.14\frac{2}{7}}$ $\mathbf{.22\frac{2}{9}}$ **.3**

$\frac{3}{4} = 4\overline{)3.00}$ with quotient **.75**; $\frac{5}{6} = 6\overline{)5.00}$ with quotient $.83\frac{2}{6} = \mathbf{.83\frac{1}{3}}$; $\frac{1}{7} = 7\overline{)1.00}$ with quotient $\mathbf{.14\frac{2}{7}}$; $\frac{2}{9} = 9\overline{)2.00}$ with quotient $\mathbf{.22\frac{2}{9}}$; $\frac{3}{10} = 10\overline{)3.0}$ with quotient **.3**

2. **.625 or $.62\frac{1}{2}$** **.5** **.4**

$\frac{5}{8} = 8\overset{\textbf{.625}}{\overline{)5.000}}$ or $8\overset{.62\frac{4}{8}}{\overline{)5.00}} = \mathbf{.62\frac{1}{2}}$ $\frac{1}{2} = 2\overset{\textbf{.5}}{\overline{)1.0}}$ $\frac{2}{5} = 5\overset{\textbf{.4}}{\overline{)2.0}}$

.35 **.11**

$\frac{7}{20} = 20\overset{\textbf{.35}}{\overline{)7.00}}$ $\frac{11}{100} = 100\overset{\textbf{.11}}{\overline{)11.00}}$

3. **.06** **.16** **.25**

$\frac{3}{50} = 50\overset{\textbf{.06}}{\overline{)3.00}}$ $\frac{4}{25} = 25\overset{\textbf{.16}}{\overline{)4.00}}$ $\frac{1}{4} = 4\overset{\textbf{.25}}{\overline{)1.00}}$

$.66\frac{2}{3}$ **$.08\frac{1}{3}$**

$\frac{2}{3} = 3\overset{\mathbf{.66\frac{2}{3}}}{\overline{)2.00}}$ $\frac{1}{12} = 12\overset{.08\frac{4}{12}}{\overline{)1.00}} = \mathbf{.08\frac{1}{3}}$

EXERCISE 7

1. **.3 .4 5.1 3.8 7.0 4.7**

2. **.36 .03 1.22 4.92 12.19 2.19**

3. **.036 .130 4.833 16.008 2.583 1.064**

EXERCISE 8

1. **1.789 1.396 1.7** 2. **.5014 .7626**

$$\begin{array}{r} .48 \\ .6 \\ +.709 \\ \hline \mathbf{1.789} \end{array} \quad \begin{array}{r} .27 \\ .726 \\ +.4 \\ \hline \mathbf{1.396} \end{array} \quad \begin{array}{r} .6 \\ .8 \\ +.3 \\ \hline \mathbf{1.7} \end{array} \quad \begin{array}{r} .0764 \\ .3 \\ +.125 \\ \hline \mathbf{.5014} \end{array} \quad \begin{array}{r} .009 \\ .28 \\ +.4736 \\ \hline \mathbf{.7626} \end{array}$$

3. **31.89 10.304 7.059** 4. **10.3625 10.6685**

$$\begin{array}{r} 4.7 \\ 26. \\ +\ 1.19 \\ \hline \mathbf{31.89} \end{array} \quad \begin{array}{r} .804 \\ 2.5 \\ +7. \\ \hline \mathbf{10.304} \end{array} \quad \begin{array}{r} .229 \\ 6. \\ +\ .83 \\ \hline \mathbf{7.059} \end{array} \quad \begin{array}{r} 1.0025 \\ 9. \\ +\ .36 \\ \hline \mathbf{10.3625} \end{array} \quad \begin{array}{r} 8. \\ .0685 \\ +2.6 \\ \hline \mathbf{10.6685} \end{array}$$

EXERCISE 9

1.	**8.643**	**11.74**	**.262**	**.331**
	9.000	12.00	.400	.360
	− .357	− .26	−.138	−.029
	8.643	**11.74**	**.262**	**.331**

2.	**7.951**	**.015**	**49.4**	**4.93**
	8.000	.020	50.0	8.93
	− .049	−.005	− .6	−4.
	7.951	**.015**	**49.4**	**4.93**

3.	**.028**	**.7695**	**.419**	**8.492**
	.036	1.0000	.700	9.000
	−.008	− .2305	−.281	− .508
	.028	**.7695**	**.419**	**8.492**

EXERCISE 10

1.	**20.8**	**2.82**	**.7**	**.054**
	2.6	.47	14	.09
	× 8	× 6	×.05	× .6
	20.8	**2.82**	.70 = **.7**	**.054**

2.	**8.4**	**.3168**	**8.184**	**.00333**
	.35	.792	.93	.37
	×24	× .4	× 8.8	× .009
	1 40	**.3168**	74 4	**.00333**
	7 0		7 44	
	8.40 = **8.4**		**8.18 4**	

3.	**208.5**	**.02242**	**.04884**	**16.92**
	4.17	.059	.0814	423
	× 50	× .38	× .6	× .04
	208.50 = **208.5**	472	**.04884**	**16.92**
		177		
		.02242		

EXERCISE 11

1.	**4.9**	**.58**	**.085**	**9.2**
	4.9	**.58**	**.085**	**9.2**
	8)39.2	3)1.74	9).765	6)55.2
	32	1 5	72	54
	7 2	24	45	1 2
	7 2	24	45	1 2

2. **2.8** **.37** **6.7**

```
     2.8          .37          6.7
12)33.6      24)8.88      52)348.4
   24           7 2          312
    9 6         1 68          36 4
    9 6         1 68          36 4
```

3. **.046** **1.87** **.43**

```
    .046         1.87          .43
7).322       16)29.92     36)15.48
   28           16           14 4
    42          13 9          1 08
    42          12 8          1 08
                 1 12
                 1 12
```

EXERCISE 12

1. **.6** **.24** 2. **9.7** **.81**

```
     .6            .24             9.7          .81
.9).5 4       .4).0 96       .2)1.9 4     .7).5 67
    5 4             8            1 8          5 6
                   16              1 4          07
                   16              1 4           7
```

3. **8.6** **37** 4. **1.09** **.08**

```
       8.6           37            1.09           .08
.06).51 6     .08)2.96     2.3)2.5 07     .54).04 32
     48             2 4           2 3             4 32
      3 6            56             2 0
      3 6            56               0
                                    2 07
                                    2 07
```

5. **16** **1.7** 6. **.8** **6.3**

```
          16              1.7               .8              6.3
.024).384       6.23)10.59 1      41.4)33.1 2     .509)3.206 7
       24                6 23             33 1 2          3 054
       144               4 36 1                             152 7
       144               4 36 1                             152 7
```

EXERCISE 13

1. **35** **300** **400** **80**

```
      3 5            3 00           4 00           8 0
2.4)84.0      .38)114.00     .09)36.00     1.6)128.0
    72            114            36             128
    12 0            0 00           0 00           0 0
    12 0
```

2. **230** **2,000** **2,155** **22**

```
       23 0            2 000         215 5           2 2
7.2)1656.0    .007)14.000     .4)862.0     8.5)187.0
    144               14          8             170
    216                0 000      06             17 0
    216                            4             17 0
      0 0                         22
                                  20
                                   2 0
                                   2 0
```

3. **400** **320** **1,300** **30**

```
        400            32 0          13 00           3 0
.015)6.000      .6)192.0     .48)624.00    6.6)198.0
      6 0           18            48            198
       000          12            144             0 0
                    12            144
                     0 0            0 00
```

EXERCISE 14 (WORD PROBLEMS)

1. (3) **74.25 kg.**

```
   165
 × .45
  8 25
 66 0
 74.25
```

2. (3) **$1.70**

```
      $ 1.70
2.8)$4.7 60
      2 8
      1 9 6
      1 9 6
         00
```

3. (4) **200,000**

```
  8.0          1,000,000
 −7.8               × .2
   .2 million  200,000.0
```

4. (4) **.288**

```
     .2875
80)23.0000
   16 0
    7 00
    6 40
      600
      560
       400
       400
```

.2875 to the nearest thousandth = **.288**

5. (2) **20,413.6**

18,450.0
+ 1,963.6
20,413.6

6. (3) **$255.00**

37.5
×6.80
30 000
225 0
255.00Ø

7. (2) **.75 yd.**

.75
13)9.75
9 1
65
65

8. (3) **103.1°**

98.6°
+ 4.5
103.1°

9. (4) **$3.83**

$2.40
×.75
12 00
1 68 0
$1.80 ØØ

$1.90
×2.3
5 7 0
3 8 0
$4.3 7 Ø

$1.80
+4.37
$6.17

$10.00
− 6.17
$ 3.83

10. (2) **3.5 min.**

16.2
−12.7
3.5

11. (3) **$27.72**

3.36
×8.25
16 80
67 2
26 88
27.72 ØØ

12. (3) **220.1 million**

41.1
71.6
58.4
+49.0
220.1

13. (2) **18.2 mi.**

18.18
22)400.00
22
180
176
4 0
2 2
1 80
1 76

18.18 to the nearest tenth = **18.2**

14. (4) **$.44**

1.68
× 4.5
84 0
6 72
7.56 Ø

$8.00
−7.56
$ .44

15. (2) **7,752.6 mi.**

46,213.4
−38,460.8
7,752.6

16. (2) **172.7 cm.**

2.54
× 68
20 32
152 4
172.72 to the nearest tenth = **172.7**

17. (2) **2.1 kg.**

2.35
1.9
1.4
+2.75
8.40

2.1
4)8.4

18. (1) **13,000,000**

8.4
2.3
+2.3
13.0

1,000,000
× 13
13,000,000

19. (2) **$6,308.25**

$6,308.25
3)$18,924.75

20. (4) **.77 m.**

2.00
−1.23
.77

21. (4) **207.9 mi.**

13.5
×15.4
5 4 0
67 5
135
207.9 0

22. (3) **24**

.282
× 84
1 128
22 56
23.688 to the nearest
unit = **24**

23. (3) **35**

35
3.80)133.00
114 0
19 00
19 00

24. (4) **40.9 mph**

4 0.88
4.5)184.0 00
180
4 0
0
4 0 0
3 6 0
4 00
3 60

40.88 to the nearest
tenth = **40.9**

25. (1) **$2.4 million**

$.75
.45
+1.2
$2.40

26. (4) **3.2 million**

7.1
−3.9
3.2

27. (3) **$3.25**

$.025
× 130
750
2 5
$3.250

28. (4) **23 l.**

3.8
× 6
22.8 to the nearest
unit = **23**

29. (2) **62¢**

$ **.62**
3.5)$2.1 70
2 1 0
70
70

30. (1) **15.3 lbs.**

62.0
−46.7
15.3

DECIMAL REVIEW

1. **.00256**

(Writing Decimals)

2. **.503**

(Comparing Decimals)

3. $\mathbf{\frac{1}{80}}$

$.0125 = \frac{125}{10{,}000} = \mathbf{\frac{1}{80}}$

(Changing Decimals to Fractions)

4. $\mathbf{.26\frac{2}{3}}$

$\frac{4}{15} = 15\overline{)4.00}$ $.26\frac{10}{15} = \mathbf{.26\frac{2}{3}}$

(Changing Fractions to Decimals)

5. **14.30**

(Rounding Off Decimals)

6. **23.945**

$$\begin{array}{r} .285 \\ 9.66 \\ +14. \\ \hline \mathbf{23.945} \end{array}$$

(Adding Decimals)

7. **4.893**

$$\begin{array}{r} 13.300 \\ -8.407 \\ \hline \mathbf{4.893} \end{array}$$

(Subtracting Decimals)

8. **2.8576**

$$\begin{array}{r} 6.08 \\ \times\ .47 \\ \hline 42\ 56 \\ 2\ 43\ 2 \\ \hline \mathbf{2.85\ 76} \end{array}$$

(Multiplying Decimals)

9. **.37**

$$\begin{array}{r} \mathbf{.37} \\ 23\overline{)8.51} \\ 6\ 9 \\ \hline 1\ 61 \\ 1\ 61 \\ \hline \end{array}$$

(Dividing Decimals by Whole Numbers)

10. **2.4**

$$\begin{array}{r} \mathbf{2.4} \\ .49_{\wedge}\overline{)1.17_{\wedge}6} \\ 98 \\ \hline 19\ 6 \\ 19\ 6 \\ \hline \end{array}$$

(Dividing Decimals by Decimals)

11. **3,000**

$$\begin{array}{r} \mathbf{3\ 000} \\ .018_{\wedge}\overline{)54.000_{\wedge}} \\ 54 \\ \hline 0\ 000 \end{array}$$

(Dividing Whole Numbers by Decimals)

12. **$1.65 million**

$$\begin{array}{r} \$4.00 \text{ million} \\ -\ 2.35 \phantom{\text{ million}} \\ \hline \mathbf{\$1.65 \text{ million}} \end{array}$$

(Decimal Word Problems)

13. **$31.20**

```
  $6.5 0
 ×  4.8
  5 2 0 0
 26 0 0
$31.2 0 0̸
```

(Decimal Word Problems)

14. **.75 yd.**

```
    .75 yd.
6)4.50
  4 2
    30
    30
```

(Decimal Word Problems)

Percents

You learned that a decimal is a kind of fraction. A percent is also a kind of fraction. Percents are often used in business. Interest rates, sales tax, and discounts are all measured in percents. For example, 7% is the sales tax rate in some states. 7% means 7 out of 100 equal parts. For every 100 cents you spend on something, you pay 7 cents in tax.

Percents are different from common fractions in two ways. One difference is that 100 is the only number that can be a denominator for percents. The other difference is that the denominator 100 is not written. Instead of writing 100, the percent sign (%) is written.

Percents are almost like two-place decimals. A decimal with two places is called hundredths. The denominator for percents is 100. Instead of two decimal places, percents use the percent sign.

The following list shows examples of percents and the decimals and fractions each percent is equal to. Study the list to learn the differences.

Percent	Decimal	Fraction
75%	.75	$\frac{75}{100} = \frac{3}{4}$
4%	.04	$\frac{4}{100} = \frac{1}{25}$
8.5%	.085	$\frac{85}{1000} = \frac{17}{200}$
90%	.9	$\frac{9}{10}$

Changing Decimals to Percents

To change a decimal to a percent, move the decimal point two places to the **right** and write a percent sign.

EXAMPLE: Change .45 to a percent.

.45 45.	.45 = 45%
Step 1	Step 2

Step 1: Move the decimal point two places to the right.
Step 2: Add a percent sign to the number. (When the decimal point comes at the end of the number, the decimal point is dropped.)

EXAMPLE: Change .087 to a percent.

.087 08.7	.087 = 8.7%
Step 1	Step 2

Step 1: Move the decimal point two places to the right.
Step 2: Add a percent sign to the number.

EXAMPLE: Change $.11\frac{1}{9}$ to a percent.

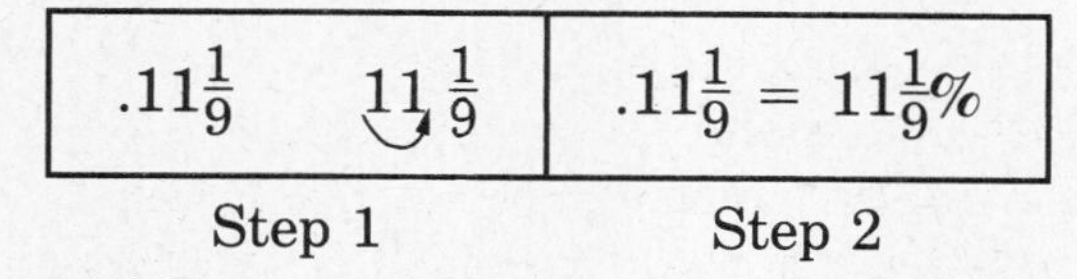

Step 1 Step 2

Step 1: Move the decimal point two places to the right.
Step 2: Add a percent sign to the number. (When the decimal point comes right before a fraction, the decimal point is dropped.)

Sometimes, you will have to put zeros after the decimal to get two places.

EXAMPLE: Change .7 to a percent.

.7 = .70	.70 70.	.7 = 70%
Step 1	Step 2	Step 3

Step 1: Put a zero at the right of .7 so that the decimal has two places.
Step 2: Move the decimal point two places to the right.
Step 3: Add the percent sign to the number.

EXERCISE 1

DIRECTIONS: Change each decimal to a percent.

1. .38 = .06 = .024 = $.08\frac{1}{3}$ =

2. .9 = .25 = .001 = .01 =

3. 4.75 = .625 = 6. = $.66\frac{2}{3}$ =

Check your answers on page 135.

Changing Percents to Decimals

To change a percent to a decimal, move the decimal point two places to the **left** and take off the percent sign.

EXAMPLE: Change 35% to a decimal.

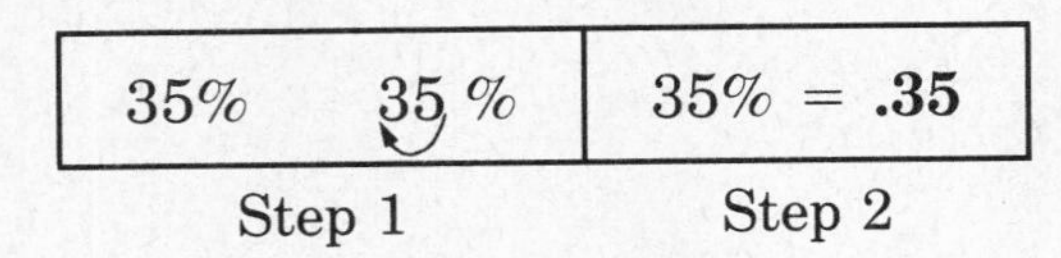

Step 1 Step 2

Step 1: Move the decimal point two places to the left. The point is understood to be at the right of the 5 in 35%.
Step 2: Take off the percent sign.

EXAMPLE: Change 20% to a decimal.

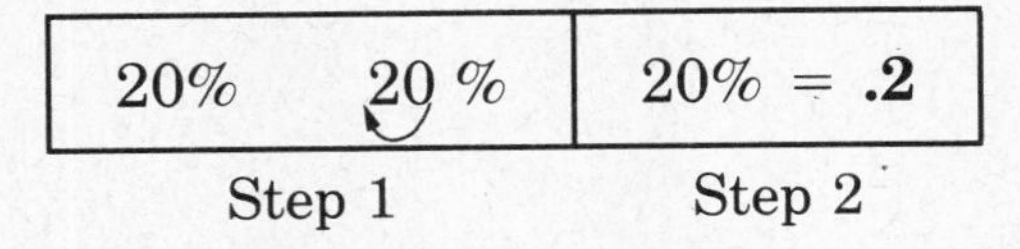

Step 1 Step 2

Step 1: Move the decimal point two places to the left.
Step 2: Take off the percent sign. (The zero at the right in .20 is taken off because it does not change the value of .2.)

Sometimes you will have to put zeros at the left of the number to get two places.

EXAMPLE: Change 6.8% to a decimal.

6.8% 06 8 %	6.8% = **.068**
Step 1	Step 2

Step 1: Move the decimal point two places to the left. Put a zero at the left of the number to make two places.
Step 2: Take off the percent sign.

EXERCISE 2

DIRECTIONS: Change each percent to a decimal.

1.	25% =	8% =	37.5% =	3% =
2.	80% =	250% =	1% =	500% =
3.	$6\frac{1}{4}$% =	11.5% =	.8% =	19% =

Check your answers on page 135.

Changing Fractions to Percents

There are two ways to change a fraction to a percent. Study the examples carefully. Then choose the method that works better for you.

One way is to find a fraction of 100%. In other words, multiply the fraction by 100%.

EXAMPLE: Change $\frac{7}{20}$ to a percent.

Step 1	Step 2
$\frac{7}{20} \times 100\% = \frac{7}{\cancel{20}_{1}} \times \frac{\overset{5}{\cancel{100}}}{1} = \frac{35}{1} = \mathbf{35}$	$\frac{7}{20} = 35\%$

Step 1: Multiply the fraction by 100%.
Step 2: Add the percent sign to the answer.

The other way to change a fraction to a percent is to change the fraction to a decimal first. Then change the decimal to a percent.

EXAMPLE: Change $\frac{4}{25}$ to a percent.

Step 1	Step 2	Step 3
$\frac{4}{25} = 25\overline{)4.00}$ = **.16** $-2\,5$ $1\,50$ $-1\,50$	.16 → 16	$\frac{4}{25} = 16\%$

Step 1: Change the fraction to a decimal.
Step 2: Move the decimal point two places to the right.
Step 3: Add the percent sign.

EXERCISE 3

DIRECTIONS: Change each fraction to a percent.

1. $\frac{2}{5} =$ $\qquad \frac{4}{25} =$ $\qquad \frac{1}{2} =$ $\qquad \frac{1}{3} =$
2. $\frac{11}{100} =$ $\qquad \frac{3}{10} =$ $\qquad \frac{5}{8} =$ $\qquad \frac{1}{16} =$
3. $\frac{4}{7} =$ $\qquad \frac{5}{6} =$ $\qquad \frac{9}{20} =$ $\qquad \frac{4}{5} =$

Check your answers on page 136.

Changing Percents to Fractions

Remember that a percent is a kind of fraction. To change a percent to a fraction, first write the digits in the percent as the numerator. Then write 100 as the denominator. Reduce the fraction if you can.

EXAMPLE: Change 24% to a fraction.

Step 1	Step 2	Step 3
$\underline{24}$	$\frac{24}{100}$	$\frac{24}{100} = \frac{6}{25}$

Step 1: Write the percent as the numerator.
Step 2: Write 100 as the denominator.
Step 3: Reduce the fraction to lowest terms.

When the percent has a decimal in it, first change the percent to a decimal. Then change the decimal to a fraction.

EXAMPLE: Change 4.5% to a fraction.

Step 1	Step 2	Step 3
4.5% 04 5 %	$.045 = \frac{45}{1{,}000}$	$\frac{45}{1{,}000} = \frac{9}{200}$

Step 1: Change the percent to a decimal. Move the decimal point two places to the left.
Step 2: Change the decimal to a fraction.
Step 3: Reduce the fraction to lowest terms.

When a percent has a fraction in it, write the digits in the percent as the numerator. Write 100 as the denominator. Then divide the numerator by the denominator. Study the next example carefully.

EXAMPLE: Change $16\frac{2}{3}\%$ to a fraction.

Step 1	Step 2	Step 3
$16\frac{2}{3}$	$\frac{16\frac{2}{3}}{100}$	$16\frac{2}{3} \div 100 = \frac{50}{3} \div \frac{100}{1} = \frac{\overset{1}{\cancel{50}}}{3} \times \frac{1}{\underset{2}{\cancel{100}}} = \frac{1}{6}$

Step 1: Write the percent as the numerator.
Step 2: Write 100 as the denominator.
Step 3: Divide.

EXERCISE 4

DIRECTIONS: Change each percent to a fraction and reduce.

1. 5% =	35% =	96% =	60% =
2. 275% =	8% =	520% =	43% =
3. 1.5% =	4.8% =	12.5% =	6.25% =
4. $4\frac{1}{6}\%$ =	$31\frac{1}{4}\%$ =	$26\frac{2}{3}\%$ =	$58\frac{1}{3}\%$ =

Check your answers on page 136.

Common Fractions, Decimals, and Percents

As you work with percent problems, you will see that sometimes it is easier to change a percent to a fraction. Other times, it is easier to change a percent to a decimal. The chart below has some of the most common percents, along with the decimals and fractions each percent is equal to. You will save a lot of time later on if you take the time now to memorize the chart.

Percent		Decimal		Fraction	Percent		Decimal		Fraction
25%	=	.25	=	$\frac{1}{4}$	20%	=	.2	=	$\frac{1}{5}$
50%	=	.5	=	$\frac{1}{2}$	40%	=	.4	=	$\frac{2}{5}$
75%	=	.75	=	$\frac{3}{4}$	60%	=	.6	=	$\frac{3}{5}$
					80%	=	.8	=	$\frac{4}{5}$
$12\frac{1}{2}\%$	=	$.12\frac{1}{2}$	=	$\frac{1}{8}$					
$37\frac{1}{2}\%$	=	$.37\frac{1}{2}$	=	$\frac{3}{8}$	10%	=	.1	=	$\frac{1}{10}$
$62\frac{1}{2}\%$	=	$.62\frac{1}{2}$	=	$\frac{5}{8}$	30%	=	.3	=	$\frac{3}{10}$
$87\frac{1}{2}\%$	=	$.87\frac{1}{2}$	=	$\frac{7}{8}$	70%	=	.7	=	$\frac{7}{10}$
					90%	=	.9	=	$\frac{9}{10}$
$33\frac{1}{3}\%$	=	$.33\frac{1}{3}$	=	$\frac{1}{3}$					
$66\frac{2}{3}\%$	=	$.66\frac{2}{3}$	=	$\frac{2}{3}$	$16\frac{2}{3}\%$	=	$.16\frac{2}{3}$	=	$\frac{1}{6}$
					$83\frac{1}{3}\%$	=	$.83\frac{1}{3}$	=	$\frac{5}{6}$

Finding a Percent of a Number

When you studied fraction word problems, you learned that a fraction immediately followed by the word *of* means to multiply. A percent immediately followed by the word *of* also means to multiply. To multiply by a percent, first change the percent to a fraction or a decimal. Then multiply by the fraction or decimal.

EXAMPLE: Find 45% of 120.

Using a fraction:

Step 1	Step 2
$45\% = \frac{45}{100} = \mathbf{\frac{9}{20}}$	$\frac{9}{20} \times 120 = \frac{9}{\cancel{20}_1} \times \frac{\cancel{120}^6}{1} = \frac{54}{1} = \mathbf{54}$

Step 1: Change the percent to a fraction.
Step 2: Multiply 120 by $\frac{9}{20}$.

Using a decimal:

Step 1	Step 2
45 % → **.45**	$\begin{array}{r} 120 \\ \times .45 \\ \hline 6\,00 \\ 48\,0 \\ \hline 54.00 \end{array} = \mathbf{54}$

Step 1: Change the percent to a decimal.
Step 2: Multiply 120 by .45.

Sometimes, decimals are easier to use than fractions. When the percent itself has a decimal, first change the percent to a decimal. Then multiply by the decimal.

EXAMPLE: Find 8.5% of 240.

Step 1	Step 2
08 5 % → **.085**	$\begin{array}{r} 240 \\ \times .085 \\ \hline 1\,200 \\ 19\,20 \\ \hline 20.400 \end{array} = \mathbf{20.4}$

Step 1: Change the percent to a decimal.
Step 2: Multiply 240 by .085.

Percents with fractions in them are hardest to use. When a percent has a fraction in it, first change the percent to a fraction. Then multiply by the fraction.

EXAMPLE: Find $26\frac{2}{3}\%$ of 90.

Step 1	Step 2	Step 3
$\dfrac{\mathbf{26\frac{2}{3}}}{\mathbf{100}}$	$26\frac{2}{3} \div 100 = \frac{80}{3} \div \frac{100}{1} =$ $\frac{\overset{4}{\cancel{80}}}{3} \times \frac{1}{\underset{5}{\cancel{100}}} = \mathbf{\frac{4}{15}}$	$\frac{4}{\underset{1}{\cancel{15}}} \times \frac{\overset{6}{\cancel{90}}}{1} = \frac{24}{1} = \mathbf{24}$

Step 1: Change the percent to a fraction. Write $26\frac{2}{3}$ as the numerator and 100 as the denominator.

Step 2: Divide $26\frac{2}{3}$ by 100.

Step 3: Multiply 90 by $\frac{4}{15}$.

EXERCISE 5

DIRECTIONS: Solve each problem. Use either fractions or decimals.

1. 80% of 65 =   75% of 120 =   24% of 150 =   60% of 35 =

2. 5% of 200 =   200% of 48 =   125% of 60 =   8% of 250 =

3. 62.5% of 40 =   10.5% of 216 =   4.2% of 500 =   6.3% of 300 =

4. 1.9% of 600 =   0.5% of 800 =   1.25% of 440 =   0.4% of 200 =

5. $66\frac{2}{3}\%$ of 90 = $37\frac{1}{2}\%$ of 120 = $8\frac{1}{3}\%$ of 60 = $87\frac{1}{2}\%$ of 400 =

6. $83\frac{1}{3}\%$ of 180 = $12\frac{1}{2}\%$ of 240 = $16\frac{2}{3}\%$ of 72 = $6\frac{1}{4}\%$ of 200 =

Check your answers on page 137.

Finding What Percent One Number Is of Another

In some percent problems, you must find a percent as an answer. When you find what percent one number is of another, you compare a **part** to a **whole.** First, make a fraction with the part as the numerator and the whole as the denominator. Then change the fraction to a percent.

EXAMPLE: 16 is what % of 64?

Step 1	Step 2	Step 3
$\frac{16}{64}$	$\frac{16}{64} = \frac{1}{4}$	$\frac{1}{\cancel{4}_{1}} \times \frac{\cancel{100}^{25}}{1} = \frac{25}{1} = \mathbf{25\%}$

Step 1: Make a fraction with the part (16) over the whole (64).
Step 2: Reduce the fraction to lowest terms.
Step 3: Change the fraction to a percent.

EXERCISE 6

DIRECTIONS: Solve each problem.

1. 18 is what % of 30? 27 is what % of 36? 72 is what % of 80?

2. 12 is what % of 36? 36 is what % of 80? 30 is what % of 45?

3. 21 is what % of 56? 8 is what % of 10? 21 is what % of 60?

4. 18 is what % of 72? 11 is what % of 66? 75 is what % of 120?

Check your answers on page 139.

Finding a Number When a Percent of It Is Known

Some percent problems seem backwards. In these problems, you have a number that is a percent of the missing number. To find the missing number, you have to divide. First, change the percent to a fraction or a decimal. Then divide the number you have by the fraction or decimal.

EXAMPLE: 40% of what number is 52?

(If you had the missing number, 40% multiplied by that number would be 52. To find the missing number, divide 52 by 40%. In algebra, you will often use this method to solve problems. It is called using *opposite operations*. Division is the opposite of multiplication.)

Using a fraction:

Step 1	Step 2
$40\% = \frac{40}{100} = \mathbf{\frac{2}{5}}$	$52 \div \frac{2}{5} = \frac{52}{1} \div \frac{2}{5} = \frac{\overset{26}{\cancel{52}}}{1} \times \frac{5}{\underset{1}{\cancel{2}}} = \frac{130}{1} = \mathbf{130}$

Step 1: Change 40% to a fraction.

Step 2: Divide 52 by $\frac{2}{5}$.

Using a decimal:

Step 1	Step 2
$40\% = 40.\% = .4$	$\begin{array}{r} \mathbf{13\,0} \\ .4\overline{)\ 52.0} \\ -4 \\ \hline 12 \\ -12 \\ \hline 0\,0 \end{array}$

Step 1: Change 40% to a decimal.
Step 2: Divide 52 by .4.

To check this problem, find 40% of 130. The answer should be 52.

EXERCISE 7

DIRECTIONS: Solve each problem.

1. 30% of what number is 24? 75% of what number is 36?
2. $16\frac{2}{3}\%$ of what number is 15? $37\frac{1}{2}\%$ of what number is 42?
3. 4.5% of what number is 1.8? 12.5% of what number is 5?
4. 60% of what number is 120? 15% of what number is 96?
5. 85% of what number is 34? $66\frac{2}{3}\%$ of what number is 52?

Check your answers on page 140.

Simple Interest

Interest is a common application of percents. Interest is money that someone pays for using someone else's money. A bank pays you interest for using your money in a savings account. You pay a bank interest for using the bank's money on a loan.

To find interest, multiply the **principal** by the **rate** and by the **time.**

The **principal** is the money borrowed or loaned.
The **rate** is the percent of interest.
The **time** is the number of years the money is borrowed or loaned.

EXAMPLE: Find the interest on $600 at 7% annual interest for one year.

Step 1	Step 2
$7\% = \frac{7}{100}$	$\frac{\overset{6}{\cancel{600}}}{1} \times \frac{7}{\underset{1}{\cancel{100}}} \times 1 = \frac{42}{1} = \mathbf{\$42}$

Step 1: Change the rate, 7%, to a fraction. (The problem also will work if you change the rate to a decimal.)
Step 2: Multiply the principal by the rate and by the time. (When the time is one year, you do not have to multiply by it. Remember, multiplying by 1 does not change the value of a number.)

When you do interest problems, follow the same rules for doing other percent problems. Use fractions when the percent itself has a fraction in it. Use decimals when the percent itself has a decimal in it.

EXERCISE 8

DIRECTIONS: Find the interest for each of the following.

1. $800 at 6% annual interest for one year.

2. $500 at 5% annual interest for one year.

3. $1,000 at 6.5% annual interest for one year.

4. $1,200 at 8.9% annual interest for one year.

5. $6,000 at 15% annual interest for one year.

6. \$400 at $3\frac{1}{4}\%$ annual interest for one year.

7. \$4,000 at $6\frac{3}{4}\%$ annual interest for one year.

8. \$1,800 at 7.5% annual interest for one year.

Check your answers on page 141.

When the time period for interest is not one year, first change the time to a fraction of a year. Then find the interest.

EXAMPLE: Find the interest on \$700 at 6% annual interest for 4 months.

Step 1	Step 2	Step 3
$6\% = \frac{6}{100}$	$\frac{4}{12} = \frac{1}{3}$	$\frac{\overset{7}{\cancel{700}}}{1} \times \frac{\overset{2}{\cancel{6}}}{\underset{1}{\cancel{100}}} \times \frac{1}{\underset{1}{\cancel{3}}} = \frac{14}{1} = \$14$

Step 1: Change 6% to a fraction.
Step 2: Change 4 months to a fraction of a year. Write 4 as the numerator and 12 as the denominator. (There are 12 months in one whole year.)
Step 3: Multiply the principal by the rate and by the time.

EXERCISE 9

DIRECTIONS: Find the interest for each of the following.

1. \$600 at 8% annual interest for 9 months.

2. \$1,500 at 5% annual interest for 1 year and 6 months.

3. \$200 at 6% annual interest for 2 years.

4. \$2,400 at 4.5% annual interest for 8 months.

5. \$3,000 at 9% annual interest for 1 year and 4 months.

6. \$700 at $6\frac{1}{2}\%$ annual interest for 2 years and 6 months.

7. $450 at 9% annual interest for 1 year and 8 months.

8. $3,200 at 12% annual interest for 2 years and 3 months.

Check your answers on page 142.

Percent Word Problems

Most applications of percent skills on the GED Test are in word problems. Here are some hints to help you identify the operations you will need to solve percent problems.

Finding a percent of a number is the most common percent problem. In these problems, you have the percent and the number you must find a percent of. You have already practiced an example of these problems with interest problems. Many examples of finding a percent of a number take two steps.

Sometimes you must find a percent of a number and then add.

EXAMPLE: Cheryl bought a shirt for $12.95.The sales tax on clothing in her state is 6%. What was the total price of the shirt, including tax?

Step 1	Step 2	Step 3	Step 4
6% = 06 % = **.06**	$12.95 × .06 **$.7770**	$.7770 **$.78**	$12.95 + .78 **$13.73**

Step 1: Change 6% to a decimal.
Step 2: Multiply the price of the shirt by .06.
Step 3: Round off the product to the nearest cent.
Step 4: Add the price of the shirt to the tax.

Another example of a percent problem in which you must add is finding a markup. A **markup** is the amount a store owner adds to the price he pays for an item. To find the price a customer pays, add the markup to the price the store owner pays.

Sometimes you must find a percent of a number and then subtract.

EXAMPLE: A coat was marked $45. The coat was put on sale for 20% off. Find the sale price of the coat.

Step 1	Step 2	Step 3
$20\% = \frac{20}{100} = \mathbf{\frac{1}{5}}$	$\frac{1}{\cancel{5}_1} \times \frac{\cancel{45}^9}{1} = \frac{9}{1} = \mathbf{\$9}$	$45 − 9 **$36**

Step 1: Change 20% to a fraction.

Step 2: Multiply the original price by $\frac{1}{5}$.
Step 3: Subtract the discount, $9, from the original price.

Finding what percent one number is of another is the easiest percent problem to recognize. The percent is not given in these problems. You have to find the percent. Sometimes you have to compare the difference between two amounts to an original amount.

EXAMPLE: Carl bought a fishing rod on sale for $30. Before the sale, the fishing rod cost $40. Find the percent of the discount on the fishing rod.

Step 1	Step 2	Step 3
$40 − 30 = **$10**	$\frac{10}{40} = \mathbf{\frac{1}{4}}$	$\frac{1}{\not{4}_1} \times \frac{\not{100}^{25}}{1} = \frac{25}{1} = \mathbf{25\%}$

Step 1: Find the difference between the prices. The difference is the discount.
Step 2: Make a fraction. Write the discount as the numerator and the original price of the fishing rod as the denominator. Reduce the fraction to lowest terms.
Step 3: Change the fraction, $\frac{1}{4}$, to a percent.

Finding a number when a percent of it is known is the hardest percent problem to recognize. Remember that in these problems you have a percent and the number you get when you multiply by the percent. You must divide to find the missing number.

EXAMPLE: The sales tax on clothing in Tony's state is 8%. Tony paid $2.36 in sales tax for a jacket. Find the price of the jacket.

Step 1	Step 2
8% = 08 % = **.08**	.08) $2.36 00 = **$ 29.50** − 1 6 76 −72 4 0 −4 0 00

Step 1: Change 8% to a decimal.
Step 2: Divide the tax, $2.36, by .08.

EXERCISE 10 (WORD PROBLEMS)

DIRECTIONS: Choose the correct answer to each problem.

1. Faye bought a sweater for \$24.80. The sales tax in her state is 6%. Find the total price of the sweater, including sales tax.

   (1) \$23.31 (2) \$24.86 (3) \$25.49
   (4) \$26.29 (5) \$30.80

2. Last year, Mrs. Cooke paid 60¢ for a loaf of bread. This year, she pays 69¢ for the same brand of bread. By what percent did the price of bread go up in a year?

   (1) 15% (2) 12% (3) 10% (4) 9% (5) $8\frac{2}{3}$%

3. Ray's Bargain Store offers 15% off the list price of every item in the store. During a special sale, there is an additional $\frac{1}{4}$ off the regular discount. During the special sale, what is the price of a suit listed at \$96?

   (1) \$87.60 (2) \$81.60 (3) \$76.00
   (4) \$68.80 (5) \$61.20

4. Gloria has \$24.50 taken from her weekly paycheck for taxes and Social Security. The deductions are 20% of Gloria's gross wages. Find Gloria's gross wage each week.

   (1) \$44.50 (2) \$122.50 (3) \$124.50
   (4) \$144.50 (5) \$490.00

5. Mr. Conrad pays a manufacturer \$12 each for dress shirts. Mr. Conrad puts a 40% markup on the shirts. How much does Mr. Conrad charge his customers for a dress shirt?

   (1) \$12.40 (2) \$14.60 (3) \$16.00
   (4) \$16.80 (5) \$18.40

6. Before Christmas, a portable radio sold for \$48. After Christmas, the radio was on sale for \$38.40. What percent was the discount on the radio?

   (1) 9.6% (2) 15% (3) 20% (4) 30% (5) 96%

7. David's Wednesday night math class usually has 24 students. Because of bad weather, one night 4 students were absent. What percent of the students were absent?

   (1) 4% (2) $16\frac{2}{3}$% (3) 20% (4) 25% (5) 40%

8. George and Margie are buying a house for $42,000. They have to make a down payment of 15%. How much is the down payment on the house?

   (1) $4,200 (2) $6,300 (3) $15,000
   (4) $25,700 (5) $27,000

9. Joan wants to buy new furniture for her living room. The furniture is listed for $520. Joan decided to pay 15% down and $24 a month for 21 months. How much more does she have to pay for the furniture by paying installments instead of paying cash?

   (1) $102 (2) $82 (3) $78 (4) $62 (5) $16

10. Colette works on a commission rate. She gets 5% of the value of the shoes she sells. One week, she made $104.50 in commission. What was the total value of the shoes she sold that week?

    (1) $522.50 (2) $1,045 (3) $2,090
    (4) $5,225 (5) $10,450

11. Fred makes $960 a month. He pays $192 a month for rent. Fred's rent is what percent of his income?

    (1) 92% (2) 80% (3) 30% (4) 25% (5) 20%

12. In 1970, Erica's house cost $24,600. In 1980, the value of her house was 16% more than the 1970 price. Find the value of Erica's house in 1980. Round off the answer to the nearest hundred.

    (1) $24,600 (2) $26,600 (3) $28,500
    (4) $29,000 (5) $30,000

13. The Consumer Price Index in 1979 was 217.4. This means that a family in 1979 had to pay 217.4% of the price they paid in 1967 for the same goods and services. For the Jones family, $8,000 covered all their expenses in 1967. How much did the family need in 1979 to cover the same expenses?

    (1) $21,740 (2) $17,392 (3) $16,740
    (4) $16,000 (5) $10,174

14. There are 225 workers at the Atlas Steel Company. 18 of the workers are women. What percent of the workers are women?

    (1) 2% (2) 8% (3) 10% (4) 12% (5) 18%

15. Nick made a down payment of $896 for a new car. The down payment was 20% of the price of the car. Find the total price of the car.

    (1) $1,792 (2) $2,896 (3) $3,584
    (4) $4,260 (5) $4,480

16. A portable TV had a list price of $120. Pete bought it on sale for $84. Find the percent of discount on the price of the TV.

    (1) 42.9% (2) 41.4% (3) 40%
    (4) 30% (5) 25%

17. Bill's gross salary is $13,500 a year. His employer deducts 21% from Bill's salary for taxes and Social Security. Find Bill's yearly net salary.

    (1) $16,335 (2) $13,479 (3) $12,835
    (4) $11,400 (5) $10,665

18. Janet owes $340 on her credit card. Every month, she has to pay a fee of 1.5% of the amount she owes. What is the monthly fee for $340?

    (1) $6.40 (2) $5.10 (3) $4.50
    (4) $3.90 (5) $3.55

19. Fred took a math test with 80 problems. He got 68 problems right. What percent of the problems did he get right?

    (1) 87% (2) 85% (3) 68% (4) 17% (5) 15%

20. Jeanne had to pay $312.50 interest on a one-year loan. The interest rate was 12.5%. How much was the loan?

    (1) $12,500 (2) $3,906 (3) $3,125
    (4) $2,500 (5) $2,188

21. Dan bought an electric drill for $24.80 and extra drill bits for $4.25. The sales tax in his state is 8%. How much did Dan pay altogether for the drill and bits, including tax?

    (1) $31.37 (2) $32.80 (3) $32.85
    (4) $37.05 (5) $38.47

22. From August to September 1980, the total number of unemployed Americans dropped from 8.0 million to 7.8 million. What was the percent of decrease in unemployment from August to September?

    (1) $1\frac{1}{2}\%$ (2) $2\frac{1}{2}\%$ (3) 4% (4) 5% (5) $5\frac{1}{2}\%$

23. Last year, the Pappas family paid $280 a month for rent. This year, they pay $315 for rent. By what percent did their rent go up?

    (1) 35% (2) 25% (3) 16% (4) 15% (5) $12\frac{1}{2}\%$

24. In a week, the Acevedo family pays $81.90 for food. Food is 35% of their budget. How much is the Acevedos' weekly budget?

    (1) $187 (2) $234 (3) $316.90
    (4) $350 (5) $365

25. Carlos wanted a camera usually priced at $180. At Thrifty Cameras, the camera was on sale for 15% off. During a Washington's Birthday sale, the camera was on sale for 10% off the regular discount. Find the Washington's Birthday sale price of the camera.

    (1) $168.30 (2) $156 (3) $155
    (4) $143 (5) $137.70

26. In Margaret's GED class, there are 15 women and 9 men. Women make up what percent of Margaret's class?

    (1) 80% (2) 75% (3) $62\frac{1}{2}$% (4) 55% (5) $37\frac{1}{2}$%

27. Mark and Heather have paid off 65% of their mortgage. So far, they have paid $15,600. Find the total amount of their mortgage.

    (1) $22,100 (2) $24,000 (3) $25,740
    (4) $36,300 (5) $101,400

28. There are 340 employees at the Central County Hospital. About 87% of them belong to a union. About how many hospital employees belong to a union?

    (1) 45 (2) 227 (3) 253 (4) 296 (5) 327

29. Ron used to make $4.50 an hour. He got a raise, and now he makes $4.86 an hour. His raise was what percent of his old hourly wage?

    (1) 3.6% (2) 5% (3) $6\frac{1}{2}$% (4) 7% (5) 8%

30. In a month, Henry sold cars for a total value of $18,500. He gets a 6% commission on every car he sells.What was Henry's total commission that month?

    (1) $3,083 (2) $1,850 (3) $1,661
    (4) $1,110 (5) $910

Check your answers on page 142.

Percent Review

These problems will give you a chance to practice the skills you learned in the percent section of this book. With each answer is the name of the section in which the skills needed for each problem are explained. For every problem you miss, review the section in which the skills for that problem are explained. Then try the problem again.

DIRECTIONS: Solve each problem.

1. Change .168 to a percent.
2. Change 9.5% to a decimal.
3. Change $\frac{7}{12}$ to a percent.
4. Change 6.4% to a fraction and reduce.
5. Find 18% of 2,400.
6. Find 6.5% of 800.
7. Find $37\frac{1}{2}\%$ of 120.
8. 36 is what % of 45?
9. 65% of what number is 39?
10. Find the interest on \$900 at $5\frac{1}{2}\%$ annual interest for 2 years and 6 months.
11. Gordon bought a shirt for \$12.95. The tax in his state is 6%. Find the total price of the shirt, including tax.
12. A 10-speed bike originally cost \$125. Jeff bought the bike on sale for 15% off. Find the sale price of the bike.
13. The Andersons spend \$95 a week on food. Their weekly budget is \$285. What percent of their budget goes for food?
14. Pete and Lee paid \$6,500 as a down payment for a house. The down payment was 20% of the price. Find the price of the house.

Check your answers on page 145.

ANSWERS & SOLUTIONS

EXERCISE 1

1. **38%** $.38 = .38 = \mathbf{38\%}$

   6% $.06 = .06 = \mathbf{6\%}$

   2.4% $.024 = .02\,4 = \mathbf{2.4\%}$

   $8\frac{1}{3}\%$ $.08\frac{1}{3} = .08\,\frac{1}{3} = \mathbf{8\frac{1}{3}\%}$

2. **90%** $.9 = .90 = \mathbf{90\%}$

   25% $.25 = .25 = \mathbf{25\%}$

   .1% $.001 = .00\,1 = \mathbf{.1\%}$

   1% $.01 = .01 = \mathbf{1\%}$

3. **475%** $4.75 = 4.75 = \mathbf{475\%}$

   62.5% $.625 = .62\,5 = \mathbf{62.5\%}$

   600% $6. = 6.00 = \mathbf{600\%}$

   $66\frac{2}{3}\%$ $.66\frac{2}{3} = 66\,\frac{2}{3} = \mathbf{66\frac{2}{3}\%}$

EXERCISE 2

1. **.25** $25\% = 25.\% = \mathbf{.25}$

   .08 $8\% = 08.\% = \mathbf{.08}$

   .375 $37.5\% = 37.5\% = \mathbf{.375}$

   .03 $3\% = 03.\% = \mathbf{.03}$

2. **.8** $80\% = 80.\% = \mathbf{.8}$

   2.5 $250\% = 2\,50.\% = \mathbf{2.5}$

   .01 $1\% = 01.\% = \mathbf{.01}$

   5 $500\% = 5\,00.\% = \mathbf{5}$

3. **$.06\frac{1}{4}$** $6\frac{1}{4}\% = 06.\frac{1}{4}\% = \mathbf{.06\frac{1}{4}}$

   .115 $11.5\% = 11.5\% = \mathbf{.115}$

   .008 $.8\% = 00.8\% = \mathbf{.008}$

   .19 $19\% = 19.\% = \mathbf{.19}$

EXERCISE 3

In the next solutions, each fraction is multiplied by 100% to get the answer. If you did the problems by changing the fractions to decimals, you should have the same answers.

1. **40%**

$\frac{2}{\underset{1}{\cancel{5}}} \times \frac{\overset{20}{\cancel{100}}}{1} = \frac{40}{1} = \mathbf{40\%}$

16%

$\frac{4}{\underset{1}{\cancel{25}}} \times \frac{\overset{4}{\cancel{100}}}{1} = \frac{16}{1} = \mathbf{16\%}$

50%

$\frac{1}{\underset{1}{\cancel{2}}} \times \frac{\overset{50}{\cancel{100}}}{1} = \frac{50}{1} = \mathbf{50\%}$

$\mathbf{33\frac{1}{3}\%}$

$\frac{1}{3} \times \frac{100}{1} = \frac{100}{3} = \mathbf{33\frac{1}{3}\%}$

2. **11%**

$\frac{11}{\underset{1}{\cancel{100}}} \times \frac{\overset{1}{\cancel{100}}}{1} = \frac{11}{1} = \mathbf{11\%}$

30%

$\frac{3}{\underset{1}{\cancel{10}}} \times \frac{\overset{10}{\cancel{100}}}{1} = \frac{30}{1} = \mathbf{30\%}$

$\mathbf{62\frac{1}{2}\%}$

$\frac{5}{\underset{2}{\cancel{8}}} \times \frac{\overset{25}{\cancel{100}}}{1} = \frac{125}{2} = \mathbf{62\frac{1}{2}\%}$

$\mathbf{6\frac{1}{4}\%}$

$\frac{1}{\underset{4}{\cancel{16}}} \times \frac{\overset{25}{\cancel{100}}}{1} = \frac{25}{4} = \mathbf{6\frac{1}{4}\%}$

3. $\mathbf{57\frac{1}{7}\%}$

$\frac{4}{7} \times \frac{100}{1} = \frac{400}{7} = \mathbf{57\frac{1}{7}\%}$

$\mathbf{83\frac{1}{3}\%}$

$\frac{5}{\underset{3}{\cancel{6}}} \times \frac{\overset{50}{\cancel{100}}}{1} = \frac{250}{3} = \mathbf{83\frac{1}{3}\%}$

45%

$\frac{9}{\underset{1}{\cancel{20}}} \times \frac{\overset{5}{\cancel{100}}}{1} = \frac{45}{1} = \mathbf{45\%}$

80%

$\frac{4}{\underset{1}{\cancel{5}}} \times \frac{\overset{20}{\cancel{100}}}{1} = \frac{80}{1} = \mathbf{80\%}$

EXERCISE 4

1. $\mathbf{\frac{1}{20}}$

$5\% = \frac{5}{100} = \mathbf{\frac{1}{20}}$

$\mathbf{\frac{7}{20}}$

$35\% = \frac{35}{100} = \mathbf{\frac{7}{20}}$

$\mathbf{\frac{24}{25}}$

$96\% = \frac{96}{100} = \mathbf{\frac{24}{25}}$

$\mathbf{\frac{3}{5}}$

$60\% = \frac{60}{100} = \mathbf{\frac{3}{5}}$

2. $\mathbf{2\frac{3}{4}}$

$275\% = \frac{275}{100} = 2\frac{75}{100} = \mathbf{2\frac{3}{4}}$

$\mathbf{\frac{2}{25}}$

$8\% = \frac{8}{100} = \mathbf{\frac{2}{25}}$

$\mathbf{5\frac{1}{5}}$

$520\% = \frac{520}{100} = 5\frac{20}{100} = \mathbf{5\frac{1}{5}}$

$\mathbf{\frac{43}{100}}$

$43\% = \mathbf{\frac{43}{100}}$

3. $\mathbf{\frac{3}{200}}$

$1.5\% = 01.5\% = .015$

$.015 = \frac{15}{1{,}000} = \mathbf{\frac{3}{200}}$

$\mathbf{\frac{6}{125}}$

$4.8\% = 04.8\% = .048$

$.048 = \frac{48}{1{,}000} = \mathbf{\frac{6}{125}}$

$\mathbf{\frac{1}{8}}$

$12.5\% = 12.5\% = .125$

$.125 = \frac{125}{1{,}000} = \mathbf{\frac{1}{8}}$

$\mathbf{\frac{1}{16}}$

$6.25\% = 06.25\% = .0625$

$.0625 = \frac{625}{10{,}000} = \mathbf{\frac{1}{16}}$

4. $\mathbf{\frac{1}{24}}$

$4\frac{1}{6}\% = \frac{4\frac{1}{6}}{100} = 4\frac{1}{6} \div 100$

$\frac{25}{6} \div \frac{100}{1} =$

$\frac{\overset{1}{\cancel{25}}}{6} \times \frac{1}{\underset{4}{\cancel{100}}} = \mathbf{\frac{1}{24}}$

$\mathbf{\frac{5}{16}}$

$31\frac{1}{4}\% = \frac{31\frac{1}{4}}{100} = 31\frac{1}{4} \div 100$

$\frac{125}{4} \div \frac{100}{1} =$

$\frac{\overset{5}{\cancel{125}}}{4} \times \frac{1}{\underset{4}{\cancel{100}}} = \mathbf{\frac{5}{16}}$

$\mathbf{\frac{4}{15}}$

$26\frac{2}{3}\% = \frac{26\frac{2}{3}}{100} = 26\frac{2}{3} \div 100$

$\frac{80}{3} \div \frac{100}{1} =$

$\frac{\overset{4}{\cancel{80}}}{3} \times \frac{1}{\underset{5}{\cancel{100}}} = \mathbf{\frac{4}{15}}$

$\mathbf{\frac{7}{12}}$

$58\frac{1}{3}\% = \frac{58\frac{1}{3}}{100} = 58\frac{1}{3} \div 100$

$\frac{175}{3} \div \frac{100}{1} =$

$\frac{\overset{7}{\cancel{175}}}{3} \times \frac{1}{\underset{4}{\cancel{100}}} = \mathbf{\frac{7}{12}}$

EXERCISE 5

1. **52**

$80\% = \frac{80}{100} = \frac{4}{5}$

$\frac{4}{\underset{1}{\cancel{5}}} \times \frac{\overset{13}{\cancel{65}}}{1} = \frac{52}{1} = \mathbf{52}$

90

$75\% = \frac{75}{100} = \frac{3}{4}$

$\frac{3}{\underset{1}{\cancel{4}}} \times \frac{\overset{30}{\cancel{120}}}{1} = \frac{90}{1} = \mathbf{90}$

36

$24\% = \frac{24}{100} = \frac{6}{25}$

$\frac{6}{\underset{1}{\cancel{25}}} \times \frac{\overset{6}{\cancel{150}}}{1} = \frac{36}{1} = \mathbf{36}$

21

$$60\% = \frac{60}{100} = \frac{3}{5}$$

$$\frac{3}{\cancel{5}_{1}} \times \frac{\cancel{35}^{7}}{1} = \frac{21}{1} = \mathbf{21}$$

2. **10**

$$5\% = \frac{5}{100} = \frac{1}{20}$$

$$\frac{1}{\cancel{20}_{1}} \times \frac{\cancel{200}^{10}}{1} = \frac{10}{1} = \mathbf{10}$$

96

$$200\% = \frac{200}{100} = 2$$

$$2 \times 48 = \mathbf{96}$$

75

$$125\% = \frac{125}{100} = \frac{5}{4}$$

$$\frac{5}{\cancel{4}_{1}} \times \frac{\cancel{60}^{15}}{1} = \frac{75}{1} = \mathbf{75}$$

20

$$8\% = \frac{8}{100} = \frac{2}{25}$$

$$\frac{2}{\cancel{25}_{1}} \times \frac{\cancel{250}^{10}}{1} = \frac{20}{1} = \mathbf{20}$$

3. **25**

$62.5\% = 62.5\% = .625$

$$\begin{array}{r} .625 \\ \times\ \ 40 \\ \hline 25.000 \end{array} = \mathbf{25}$$

22.68

$10.5\% = 10.5\% = .105$

$$\begin{array}{r} 216 \\ \times .105 \\ \hline 1\ 080 \\ 21\ 60\ \ \\ \hline 22.680 \end{array} = \mathbf{22.68}$$

21

$4.2\% = 04.2\% = .042$

$$\begin{array}{r} .042 \\ \times\ 500 \\ \hline 21.000 \end{array} = \mathbf{21}$$

18.9

$6.3\% = 06.3\% = .063$

$$\begin{array}{r} .063 \\ \times\ 300 \\ \hline 18.900 \end{array} = \mathbf{18.9}$$

4. **11.4**

$1.9\% = 01.9\% = .019$

$$\begin{array}{r} .019 \\ \times\ 600 \\ \hline 11.400 \end{array} = \mathbf{11.4}$$

4

$0.5\% = 00.5\% = .005$

$$\begin{array}{r} .005 \\ \times\ 800 \\ \hline 4.000 \end{array} = \mathbf{4}$$

5.5

$1.25\% = 01.25\% = .0125$

$$\begin{array}{r} .0125 \\ \times 440 \\ \hline 5000 \\ 5\ 00\ \ \ \ \\ \hline 5.5000 \end{array} = \mathbf{5.5}$$

.8

$0.4\% = 00.4\% = .004$

$$\begin{array}{r} .004 \\ \times\ 200 \\ \hline .800 \end{array} = \mathbf{.8}$$

5. **60**

$66\frac{2}{3}\% = \frac{2}{3}$

$\frac{2}{\underset{1}{\cancel{3}}} \times \frac{\overset{30}{\cancel{90}}}{1} = \frac{60}{1} = \mathbf{60}$

45

$37\frac{1}{2}\% = \frac{3}{8}$

$\frac{3}{\underset{1}{\cancel{8}}} \times \frac{\overset{15}{\cancel{120}}}{1} = \frac{45}{1} = \mathbf{45}$

5

$8\frac{1}{3}\% = \frac{8\frac{1}{3}}{100} = 8\frac{1}{3} \div 100 =$

$\frac{25}{3} \div \frac{100}{1} = \frac{\overset{1}{\cancel{25}}}{3} \times \frac{1}{\underset{4}{\cancel{100}}} = \frac{1}{12}$

$\frac{1}{\underset{1}{\cancel{12}}} \times \frac{\overset{5}{\cancel{60}}}{1} = \frac{5}{1} = \mathbf{5}$

350

$87\frac{1}{2}\% = \frac{7}{8}$

$\frac{7}{\underset{1}{\cancel{8}}} \times \frac{\overset{50}{\cancel{400}}}{1} = \frac{350}{1} = \mathbf{350}$

6. **150**

$83\frac{1}{3}\% = \frac{5}{6}$

$\frac{5}{\underset{1}{\cancel{6}}} \times \frac{\overset{30}{\cancel{180}}}{1} = \frac{150}{1} = \mathbf{150}$

30

$12\frac{1}{2}\% = \frac{1}{8}$

$\frac{1}{\underset{1}{\cancel{8}}} \times \frac{\overset{30}{\cancel{240}}}{1} = \frac{30}{1} = \mathbf{30}$

12

$16\frac{2}{3}\% = \frac{1}{6}$

$\frac{1}{\underset{1}{\cancel{6}}} \times \frac{\overset{12}{\cancel{72}}}{1} = \frac{12}{1} = \mathbf{12}$

$\mathbf{12\frac{1}{2}}$

$6\frac{1}{4}\% = \frac{6\frac{1}{4}}{100} = 6\frac{1}{4} \div 100 =$

$\frac{25}{4} \div \frac{100}{1} = \frac{\overset{1}{\cancel{25}}}{4} \times \frac{1}{\underset{4}{\cancel{100}}} = \frac{1}{16}$

$\frac{1}{\underset{2}{\cancel{16}}} \times \frac{\overset{25}{\cancel{200}}}{1} = \frac{25}{2} = \mathbf{12\frac{1}{2}}$

EXERCISE 6

1. **60%**

$\frac{18}{30} = \frac{3}{5}$

$\frac{3}{\underset{1}{\cancel{5}}} \times \frac{\overset{20}{\cancel{100}}}{1} = \frac{60}{1} = \mathbf{60\%}$

75%

$\frac{27}{36} = \frac{3}{4}$

$\frac{3}{\underset{1}{\cancel{4}}} \times \frac{\overset{25}{\cancel{100}}}{1} = \frac{75}{1} = \mathbf{75\%}$

90%

$\frac{72}{80} = \frac{9}{10}$

$\frac{9}{\underset{1}{\cancel{10}}} \times \frac{\overset{10}{\cancel{100}}}{1} = \frac{90}{1} = \mathbf{90\%}$

2. **$33\frac{1}{3}\%$** — **45%** — **$66\frac{2}{3}\%$**

$\frac{12}{36} = \frac{1}{3}$

$\frac{1}{3} \times \frac{100}{1} = \frac{100}{3} = \mathbf{33\frac{1}{3}\%}$

$\frac{36}{80} = \frac{9}{20}$

$\frac{9}{\cancel{20}_{1}} \times \frac{\cancel{100}^{5}}{1} = \frac{45}{1} = \mathbf{45\%}$

$\frac{30}{45} = \frac{2}{3}$

$\frac{2}{3} \times \frac{100}{1} = \frac{200}{3} = \mathbf{66\frac{2}{3}\%}$

3. **$37\frac{1}{2}\%$** — **80%** — **35%**

$\frac{21}{56} = \frac{3}{8}$

$\frac{3}{\cancel{8}_{2}} \times \frac{\cancel{100}^{25}}{1} = \frac{75}{2} = \mathbf{37\frac{1}{2}\%}$

$\frac{8}{10} = \frac{4}{5}$

$\frac{4}{\cancel{5}_{1}} \times \frac{\cancel{100}^{20}}{1} = \frac{80}{1} = \mathbf{80\%}$

$\frac{21}{60} = \frac{7}{20}$

$\frac{7}{\cancel{20}_{1}} \times \frac{\cancel{100}^{5}}{1} = \frac{35}{1} = \mathbf{35\%}$

4. **25%** — **$16\frac{2}{3}\%$** — **$62\frac{1}{2}\%$**

$\frac{18}{72} = \frac{1}{4}$

$\frac{1}{\cancel{4}_{1}} \times \frac{\cancel{100}^{25}}{1} = \frac{25}{1} = \mathbf{25\%}$

$\frac{11}{66} = \frac{1}{6}$

$\frac{1}{\cancel{6}_{3}} \times \frac{\cancel{100}^{50}}{1} = \frac{50}{3} = \mathbf{16\frac{2}{3}\%}$

$\frac{75}{120} = \frac{5}{8}$

$\frac{5}{\cancel{8}_{2}} \times \frac{\cancel{100}^{25}}{1} = \frac{125}{2} = \mathbf{62\frac{1}{2}\%}$

EXERCISE 7

1. **80** — **48**

$30\% = \frac{3}{10}$

$24 \div \frac{3}{10} = \frac{\cancel{24}^{8}}{1} \times \frac{10}{\cancel{3}_{1}} = \frac{80}{1} = \mathbf{80}$

$75\% = \frac{3}{4}$

$36 \div \frac{3}{4} = \frac{\cancel{36}^{12}}{1} \times \frac{4}{\cancel{3}_{1}} = \frac{48}{1} = \mathbf{48}$

2. **90** — **112**

$16\frac{2}{3}\% = \frac{1}{6}$

$15 \div \frac{1}{6} = \frac{15}{1} \times \frac{6}{1} = \frac{90}{1} = \mathbf{90}$

$37\frac{1}{2}\% = \frac{3}{8}$

$42 \div \frac{3}{8} = \frac{\cancel{42}^{14}}{1} \times \frac{8}{\cancel{3}_{1}} = \mathbf{112}$

3. **40** — **40**

4.5% = 04.5% = .045

$$\begin{array}{r} \mathbf{40} \\ .045\overline{)1.800} \\ \underline{1\ 80} \\ 00 \end{array}$$

12.5% = 12.5% = .125

$$\begin{array}{r} \mathbf{40} \\ .125\overline{)5.000} \\ \underline{5\ 00} \\ 00 \end{array}$$

4. **200**

$60\% = \frac{3}{5}$

$120 \div \frac{3}{5} = \frac{\overset{40}{\cancel{120}}}{1} \times \frac{5}{\underset{1}{\cancel{3}}} = \frac{200}{1} = \mathbf{200}$

640

$15\% = \frac{15}{100} = \frac{3}{20}$

$96 \div \frac{3}{20} = \frac{\overset{32}{\cancel{96}}}{1} \times \frac{20}{\underset{1}{\cancel{3}}} = \mathbf{640}$

5. **40**

$85\% = \frac{85}{100} = \frac{17}{20}$

$34 \div \frac{17}{20} = \frac{\overset{2}{\cancel{34}}}{1} \times \frac{20}{\underset{1}{\cancel{17}}} = \frac{40}{1} = \mathbf{40}$

78

$66\frac{2}{3}\% = \frac{2}{3}$

$52 \div \frac{2}{3} = \frac{\overset{26}{\cancel{52}}}{1} \times \frac{3}{\underset{1}{\cancel{2}}} = \frac{78}{1} = \mathbf{78}$

EXERCISE 8

1. **\$48**

$\frac{\overset{8}{\cancel{\$800}}}{1} \times \frac{6}{\underset{1}{\cancel{100}}} = \mathbf{\$48}$

2. **\$25**

$\frac{\overset{5}{\cancel{\$500}}}{1} \times \frac{5}{\underset{1}{\cancel{100}}} = \mathbf{\$25}$

3. **\$65**

$6.5\% = .065$

$$\begin{array}{r} .065 \\ \times 1{,}000 \\ \hline \mathbf{\$65.00\cancel{0}} \end{array}$$

4. **\$106.80**

$8.9\% = .089$

$$\begin{array}{r} \$1{,}200 \\ \times .089 \\ \hline 10\ 800 \\ 96\ 00 \\ \hline \mathbf{\$106.80\cancel{0}} \end{array}$$

5. **\$900**

$\frac{\overset{60}{\cancel{\$6{,}000}}}{1} \times \frac{15}{\underset{1}{\cancel{100}}} = \mathbf{\$900}$

6. **\$13**

$\frac{\$400}{100} \times 3\frac{1}{4} =$

$\frac{\overset{\overset{1}{\cancel{4}}}{\cancel{400}}}{\underset{1}{\cancel{100}}} \times \frac{13}{\underset{1}{\cancel{4}}} = \mathbf{\$13}$

7. **\$270**

$\frac{\$4{,}000}{100} \times 6\frac{3}{4} =$

$\frac{\overset{\overset{10}{\cancel{40}}}{\cancel{4{,}000}}}{\underset{1}{\cancel{100}}} \times \frac{27}{\underset{1}{\cancel{4}}} = \mathbf{\$270}$

8. **\$135**

$7.5\% = .075$

$$\begin{array}{r} \$1{,}800 \\ \times .075 \\ \hline 9\ 000 \\ 126\ 00 \\ \hline \mathbf{\$135.00\cancel{0}} \end{array}$$

EXERCISE 9

1. $36

9 mos. $= \frac{9}{12} = \frac{3}{4}$ yr.

$$\frac{\overset{6}{\cancel{\$600}}}{1} \times \frac{\overset{2}{\cancel{8}}}{\underset{1}{\cancel{100}}} \times \frac{3}{\underset{1}{\cancel{4}}} = \mathbf{\$36}$$

2. $112.50

1 yr. 6 mos. $= 1\frac{6}{12} = 1\frac{1}{2} = \frac{3}{2}$ yrs.

$$\frac{\overset{15}{\cancel{\$1{,}500}}}{1} \times \frac{5}{\underset{1}{\cancel{100}}} \times \frac{3}{2} = \frac{225}{2} = \mathbf{\$112.50}$$

3. $24

$$\frac{\overset{2}{\cancel{\$200}}}{1} \times \frac{6}{\underset{1}{\cancel{100}}} \times \frac{2}{1} = \mathbf{\$24}$$

4. $72

$4.5\% = .045$ 8 mos. $= \frac{8}{12} = \frac{2}{3}$ yr.

$$\begin{array}{r} \$2{,}400 \\ \times .045 \\ \hline 12\ 000 \\ 96\ 00 \\ \hline \$108.00\cancel{0} \end{array}$$

$$\frac{\overset{36}{\cancel{\$108}}}{1} \times \frac{2}{\underset{1}{\cancel{3}}} = \mathbf{\$72}$$

5. $360

1 yr. 4 mos. $= 1\frac{4}{12} = 1\frac{1}{3} = \frac{4}{3}$ yrs.

$$\frac{\overset{30}{\cancel{\$3{,}000}}}{1} \times \frac{\overset{3}{\cancel{9}}}{\underset{1}{\cancel{100}}} \times \frac{4}{\underset{1}{\cancel{3}}} = \mathbf{\$360}$$

6. $113.75

2 yrs. 6 mos. $= 2\frac{6}{12} = 2\frac{1}{2} = \frac{5}{2}$ yrs.

$$\frac{\overset{7}{\cancel{\$700}}}{1} \times \frac{13}{\underset{2}{\cancel{200}}} \times \frac{5}{2} = \frac{455}{4} = \mathbf{\$113.75}$$

7. $67.50

1 yr. 8 mos. $= 1\frac{8}{12} = 1\frac{2}{3} = \frac{5}{3}$ yrs.

$$\frac{\overset{9}{\cancel{\$450}}}{1} \times \frac{\overset{3}{\cancel{9}}}{\underset{2}{\cancel{100}}} \times \frac{5}{\underset{1}{\cancel{3}}} = \frac{135}{2} = \mathbf{\$67.50}$$

8. $864

2 yrs. 3 mos. $= 2\frac{3}{12} = 2\frac{1}{4} = \frac{9}{4}$ yrs.

$$\frac{\overset{32}{\cancel{\$3{,}200}}}{1} \times \frac{\overset{3}{\cancel{12}}}{\underset{1}{\cancel{100}}} \times \frac{9}{\underset{1}{\cancel{4}}} = \mathbf{\$864}$$

EXERCISE 10 (WORD PROBLEMS)

1. (4) **$26.29**

$$\begin{array}{r} \$24.80 \\ \times .06 \\ \hline \$1.48\ 80 \end{array} \qquad \begin{array}{r} \$24.80 \\ +\ 1.49 \\ \hline \mathbf{\$26.29} \end{array}$$

to the nearest cent = $1.49

2. (1) **15%**

$$\begin{array}{r} 69¢ \\ -60 \\ \hline 9¢ \end{array} \qquad \frac{9}{60} = \frac{3}{20}$$

$$\frac{3}{\underset{1}{\cancel{20}}} \times \frac{\overset{5}{\cancel{100\%}}}{1} = \mathbf{15\%}$$

3. (5) **$61.20**

$$\begin{array}{r} 96 \\ \times .15 \\ \hline 4\ 80 \\ 9\ 6 \\ \hline 14.40 \end{array} \qquad \begin{array}{r} 96.00 \\ -14.40 \\ \hline 81.60 \end{array} \qquad \frac{1}{\underset{1}{\cancel{4}}} \times \frac{\overset{20.40}{\cancel{81.60}}}{1} = 20.40 \qquad \begin{array}{r} \$81.60 \\ -20.40 \\ \hline \mathbf{\$61.20} \end{array}$$

4. **(2) $122.50** 20% = .2

$$\begin{array}{r} \mathbf{\$12\,2.50} \\ .2\overline{)\$24.5\,00} \end{array}$$

5. **(4) $16.80** 40% = .4

$$\begin{array}{r} \$12 \\ \times .4 \\ \hline \$4.8 \end{array} \qquad \begin{array}{r} \$12.00 \\ +\ 4.80 \\ \hline \mathbf{\$16.80} \end{array}$$

6. **(3) 20%**

$$\begin{array}{r} \$48.00 \\ -38.40 \\ \hline \$\ 9.60 \end{array} \qquad \frac{9.60}{48.00} = \frac{1}{5}$$

$$\frac{1}{\not{5}_1} \times \frac{\not{100\%}^{20}}{1} = \mathbf{20\%}$$

7. **(2)** $\mathbf{16\frac{2}{3}\%}$

$$\frac{4}{24} = \frac{1}{6}$$

$$\frac{1}{\not{6}_3} \times \frac{\not{100\%}^{50}}{1} = \frac{50}{3} = \mathbf{16\frac{2}{3}\%}$$

8. **(2) $6,300** 15% = .15

$$\begin{array}{r} \$42,000 \\ \times \quad .15 \\ \hline 2\,100\,00 \\ 4\,200\,0 \\ \hline \mathbf{\$6,300.00} \end{array}$$

9. **(4) $62**

$$\begin{array}{r} \$520 \\ \times .15 \\ \hline 26\,00 \\ 52\,0 \\ \hline \$78.00 \end{array} \qquad \begin{array}{r} \$24 \\ \times 21 \\ \hline 24 \\ 48 \\ \hline \$504 \end{array} \qquad \begin{array}{r} \$504 \\ +78 \\ \hline \$582 \end{array} \qquad \begin{array}{r} \$582 \\ -520 \\ \hline \mathbf{\$\ 62} \end{array}$$

10. **(3) $2,090** 5% = .05

$$\begin{array}{r} \mathbf{\$\ 20\,90} \\ .05\overline{)\$104.50} \end{array}$$

11. **(5) 20%**

$$\frac{192}{960} = \frac{1}{5} \qquad \frac{1}{\not{5}_1} \times \frac{\not{100\%}^{20}}{1} = \mathbf{20\%}$$

12. **(3) $28,500**

$$\begin{array}{r} \$24,600 \\ \times .16 \\ \hline 1\,476\,00 \\ 2\,460\,0 \\ \hline 3,936.00 \end{array} \qquad \begin{array}{r} 24,600 \\ +3,936 \\ \hline \end{array}$$

28,536 to the nearest hundred = **$28,500**

13. **(2) $17,392**

217.4% = 2.174

$$\begin{array}{r} 2.174 \\ \times \$8,000 \\ \hline \mathbf{\$17,392.00\not{0}} \end{array}$$

14. **(2) 8%**

$$\frac{18}{225} = \frac{2}{25}$$

$$\frac{2}{\not{25}_1} \times \frac{\not{100\%}^{4}}{1} = \mathbf{8\%}$$

15. **(5) $4,480** 20% = .2

$$\begin{array}{r} \mathbf{\$448\ 0} \\ .2\overline{)\$896.0} \end{array}$$

16. **(4) 30%**

$$\begin{array}{r} \$120 \\ -84 \\ \hline \$\ 36 \end{array} \qquad \frac{36}{120} = \frac{3}{10}$$

$$\frac{3}{\not{10}_1} \times \frac{\not{100\%}^{10}}{1} = \mathbf{30\%}$$

17. **(5) $10,665**

$$\begin{array}{r} \$13,500 \\ \times \quad .21 \\ \hline 135\,00 \\ 2700\,0 \\ \hline \$2,835.00 \end{array} \qquad \begin{array}{r} \$13,500 \\ -2,835 \\ \hline \mathbf{\$10,665} \end{array}$$

18. **(2) $5.10** 1.5% = .015

$$\begin{array}{r} \$340 \\ \times .015 \\ \hline 1\,700 \\ 3\,40 \\ \hline \mathbf{\$5.10\not{0}} \end{array}$$

19. **(2) 85%**

$$\frac{68}{80} = \frac{17}{20}$$

$$\frac{17}{\not{20}_1} \times \frac{\not{100\%}^{5}}{1} = \mathbf{85\%}$$

20. **(4) $2,500** $12.5\% = .125$

$$\begin{array}{r} \$\ \ 2\,500 \\ .125\overline{)\$312.500} \\ \underline{250} \\ 62\ 5 \\ \underline{62\ 5} \\ 000 \end{array}$$

21. **(1) $31.37**

$$\begin{array}{r} \$24.80 \\ \underline{+\ 4.25} \\ \$29.05 \\ \underline{\times\ \ .08} \\ \$2.32\ 40 \end{array}$$

to the nearest cent = $2.32

$$\begin{array}{r} \$29.05 \\ \underline{+\ 2.32} \\ \mathbf{\$31.37} \end{array}$$

22. **(2) $2\frac{1}{2}\%$**

$$\begin{array}{r} 8.0 \\ \underline{-7.8} \\ .2 \end{array} \qquad \frac{.2}{8.0} = \frac{1}{40}$$

$$\frac{1}{\cancel{40}_{2}} \times \frac{\cancel{100\%}^{5}}{1} = \frac{5}{2} = \mathbf{2\tfrac{1}{2}\%}$$

23. **(5) $12\frac{1}{2}\%$**

$$\begin{array}{r} \$315 \\ \underline{-280} \\ \$\ 35 \end{array} \qquad \frac{35}{280} = \frac{1}{8}$$

$$\frac{1}{\cancel{8}_{2}} \times \frac{\cancel{100\%}^{25}}{1} = \frac{25}{2} = \mathbf{12\tfrac{1}{2}\%}$$

24. **(2) $234** $35\% = .35$

$$\begin{array}{r} \$\ 2\,34 \\ .35\overline{)\$81.90} \\ \underline{70} \\ 11\ 9 \\ \underline{10\ 5} \\ 1\ 40 \\ \underline{1\ 40} \end{array}$$

25. **(5) $137.70** $15\% = .15$ $10\% = .1$

$$\begin{array}{r} \$180 \\ \underline{\times .15} \\ 9\ 00 \\ \underline{18\ 0\ \ } \\ \$27.00 \end{array} \qquad \begin{array}{r} \$180 \\ \underline{-\ 27} \\ \$153 \end{array} \qquad \begin{array}{r} \$153 \\ \underline{\times\ \ .1} \\ \$15.3 \end{array} \qquad \begin{array}{r} \$153.00 \\ \underline{-\ 15.30} \\ \mathbf{\$137.70} \end{array}$$

26. **(3) $62\frac{1}{2}\%$**

$$\begin{array}{r} 15 \text{ women} \\ \underline{+\ \ 9 \text{ men}\ \ \ } \\ 24 \text{ total} \end{array} \qquad \frac{15}{24} = \frac{5}{8}$$

$$\frac{5}{\cancel{8}_{2}} \times \frac{\cancel{100\%}^{25}}{1} = \frac{125}{2} = \mathbf{62\tfrac{1}{2}\%}$$

27. **(2) $24,000** $65\% = .65$

$$\begin{array}{r} \$\ \ \ 240\ 00 \\ .65\overline{)\$15,600.00} \\ \underline{13\ 0} \\ 2\ 60 \\ \underline{2\ 60} \\ 00\ 00 \end{array}$$

28. **(4) 296**

$$\begin{array}{r} 340 \\ \underline{\times .87} \\ 23\ 80 \\ \underline{272\ 0\ \ } \\ 295.80 \end{array}$$

295.80 to the nearest unit = **296**

29. **(5) 8%**

$$\begin{array}{r} \$4.86 \\ \underline{-\ 4.50} \\ \$\ .36 \end{array} \qquad \frac{.36}{4.50} = \frac{2}{25}$$

$$\frac{2}{\cancel{25}_{1}} \times \frac{\cancel{100\%}^{4}}{1} = \mathbf{8\%}$$

30. **(4) $1,110**

$$\begin{array}{r} \$18,500 \\ \underline{\times\ \ \ \ \ .06} \\ \mathbf{\$1,110.00} \end{array}$$

PERCENT REVIEW

1. **16.8%**

$.168 = .16.8 = \mathbf{16.8\%}$

(Changing Decimals to Percents)

2. **.095**

$9.5\% = 09.5\% = \mathbf{.095}$

(Changing Percents to Decimals)

3. $\mathbf{58\frac{1}{3}\%}$

$\frac{7}{\cancel{12}_{3}} \times \frac{\cancel{100}^{25}}{1} = \frac{175}{3} = \mathbf{58\frac{1}{3}\%}$

(Changing Fractions to Percents)

4. $\mathbf{\frac{8}{125}}$

$6.4\% = .064 = \frac{64}{1,000} = \frac{8}{125}$

(Changing Percents to Fractions)

5. **432**

$18\% = \frac{18}{100}$

$\frac{18}{\cancel{100}_{1}} \times \frac{\cancel{2,400}^{24}}{1} = \mathbf{432}$

(Finding a Percent of a Number)

6. **52**

$6.5\% = .065$

$$\begin{array}{r} .065 \\ \times\ 800 \\ \hline 52.000 = \mathbf{52} \end{array}$$

(Finding a Percent of a Number)

7. **45**

$37\frac{1}{2}\% = \frac{3}{8}$

$\frac{3}{\cancel{8}_{1}} \times \frac{\cancel{120}^{15}}{1} = \mathbf{45}$

(Finding a Percent of a Number)

8. **80%**

$\frac{36}{45} = \frac{4}{5} = \mathbf{80\%}$

(Finding What Percent One Number Is of Another)

9. **60**

$65\% = .65$

$$\begin{array}{r} \mathbf{60} \\ .65\overline{)39.00} \\ \underline{39\ 0} \\ 00 \end{array}$$

(Finding a Number When a Percent of It Is Known)

10. **$123.75**

2 yrs. 6 mos. $= 2\frac{6}{12} = 2\frac{1}{2} = \frac{5}{2}$ yrs.

$\frac{\cancel{\$900}^{9}}{1} \times \frac{11}{\cancel{200}_{2}} \times \frac{5}{2} = \frac{495}{4} = \mathbf{\$123.75}$

(Simple Interest)

11. **$13.73**

$6\% = .06$

$$\begin{array}{r} \$12.95 \\ \times .06 \\ \hline \$.77\ 70 \end{array}$$

to the nearest cent = $.78

$$\begin{array}{r} \$12.95 \\ +.78 \\ \hline \mathbf{\$13.73} \end{array}$$

(Percent Word Problems)

12. **$106.25**

$15\% = .15$

$$\begin{array}{r} \$125 \\ \times .15 \\ \hline 6\ 25 \\ 12\ 5 \\ \hline \$18.75 \end{array} \qquad \begin{array}{r} \$125.00 \\ -\ 18.75 \\ \hline \mathbf{\$106.25} \end{array}$$

(Percent Word Problems)

13. $\mathbf{33\frac{1}{3}\%}$

$\frac{95}{285} = \frac{1}{3} = \mathbf{33\frac{1}{3}\%}$

(Percent Word Problems)

14. **$32,500**

$20\% = .2$

$$\begin{array}{r} \mathbf{\$3\,2{,}50\,0} \\ .2\overline{)\$6{,}5\,00.0} \end{array}$$

(Percent Word Problems)

Measurements

AMERICAN STANDARD SYSTEM

The list below gives units of measurement for length, weight, time, and liquid measure. These units of length, weight, and liquid measure are the units commonly used in the U.S. After each unit is an abbreviation or symbol for that unit. The list also tells how many smaller units each bigger unit is equal to. Memorize any units on the list you do not already know.

Length

1 foot (ft. or ′) = 12 inches (in. or ″)
1 yard (yd.) = 3 ft. or 36 in.
1 mile (mi.) = 5,280 ft. or 1,760 yds.

Weight

1 pound (1 lb.) = 16 ounces (oz.)
1 ton (t.) = 2,000 lbs.

Time

1 minute (min.) = 60 seconds (sec.)
1 hour (hr.) = 60 min.
1 day (da.) = 24 hrs.
1 week (wk.) = 7 days
1 year (yr.) = 365 days or 12 months (mos.) or 52 wks.

Liquid Measure

1 pint (pt.) = 16 oz.
1 quart (qt.) = 2 pts. or 32 oz.
1 gallon (gal.) = 4 qts.

To convert a bigger unit to smaller units, **multiply** the number of bigger units by the number of smaller units that are in one of the bigger units.

EXAMPLE: 4 ft. = ______ in.

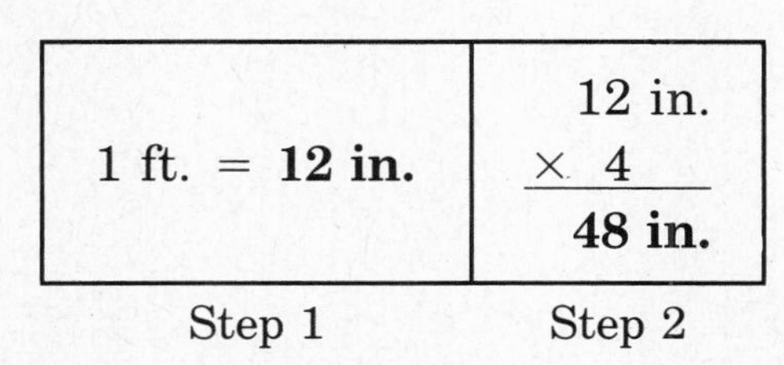

Step 1: Find the number of inches in one foot. There are 12 inches in one foot.

Step 2: Multiply the number of inches in one foot (12) by the number of feet in the problem (4).

To convert smaller units to bigger units, **divide** the number of smaller units you have by the number of smaller units that are in one of the bigger units.

EXAMPLE: 32 oz. = ______ pt.

1 pt. = **16 oz.**	$16\overline{)\,32}$ **2 pts.** -32
Step 1	Step 2

Step 1: Find the number of ounces in one pint. There are 16 ounces in one pint.
Step 2: Divide the number of ounces you have (32) by the number of ounces that are in one pint (16).

Sometimes, your answer will be a fraction.

EXAMPLE: 1,500 lbs. = ______ t.

1 t. = **2,000 lbs.**	**2,000$\overline{)1{,}500}$** = $\mathbf{\frac{1{,}500}{2{,}000}}$ = $\mathbf{\frac{3}{4}}$ **t.**
Step 1	Step 2

Step 1: Find the number of pounds in one ton. There are 2,000 pounds in one ton.
Step 2: Divide the number of pounds you have (1,500) by the number of pounds there are in one ton (2,000). 2,000 does not divide into 1,500. Write a fraction with 1,500 as the top number and 2,000 as the bottom number. (Remember that a fraction means the same as "divided by.") Reduce the fraction to lowest terms.

EXERCISE 1

DIRECTIONS: Change the given unit to the other unit shown in the problem.

1. 3 min. = ______ sec. 64 oz. = ______ lb. 3 qts. = ______ gal.
2. 48 in. = ______ yd. 40 sec. = ______ min. 42 da. = ______ wk.
3. 1,200 lbs. = ______ t. 9 ft. = ______ in. 6 lbs. = ______ oz.
4. 5 gals. = ______ qt. 20 mos. = ______ yr. 1,320 ft. = ______ mi.
5. 200 in. = ______ yd. 50 min. = ______ hr. 2 mi. = ______ ft.

6. 7 da. = ______ hr. 9 gals. = ______ qt. 144 in. = ______ yd.

7. 40 mos. = ______ yr. 8,500 lbs. = ______ t. 13 wks. = ______ yr.

Check your answers on page 156.

Metric System

The metric system often is used to measure things. There are three basic units in the metric system:

1. The **meter** (m.) is the basic unit of length. A meter is a little longer than one yard.
2. The **gram** (g.) is the basic unit of weight. A gram is about $\frac{1}{30}$ of an ounce.
3. The **liter** (l.) is the basic unit of liquid measure. A liter is a little more than one quart.

Bigger and smaller metric units are made by putting a *prefix* before the basic units (meter, gram, and liter). The next list gives the most common metric prefixes and their meanings.

Prefix	**Meaning**
kilo- (k)	1,000
deci- (d)	$\frac{1}{10}$
centi- (c)	$\frac{1}{100}$
milli- (m)	$\frac{1}{1{,}000}$

EXAMPLE: One kilometer (km.) = ______ meters (m.)

Find the prefix *kilo-*. *Kilo-* means 1,000.

One kilometer = 1,000 meters

EXAMPLE: One centigram (cg.) = ______ gram (g.)

Find the prefix *centi-*. *Centi-* means $\frac{1}{100}$.

One centigram = $\frac{1}{100}$ gram

EXERCISE 2

DIRECTIONS: Fill in each blank.

1. 1 millimeter (mm.) = ______ meter 1 centigram (cg.) = ______ gram

2. 1 deciliter (dl.) = ______ liter 1 milliliter (ml.) = ______ liter

3. 1 centimeter (cm.) = ______ meter   1 kilogram (kg.) = ______ gram

4. 1 kiloliter (kl.) = ______ liter   1 decimeter (dm.) = ______ meter

5. 1 milligram (mg.) = ______ gram   1 decigram (dg.) = ______ gram

Check your answers on page 157.

Changing between the American standard system and the metric system is hard because the two systems are so different. You do not have to memorize the differences between the two systems for the GED Test. However, you should be able to solve a problem if you are asked to change from one system to the other. Here is an example of this type of problem.

EXAMPLE: 1 mile = 1.6 kilometers. Change 5 miles to kilometers.

$$\begin{array}{r} 1.6 \\ \times 5 \\ \hline 8.0 \end{array} = \mathbf{8\ km.}$$

Multiply 5 miles by the number of kilometers in one mile (1.6).

EXERCISE 3

DIRECTIONS: Solve each problem.

1. 1 quart = .946 liter. Convert 4 quarts to liters.

2. 1 pound = .453 kilogram. Convert 100 pounds to kilograms.

3. 1 meter = 39.37 inches. Convert 3 meters to inches.

4. 1 liter = 1.05 quarts. Convert 8 liters to quarts.

5. 1 inch = 2.54 centimeters. Convert 12 inches to centimeters.

6. 1 gram = .035 ounce. Convert 25 grams to ounces.

7. 1 gallon = 3.78 liters. Convert 50 gallons to liters.

8. 1 kilogram = 2.2 pounds. Convert 65 kilograms to pounds.

Check your answers on page 157.

BASIC OPERATIONS WITH MEASUREMENTS

To add, subtract, multiply, or divide measurements, remember to convert units when you carry and borrow.

In addition problems, first add the units separately. Then convert smaller units to bigger units.

EXAMPLE: Add 2 lbs. 9 oz. and 3 lbs. 11 oz.

Steps 1-2	Step 3	Step 4
2 lbs. 9 oz. **+3 lbs. 11 oz.** **5 lbs. 20 oz.**	**1 lb. 4 oz.** 16)20	5 lbs. +1 lb. 4 oz. **6 lbs. 4 oz.**

Step 1: Set up the problem. Put pounds under pounds and ounces under ounces.
Step 2: Add the pounds and ounces separately.
Step 3: Convert the ounces to pounds.
Step 4: Add the converted ounces to the other pounds to get the final answer.

In subtraction problems, borrow bigger units when you do not have enough smaller units to subtract from.

EXAMPLE: Subtract 4 ft. 10 in. from 7 ft. 3 in.

Step 1	Steps 2-3	Step 4
7 ft. 3 in. **−4 ft. 10 in.**	(1 ft. = 12 in.) 7 ft. 3 in. = **6 ft. 15 in.**	6 ft. 15 in. −4 ft. 10 in. **2 ft. 5 in.**

Step 1: Set up the problem. Put feet under feet and inches under inches.
Step 2: Try to subtract the inches. You cannot subtract 10 from 3. Therefore, you must borrow from the feet column. Subtract 1 foot from the 7 feet. Then convert the foot to inches.
Step 3: Add the 12 inches to the 3 inches.
Step 4: Subtract the inches and feet separately.

In multiplication problems, first multiply the units separately. Then convert smaller units to bigger units.

EXAMPLE: Multiply 4 min. 15 sec. by 5.

Step 1	Step 2	Step 3
4 min. 15 sec. × 5 × 5 **20 min. 75 sec.**	**1 min. 15 sec.** 60) 75 −60 15	20 min. + 1 min. 15 sec. **21 min. 15 sec.**

Step 1: Set up the problem. Multiply the minutes and seconds separately.
Step 2: Convert the seconds to minutes.
Step 3: Add the converted minutes to the other minutes to get the final answer.

In division problems, convert any remaining bigger units to smaller units before you bring down the other smaller units in the problem.

EXAMPLE: Divide 8 gals. 1 qt. by 3.

Step 1	Step 2	Steps 3-4
2 gals. 3) 8 gals. 1 qt. −6 2 gals.	(1 gal. = 4 qts.) 2 × 4 = **8 qts.**	2 gals. **3 qts.** 3) 8 gals. 1 qt. −6 2 gals. = **8 qts.** **+ 1 qt.** **9 qts.** − 9

Step 1: Set up the problem. Divide 8 gallons by 3.
Step 2: The 3 does not divide into 2. Therefore, change the remaining 2 gallons to quarts.
Step 3: Add the 8 quarts to the other quart.
Step 4: Divide the 9 quarts by 3.

EXERCISE 4

DIRECTIONS: Solve each problem.

1. Add 5 yds. 2 ft. and 6 yds. 1 ft.

2. Multiply 3 t. 500 lbs. by 8.

3. Divide 13 lbs. 2 oz. by 6.

4. Subtract 7 gals. 3 qts. from 10 gals. 1 qt.

5. Multiply 4 da. 10 hrs. by 3.

6. Add 5 hrs. 20 min., 3 hrs. 50 min., and 4 hrs. 35 min.

7. Divide 10 ft. 6 in. by 6.

8. Subtract 5 wks. 3 da. from 12 wks.

9. Divide 23 hrs. 34 min. by 7.

10. Add 3 t. 1,000 lbs., 4 t. 1,200 lbs., and 2 t. 1,450 lbs.

11. Subtract 5 min. 47 sec. from 9 min. 15 sec.

12. Multiply 6 ft. 3 in. by 9.

13. Subtract 8 lbs. 9 oz. from 15 lbs. 4 oz.

14. Multiply 3 qts. 1 pt. by 12.

15. Divide 8 yds. 1 ft. by 5.

16. Add 5 gals. 2 qts., 9 gals. 3 qts., and 6 gals. 3 qts.

17. Divide 8 yrs. 6 mos. by 3.

18. Subtract 8 t. 1,600 lbs. from 10 t. 500 lbs.

19. Add 2 ft. 9 in., 3 ft. 11 in., and 5 ft. 6 in.

20. Multiply 5 wks. 6 da. by 4.

Check your answers on page 158.

Measurement Word Problems

The next problems will give you a chance to apply your measurement skills to word problems. Some problems may take more than one operation to find an answer. Outline for yourself the steps you need to take to find each answer.

EXERCISE 5 (WORD PROBLEMS)

DIRECTIONS: Choose the correct answer to each problem.

1. The measurement around each window in Colette's house is 136 inches. Colette wants to buy weather stripping to go around 6 windows. The weather stripping is sold by the yard. How many yards should Colette buy?

   (1) 9 yds. (2) $14\frac{5}{6}$ yds. (3) $22\frac{2}{3}$ yds.
   (4) 68 yds. (5) 136 yds.

2. Phil bought two packages of ground beef. One package weighed 1 pound 11 ounces. The other weighed 2 pounds 8 ounces. What was the combined weight of the two packages?

   (1) 2 lbs. 15 oz. (2) 3 lbs. 17 oz. (3) 3 lbs. 18 oz.
   (4) 4 lbs. 3 oz. (5) 4 lbs. 9 oz.

3. Ann bought a 16-ounce package of margarine for $1.79. To the nearest cent, what was the price of one ounce of margarine?

   (1) $1.79 (2) $.39 (3) $.22
   (4) $.17 (5) $.11

4. Kathryn bought an 8-ounce package of margarine for 99¢. To the nearest cent, what was the price of one ounce of margarine?

   (1) $.99 (2) $.50 (3) $.49 (4) $.12 (5) $.11

5. A flight leaves New York at 9:40 A.M. local time. The flight arrives in San Francisco at 1:07 P.M. local time. When the time in New York is 9:00 A.M., the time in San Francisco is 6:00 A.M. How long does the flight from New York to San Francisco take?

   (1) 3 hrs. 27 min. (2) 4 hrs. 27 min. (3) 6 hrs. 27 min.
   (4) 8 hrs. 7 min. (5) 10 hrs. 7 min.

6. Nick bought a piece of lumber 8 feet long. From it he cut a piece 6 feet 7 inches long. How long was the piece left over?

   (1) 5 in. (2) 1 ft. 5 in. (3) 2 ft. 5 in.
   (4) 2 ft. 7 in. (5) 3 ft. 7 in.

7. In 1934, Glenn Cunningham ran a mile in 4 min. 7 sec. In 1979, Sebastian Coe ran a mile in 3 min. 49 sec. Coe's time was how much faster than Cunningham's time?

   (1) 18 sec. (2) 21 sec. (3) 58 sec.
   (4) 1 min. 18 sec. (5) 1 min. 21 sec.

8. The Coljer Company is shipping a load weighing 7 t. 200 lbs. on 5 trucks. If the weight is divided equally, how much will each truck carry?

   (1) 2 t. 840 lbs. (2) 2 t. 520 lbs. (3) 1 t. 840 lbs.
   (4) 1 t. 520 lbs. (5) 1 t. 200 lbs.

9. Fred mailed 3 packages, each weighing 2 lbs. 9 oz. What was the total weight of the packages?

   (1) $13\frac{2}{3}$ oz. (2) 7 lbs. 11 oz. (3) 8 lbs. 3 oz.
   (4) 8 lbs. 7 oz. (5) 8 lbs. 14 oz.

10. The Allen family uses 3 qts. 1 pt. of milk every day. In a week, how much milk do the Allens use?

    (1) 22 qts. 3 pts. (2) 24 qts. 1 pt. (3) 31 qts. 3 pts.
    (4) 32 qts. 1 pt. (5) 33 qts. 1 pt.

11. Fran worked 2 hrs. 25 min. on Monday painting her apartment. Tuesday, she worked 1 hr. 40 min. Wednesday, she worked 3 hrs. 10 min. Altogether, how much time did Fran spend painting her apartment?

    (1) 7 hrs. 15 min. (2) 7 hrs. 25 min. (3) 8 hrs. 15 min.
    (4) 8 hrs. 25 min. (5) 8 hrs. 45 min.

12. A recipe calls for 2 lbs. 9 oz. of ground beef. The recipe is enough for 6 people. How much ground beef would be needed for 12 people?

    (1) 15 lbs. 6 oz. (2) 10 lbs. 3 oz. (3) 5 lbs. 2 oz.
    (4) 5 lbs. 1 oz. (5) 3 lbs. 2 oz.

13. The Rigbys burn coal to heat their house. In September, they had 600 pounds of coal. By January, they had burned 250 pounds of the coal. What fraction of the coal was left by January?

    (1) $\frac{5}{12}$ (2) $\frac{1}{2}$ (3) $\frac{7}{12}$ (4) $\frac{2}{3}$ (5) $\frac{3}{4}$

14. Pete wants to cut a piece of lumber 9 ft. 11 in. long into 7 equal pieces. How long will each piece be?

    (1) 2 ft. 4 in. (2) 2 ft. 2 in. (3) 1 ft. 9 in.
    (4) 1 ft. 7 in. (5) 1 ft. 5 in.

15. Sue works part-time as a waitress. Thursday, she worked 2 hrs. 30 min. Friday, she worked 3 hrs. 15 min. Saturday, she worked 4 hrs. 45 min. What was the average time she worked each day?

    (1) 3 hrs. 30 min. (2) 3 hrs. $43\frac{1}{3}$ min. (3) 4 hrs. 30 min.
    (4) 6 hrs. 20 min. (5) 10 hrs. 30 min.

16. Al left his house at 8:30 in the morning. He returned home at 4:20 in the afternoon. How long was Al out?

(1) 4 hrs. 10 min. (2) 4 hrs. 50 min. (3) 7 hrs. 10 min.
(4) 7 hrs. 50 min. (5) 8 hrs. 30 min.

17. Bob, Jeff, and Gordon caught a total of 14 lbs. 7 oz. of fish. They shared the fish equally. How many pounds did each man get?

(1) 4 lbs. 13 oz. (2) 4 lbs. 27 oz. (3) 6 lbs. 3 oz.
(4) 6 lbs. 13 oz. (5) 8 lbs. 3 oz.

18. The total measurement around John's roof is 98 feet. John wants to put a new gutter around his roof. The gutter is sold by the yard. How many yards should John buy?

(1) $9\frac{4}{5}$ yds. (2) $19\frac{3}{5}$ yds. (3) $32\frac{2}{3}$ yds.
(4) $65\frac{1}{3}$ yds. (5) 294 yds.

19. The truck Jack drives weighs 1,900 lbs. When he filled the truck with a load, it weighed 3 t. 650 lbs. Find the weight of the load.

(1) 1 t. 1,750 lbs. (2) 2 t. 750 lbs. (3) 2 t. 1,750 lbs.
(4) 2 t. 1,850 lbs. (5) 2 t. 1,950 lbs.

20. The Verrazano Narrows Bridge is 4,260 feet long. The length of the bridge is about what fraction of a mile? (1 mile = 5,280 ft.)

(1) $1\frac{1}{2}$ mi. (2) $\frac{1}{2}$ mi. (3) $\frac{2}{3}$ mi.
(4) $\frac{3}{4}$ mi. (5) $\frac{4}{5}$ mi.

Check your answers on page 160.

ANSWERS & SOLUTIONS

EXERCISE 1

1. **180 sec.** **4 lbs.** $\frac{3}{4}$ **gals.**

$$\begin{array}{r} 60 \\ \times 3 \\ \hline \textbf{180 sec.} \end{array} \qquad 16\overline{)64} = \textbf{4 lbs.}$$

2. $1\frac{1}{3}$ **yds.** $\frac{2}{3}$ **min.** **6 wks.**

$$36\overline{)48} = 1\frac{12}{36} = 1\frac{1}{3} \text{ yds.} \quad \begin{array}{r} 48 \\ 36 \\ \hline 12 \end{array} \qquad \frac{40}{60} = \frac{2}{3} \text{ min.} \qquad 7\overline{)42} = \textbf{6 wks.}$$

3. **$\frac{3}{5}$ t.** **108 in.** **96 oz.**

$\frac{1{,}200}{2{,}000} = \frac{3}{5}$ **t.**

$\begin{array}{r} 12 \\ \times 9 \\ \hline \end{array}$ **108 in.**

$\begin{array}{r} 16 \\ \times 6 \\ \hline \end{array}$ **96 oz.**

4. **20 qts.** **$1\frac{2}{3}$ yrs.** **$\frac{1}{4}$ mi.**

$\begin{array}{r} 5 \\ \times 4 \\ \hline \end{array}$ **20 qts.**

$12\overline{)20}$ $= 1\frac{8}{12}$; 12; remainder 8; $1\frac{8}{12} = 1\frac{2}{3}$ **yrs.**

$\frac{1{,}320}{5{,}280} = \frac{1}{4}$ **mi.**

5. **$5\frac{5}{9}$ yds.** **$\frac{5}{6}$ hr.** **10,560 ft.**

$36\overline{)200}$ $= 5\frac{20}{36}$; 180; remainder 20; $5\frac{20}{36} = 5\frac{5}{9}$ **yds.**

$\frac{50}{60} = \frac{5}{6}$ **hr.**

$\begin{array}{r} 5{,}280 \\ \times \quad 2 \\ \hline \end{array}$ **10,560 ft.**

6. **168 hrs.** **36 qts.** **4 yds.**

$\begin{array}{r} 24 \\ \times 7 \\ \hline \end{array}$ **168 hrs.**

$\begin{array}{r} 9 \\ \times 4 \\ \hline \end{array}$ **36 qts.**

$36\overline{)144}$ = **4 yds.**; 144

7. **$3\frac{1}{3}$ yrs.** **$4\frac{1}{4}$ t.** **$\frac{1}{4}$ yr.**

$12\overline{)40}$ $= 3\frac{4}{12}$; 36; remainder 4; $3\frac{4}{12} = 3\frac{1}{3}$ **yrs.**

$2{,}000\overline{)8{,}500}$ $= 4\frac{500}{2{,}000}$; $8{,}000$; remainder 500; $4\frac{500}{2{,}000} = 4\frac{1}{4}$ **t.**

$\frac{13}{52} = \frac{1}{4}$ **yr.**

EXERCISE 2

1. **$\frac{1}{1{,}000}$ meter** **$\frac{1}{100}$ gram**
2. **$\frac{1}{10}$ liter** **$\frac{1}{1{,}000}$ liter**
3. **$\frac{1}{100}$ meter** **1,000 grams**
4. **1,000 liters** **$\frac{1}{10}$ meter**
5. **$\frac{1}{1{,}000}$ gram** **$\frac{1}{10}$ gram**

EXERCISE 3

1. **3.784 l.**

$\begin{array}{r} .946 \\ \times \quad 4 \\ \hline \end{array}$ **3.784 l.**

2. **45.3 kg.**

$\begin{array}{r} .453 \\ \times 100 \\ \hline 45.300 \end{array}$ = **45.3 kg.**

3. **118.11 in.**

$\begin{array}{r} 39.37 \\ \times \quad 3 \\ \hline \end{array}$ **118.11 in.**

4. **8.4 qts.**

```
 1.05
×  8
 8.40 = 8.4 qts.
```

5. **30.48 cm.**

```
  2.54
×  12
  5 08
 25 4
 30.48 cm.
```

6. **.875 oz.**

```
 .035
 ×25
  175
  70
 .875 oz.
```

7. **189 l.**

```
   3.78
 ×  50
 189.00 = 189 l.
```

8. **143 lbs.**

```
   65
 ×2.2
  13 0
 130
 143.0 = 143 lbs.
```

EXERCISE 4

1. **12 yds.**

```
  5 yds. 2 ft.
 +6 yds. 1 ft.
 11 yds. 3 ft. = 12 yds.
```

2. **26 t.**

```
  3 t.      500 lbs.
 ×8       ×   8
 24 t.    4,000 lbs. = 26 t.
```

3. **2 lbs. 3 oz.**

```
   2 lbs.        3 oz.
 6)13 lbs.       2 oz.
   12
    1 lb. =     16 oz.
              +  2
                18 oz.
                18
```

4. **2 gals. 2 qts.**

```
  10 gals. 1 qt.  =   9 gals. 5 qts.
 −  7 gals. 3 qts.   −7 gals. 3 qts.
                      2 gals. 2 qts.
```

5. **13 da. 6 hrs.**

```
  4 da.    10 hrs.
 ×3        ×3
 12 da.    30 hrs. = 13 da. 6 hrs.
```

6. **13 hrs. 45 min.**

```
   5 hrs.  20 min.
   3 hrs.  50 min.
  +4 hrs.  35 min.
  12 hrs. 105 min. = 13 hrs. 45 min.
```

7. **1 ft. 9 in.**

```
   1 ft.       9 in.
 6)10 ft.      6 in.
    6
    4 ft. =   48 in.
            +  6
              54 in.
              54
```

8. **6 wks. 4 da.**

```
  12 wks.          =   11 wks. 7 da.
 −  5 wks. 3 da.     −  5 wks. 3 da.
                        6 wks. 4 da.
```

9. **3 hrs. 22 min.**

3 hrs. 22 min.
7)23 hrs. 34 min.
21
2 hrs. = 120 min.
+ 34
154 min.
14
14
14

10. **10 t. 1,650 lbs.**

3 t. 1,000 lbs.
4 t. 1,200 lbs.
+2 t. 1,450 lbs.
9 t. 3,650 lbs. = **10 t. 1,650 lbs.**

11. **3 min. 28 sec.**

9 min. 15 sec. = 8 min. 75 sec.
−5 min. 47 sec. −5 min. 47 sec.
3 min. 28 sec.

12. **56 ft. 3 in.**

6 ft. 3 in.
×9 ×9
54 ft. 27 in. = **56 ft. 3 in.**

13. **6 lbs. 11 oz.**

15 lbs. 4 oz. = 14 lbs. 20 oz.
− 8 lbs. 9 oz. − 8 lbs. 9 oz.
6 lbs. 11 oz.

14. **42 qts.**

3 qts. 1 pt.
×12 ×12
36 qts. 12 pts. = **42 qts.**

15. **1 yd. 2 ft.**

1 yd. 2 ft.
5)8 yds. 1 ft.
5 yds.
3 yds. = 9 ft.
+1
10 ft.
10

16. **22 gals.**

5 gals. 2 qts.
9 gals. 3 qts.
+6 gals. 3 qts.
20 gals. 8 qts. = **22 gals.**

17. **2 yrs. 10 mos.**

2 yrs. 10 mos.
3)8 yrs. 6 mos.
6
2 yrs. = 24 mos.
+ 6
30 mos.
30

18. **1 t. 900 lbs.**

10 t. 500 lbs. = 9 t. 2,500 lbs.
− 8 t. 1,600 lbs. −8 t. 1,600 lbs.
1 t. 900 lbs.

19. **12 ft. 2 in.**

2 ft. 9 in.
3 ft. 11 in.
+5 ft. 6 in.
10 ft. 26 in. = **12 ft. 2 in.**

20. **23 wks. 3 da.**

5 wks. 6 da.
×4 ×4
20 wks. 24 da. = **23 wks. 3 da.**

EXERCISE 5 (WORD PROBLEMS)

1. (3) **$22\frac{2}{3}$ yds.**

$$136 \times 6 = 816 \text{ in.}$$

$$816 \div 36 = 22\frac{24}{36} = \mathbf{22\frac{2}{3}\ yds.}$$

```
   22 24/36
36)816
   72
    96
    72
    24
```

2. (4) **4 lbs. 3 oz.**

```
  1 lb.   11 oz.
+2 lbs.    8 oz.
  3 lbs.  19 oz. = 4 lbs. 3 oz.
```

(1 lb. = 16 oz.)

3. (5) **$.11**

```
    $ .111
16)$1.790
    1 6
      19
      16
       30
       16
```

to the nearest cent = **$.11**

4. (4) **$.12**

```
   $.123
8)$.990
   8
   19
   16
    30
    24
```

to the nearest cent = **$.12**

5. (3) **6 hrs. 27 min.**

Call 9:40 9 hrs. 40 min.
Call 1:07 13 hrs. 7 min. (Add 1 hr. 7 min. to 12 hrs.)

```
  13 hrs.  7 min. =   12 hrs. 67 min.
-  9 hrs. 40 min.   -  9 hrs. 40 min.
                       3 hrs. 27 min.
                      +3 hrs.          (time difference)
                       6 hrs. 27 min.
```

6. (2) **1 ft. 5 in.**

```
  8 ft.        =   7 ft. 12 in.
- 6 ft. 7 in.    - 6 ft.  7 in.
                   1 ft.  5 in.
```

7. (1) **18 sec.**

```
  4 min.  7 sec. =   3 min. 67 sec.
- 3 min. 49 sec.   - 3 min. 49 sec.
                            18 sec.
```

8. (3) **1 t. 840 lbs.**

```
   1 t.         840 lbs.
5)7 t.          200 lbs.
  5
  2 t. =   4,000 lbs.
          +  200
           4 200
           4 0
             20
             20
             00
```

9. (2) **7 lbs. 11 oz.**

```
  2 lbs.     9 oz.
 ×3         ×3
  6 lbs.    27 oz. = 7 lbs. 11 oz.
```

(1 lb. = 16 oz.)

10. **(2) 24 qts. 1 pt.**

$$\begin{array}{rr} 3 \text{ qts.} & 1 \text{ pt.} \\ \times 7 & \times 7 \\ \hline 21 \text{ qts.} & 7 \text{ pts.} \end{array} = \textbf{24 qts. 1 pt.}$$

11. **(1) 7 hrs. 15 min.**

$$\begin{array}{r} 2 \text{ hrs. } 25 \text{ min.} \\ 1 \text{ hr. } 40 \text{ min.} \\ +3 \text{ hrs. } 10 \text{ min.} \\ \hline 6 \text{ hrs. } 75 \text{ min.} \end{array} = \textbf{7 hrs. 15 min.}$$

12. **(3) 5 lbs. 2 oz.**

$$6\overline{)12} = 2 \qquad \begin{array}{rr} 2 \text{ lbs.} & 9 \text{ oz.} \\ \times 2 & \times 2 \\ \hline 4 \text{ lbs.} & 18 \text{ oz.} \end{array} = \textbf{5 lbs. 2 oz.}$$

13. **(3)** $\mathbf{\frac{7}{12}}$

$$\begin{array}{r} 600 \text{ lbs.} \\ -250 \text{ lbs.} \\ \hline 350 \text{ lbs.} \end{array} \qquad \frac{350 \text{ lbs.}}{600 \text{ lbs.}} = \mathbf{\frac{7}{12}}$$

14. **(5) 1 ft. 5 in.**

$$\begin{array}{rll} & \textbf{1 ft.} & \textbf{5 in.} \\ 7) & 9 \text{ ft.} & 11 \text{ in.} \\ & \underline{7} & \\ & 2 \text{ ft.} = & 24 \text{ in.} \\ & & +\underline{11} \\ & & 35 \text{ in.} \\ & & \underline{35} \end{array}$$

15. **(1) 3 hrs. 30 min.**

$$\begin{array}{r} 2 \text{ hrs. } 30 \text{ min.} \\ 3 \text{ hrs. } 15 \text{ min.} \\ +4 \text{ hrs. } 45 \text{ min.} \\ \hline 9 \text{ hrs. } 90 \text{ min.} \end{array} = 10 \text{ hrs. } 30 \text{ min.}$$

$$\begin{array}{rll} & \textbf{3 hrs.} & \textbf{30 min.} \\ 3) & 10 \text{ hrs.} & 30 \text{ min.} \\ & \underline{9} & \\ & 1 \text{ hr.} = & 60 \text{ min.} \\ & & +\underline{30} \\ & & 90 \\ & & \underline{9} \\ & & 00 \end{array}$$

16. **(4) 7 hrs. 50 min.**

Call 8:30 8 hrs. 30 min.
Call 4:20 16 hrs. 20 min. (Add 4 hrs. 20 min. to 12 hrs.)

$$\begin{array}{r} 16 \text{ hrs. } 20 \text{ min.} \\ -\ 8 \text{ hrs. } 30 \text{ min.} \\ \hline \end{array} = \begin{array}{r} 15 \text{ hrs. } 80 \text{ min.} \\ -\ 8 \text{ hrs. } 30 \text{ min.} \\ \hline \textbf{7 hrs. 50 min.} \end{array}$$

17. **(1) 4 lbs. 13 oz.**

$$\begin{array}{rll} & \textbf{4 lbs.} & \textbf{13 oz.} \\ 3) & 14 \text{ lbs.} & 7 \text{ oz.} \\ & \underline{12} & \\ & 2 \text{ lbs.} = & 32 \text{ oz.} \\ & & +\underline{\ 7} \\ & & 39 \\ & & \underline{39} \end{array}$$

18. **(3)** $\mathbf{32\frac{2}{3}}$ **yds.**

$$\begin{array}{rl} & \mathbf{32\frac{2}{3}} \textbf{ yds.} \\ 3) & 98 \\ & \underline{9} \\ & 08 \\ & \underline{\ 6} \\ & \ 2 \end{array}$$

19. **(2) 2 t. 750 lbs.**

$$\begin{array}{r} 3 \text{ t. } \ 650 \text{ lbs.} \\ -\quad 1{,}900 \text{ lbs.} \\ \hline \end{array} = \begin{array}{r} 2 \text{ t. } 2{,}650 \text{ lbs.} \\ -\quad 1{,}900 \text{ lbs.} \\ \hline \textbf{2 t. \ 750 lbs.} \end{array}$$

20. **(5)** $\mathbf{\frac{4}{5}}$ **mi.**

$$\frac{4{,}260}{5{,}280} = \text{about } \frac{4{,}000}{5{,}000} = \mathbf{\frac{4}{5}}$$

Tables and Graphs

On the GED Mathematics Test, you will find some questions that test your ability to read and use tables and graphs. Using tables and graphs is an important skill to learn not only for the GED Test but also for everyday life. You probably use tables and graphs to figure your income tax, work out your budget, or compare the costs of different items. Tables and graphs often appear in newspapers. They are used to present numerical information about the government, the economy, and society.

In this part of the book, you will study the different ways that information is presented in tables and graphs. There are many word problems in this section. The word problems will help to sharpen your skills in reading and using information in tables and graphs.

TABLES

A **table** is a set of numbers in rows and columns. A table is a way of presenting numerical information.

The table below shows how much it costs to run some common electrical appliances. The first column of numbers shows the cost of running the appliances in the summer. The second column shows the winter cost.

What It Costs to Run Electrical Appliances

appliance		in summer	in winter
Refrigerator 12 cu. ft., manual defrost	cents per day	25	22
Toaster	cents per slice	0.3	0.3
Television			
B & W	cents per hour	2.9	2.6
Color	cents per hour	3.9	3.4
Light bulb			
60-watt	cents per hour	0.7	0.6
100-watt	cents per hour	1.2	1.0
Iron	cents per hour	6.7	5.9
Clock	cents per month	23.4	21

The GED Test often contains questions about information in tables.

EXAMPLE: Find the cost of burning a 60-watt bulb for 10 hours during the summer.

Step 1: Find the "60-watt" under "light bulb." Read across the line to the number under the column called "in summer." The number is 0.7 cents per hour.

Step 2: Multiply 0.7 by 10 hours: $0.7 \times 10 = 7¢$.

EXERCISE 1

Use the table called "What It Costs to Run Electrical Appliances" to answer the next questions.

1. How much does it cost to run a color television for one hour during the winter?

2. How much does it cost to use an iron for an hour during the summer?

3. How much does it cost to burn a 100-watt bulb for one hour during the summer?

4. How much does it cost to run an electric clock for a month during the winter?

5. How much does it cost to watch a black-and-white television for four hours during the summer? Round off the answer to the nearest cent.

6. During the winter, the Duggans burn four 100-watt bulbs and three 60-watt bulbs for an average of 5 hours each day. To the nearest cent, how much does it cost each day to burn the light bulbs?

7. How much more does it cost to run an electric clock for a month in the summer than for a month in the winter?

8. Mrs. Adams toasts 6 slices of toast every day, 7 days a week for her family. To the nearest cent, how much does the electricity for the toaster cost each week?

9. How much does it cost to run a 12 cubic foot refrigerator for 30 days during the summer?

10. Jack watches his color television for an average of 3.5 hours a day all year round. To the nearest cent, how much more does it cost Jack to watch TV on a day in the summer than on a day in the winter?

Check your answers on page 181.

The next table shows the cost of sending packages up to 5 pounds in weight by parcel post. To find the cost, you must know the weight of the package and the zone the package is going to.

Parcel Post Rate Schedule

Weight		Zones				
from	**to**	**Local**	**1 & 2**	**3**	**4**	**5**
1 lb.	2 lbs.	$1.15	$1.35	$1.39	$1.56	$1.72
2 lbs.	3 lbs.	1.23	1.45	1.53	1.73	1.86
3 lbs.	4 lbs.	1.29	1.50	1.65	1.82	2.00
4 lbs.	5 lbs.	1.36	1.66	1.77	1.92	2.14

EXAMPLE: Find the cost of sending a package weighing 3 lbs. 5 oz. to Zone 4.

$1.82

Step 1: Find the row that starts with 3 lbs. under "weight."
Step 2: Read across the line to the column under the number 4.

EXAMPLE: How much more does it cost to send a package weighing 2 lbs. 14 oz. to Zone 3 than to a local address?

Step 1: Find the row that starts with 2 lbs. under "weight."
Step 2: Read across the line to the column under the number 3. The cost to Zone 3 is $1.53.
Step 3: Next, read across to the column under the word "Local." The cost to a local address is $1.23.
Step 4: Subtract the local cost from the Zone 3 cost: $1.53 − $1.23 = $.30. The difference in cost is $.30.

EXERCISE 2

DIRECTIONS: Use the "Parcel Post Rate Schedule" to answer the next questions.

1. Find the cost of sending a package weighing 2 lbs. 4 oz. to Zone 5.

2. How much does it cost to send a package weighing 4 lbs. 2 oz. to a local address?

3. Find the cost of sending a package weighing 1 lb. 10 oz. to Zone 2.

4. How much does it cost to send a package weighing 3 lbs. 9 oz. to Zone 3?

5. Colette sent a package weighing 1 lb. 6 oz. and another weighing 1 lb. 10 oz. both to Zone 4. Find the total cost of sending the two packages.

6. How much more does it cost to send a package weighing exactly 3 lbs. to Zone 5 than to a local address?

7. Jeff sent a package weighing 2 lbs. 3 oz. to Zone 1, another weighing 2 lbs. 9 oz. to Zone 4, and a third weighing 3 lbs. 1 oz. to Zone 3. Find the total cost of sending the three packages.

8. Frank has a choice of sending 1 package weighing 4 lbs. 3 oz. to Zone 2 or two packages, each weighing 2 lbs. 9 oz., to Zone 2. How much cheaper is it to send one big package?

9. How much more does it cost to send a package weighing 2 lbs. 1 oz. to Zone 5 than to send a package weighing 1 lb. 11 oz. to Zone 5?

10. By what percent does the cost of sending a package weighing 3 lbs. 6 oz. increase from Zone 2 to Zone 5?

Check your answers on page 182.

Pictographs

A **pictograph** uses a symbol to represent a number of things. Somewhere near the pictograph, a key tells what each symbol stands for. The amounts shown in a pictograph are always rounded off.

The pictograph below shows the number of eggs produced in a year in four different states. The key below the pictograph tells you that each egg stands for one billion eggs.

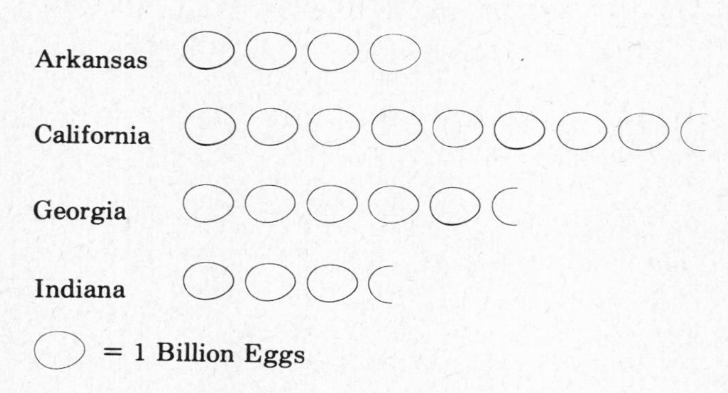

EXAMPLE: How many eggs does Indiana produce in a year?

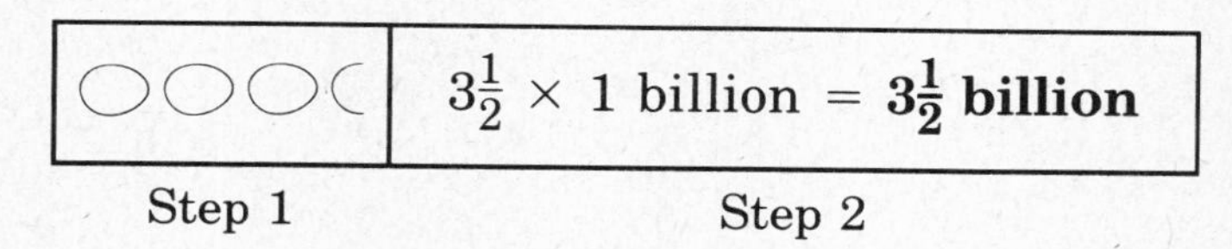

Step 1: Count the number of eggs at the right of the word "Indiana" in the pictograph. There are $3\frac{1}{2}$ eggs.

Step 2: Each egg stands for 1 billion. Multiply $3\frac{1}{2}$ by 1 billion to get the answer.

EXERCISE 3

DIRECTIONS: Use the pictograph about egg production to answer these questions.

1. How many eggs does Arkansas produce in a year?

2. How many eggs does California produce in a year?

3. How many eggs does Georgia produce in a year?

4. Together, the four states in the pictograph produce how many eggs in a year?

5. In a year, the top producer of eggs produces how many more eggs than the second-place producer?

6. The combined egg production in Arkansas and Indiana is how much less than the egg production in California?

7. In a year, North Carolina produces 3,081,000,000 eggs. To the nearest half egg, how many egg symbols would you need to show North Carolina's yearly egg production?

8. In a year, the total egg production in the U.S. was 67 billion eggs. The combined production of the four states in the pictograph was about what fraction of the total U.S. production?

9. Ohio produced about 2,140,000,000 eggs in a year. The egg production in Ohio is about what percent of the production in Arkansas?

10. The yearly egg production in Indiana is what percent of the egg production in Arkansas?

Check your answers on page 182.

The next pictograph shows the number of barrels of oil produced each day in six countries near the Persian Gulf.

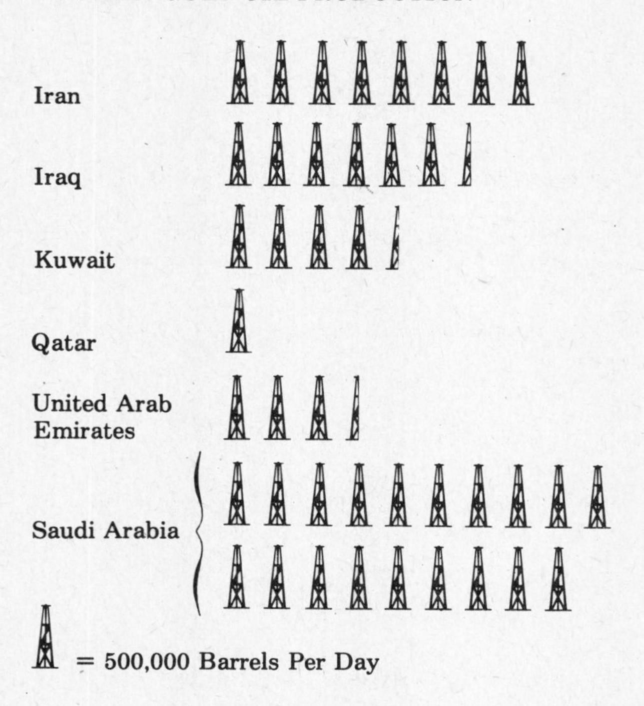

EXAMPLE: How many more barrels of oil per day does Iraq produce than Kuwait?

Step 1: Count the number of symbols across from the word "Iraq." There are about $6\frac{1}{2}$ symbols.

Step 2: Count the number of symbols across from the word "Kuwait." There are about $4\frac{1}{2}$ symbols.

Step 3: Subtract the number of symbols across from Kuwait from the number of symbols across from Iraq: $6\frac{1}{2} - 4\frac{1}{2} = 2$.

Step 4: Multiply the number of barrels that each symbol stands for by the number of symbols left: $500{,}000 \times 2 = 1{,}000{,}000$. Iraq produces 1,000,000 barrels of oil per day more than Kuwait.

EXERCISE 4

DIRECTIONS: Use the pictograph on Persian Gulf oil production to answer the next questions.

1. Each picture of an oil well represents how many barrels of oil?

   (1) 1 (2) 1,000 (3) 50,000
   (4) 500,000 (5) 5,000,000

2. How many barrels of oil does Saudi Arabia produce in a day?

   (1) 9,500 (2) 95,000 (3) 950,000
   (4) 9,500,000 (5) 95,000,000

3. The daily oil production in Qatar is what fraction of the oil production in Iran?

   (1) $\frac{1}{8}$ (2) $\frac{1}{5}$ (3) $\frac{1}{4}$ (4) $\frac{1}{3}$ (5) $\frac{1}{2}$

4. A barrel of crude oil costs about $30. Find the value of the daily oil production in Saudi Arabia.

   (1) $28,500,000 (2) $30,000,000 (3) $285,000,000
   (4) $300,000,000 (5) $950,000,000

5. The combined daily oil production of which three countries is equal to the daily production in Saudi Arabia?

   (1) Iran, Iraq, and Kuwait
   (2) Iraq, Kuwait, and Qatar
   (3) Iran, Kuwait, and the United Arab Emirates
   (4) Iraq, Kuwait, and the United Arab Emirates
   (5) Kuwait, Qatar, and the United Arab Emirates

6. One barrel holds 42 gallons of oil. Find the daily oil production of Qatar in gallons.

   (1) 210 million gals. (2) 21 million gals.
   (3) 12 million gals. (4) 2.1 million gals.
   (5) 1.2 million gals.

Check your answers on page 183.

Circle Graphs

A circle graph shows the parts of a whole amount. Each "piece" of a circle represents a percent of a whole or a dollar amount of a whole.

The circle graph below shows how one family spends the money it makes each month. The percents in the graph add up to 100%, or its whole monthly budget.

ROSA FAMILY'S MONTHLY BUDGET

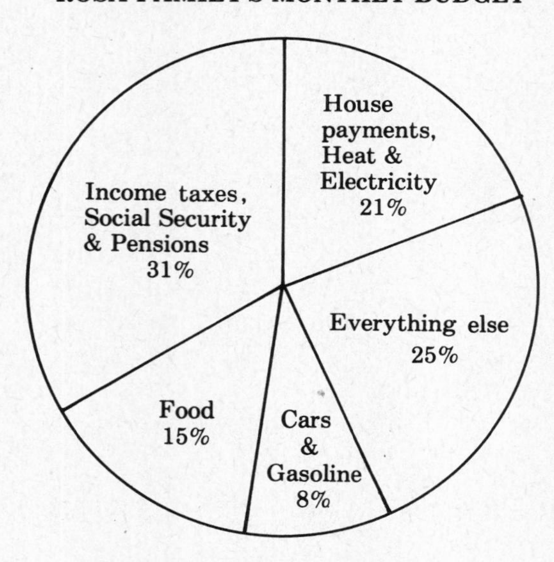

EXAMPLE: The Rosa family has a combined income, before expenses, of $1,600 a month. How much of the Rosa's income goes for food each month?

Step 1	Step 2
food = **15%**	$1,600 × 15% = $1,600 × .15 80 00 160 0 **$240.00**

Step 1: Find the percent of the budget spent on food. Food represents 15% of the budget.

Step 2: Find 15% of $1,600.

EXERCISE 5

DIRECTIONS: Use the graph of the Rosa family's budget to answer the next questions.

1. Cars and gasoline represent what percent of the Rosas' budget?
2. House payments, heat, and electricity represent what percent of the Rosas' budget?
3. Income taxes, Social Security, and pensions represent what percent of the Rosas' budget?
4. Food represents what fraction of the Rosas' budget?
5. The Rosas' combined income, before expenses, is $1,600. How much of their income goes for income taxes, Social Security, and pensions?
6. What fraction of the Rosas' budget goes for cars and gasoline?
7. The Rosas' combined income is $1,600 a month. How much do they spend on house payments, heat, and electricity in a year?
8. Out of every dollar they earn, how much do the Rosas spend on food?
9. With a monthly budget of $1,600, how much money do the Rosas have each month for "everything else"?
10. The Rosas try to save $160 every month. This amount represents what percent of their monthly income?

Check your answers on page 183.

The next circle graph represents another family's monthly budget. Each piece of the graph represents a dollar amount.

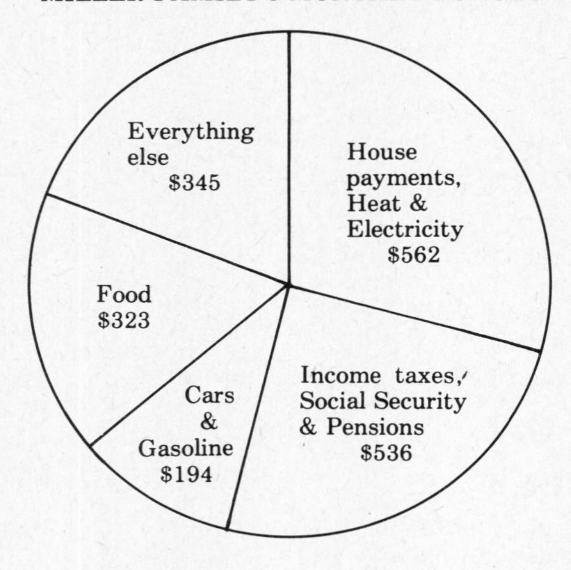

EXAMPLE: Together, how much does the Miller family spend every month on food, house payments, heat, and electricity?

Step 1: Find the amount of money that the Millers spend on food. They spend $323 every month on food.

Step 2: Find the amount of money that the Millers spend on house payments, heat, and electricity. They spend $562 every month on house payments, heat, and electricity.

Step 3: Add the amounts: $323 + $562 = $885. The Millers spend $885 every month on food, house payments, heat, and electricity.

EXERCISE 6

DIRECTIONS: Use the graph of the Miller family's budget to answer the next questions.

1. How much do the Millers spend each month on food?

2. How much do the Millers spend each month on cars and gasoline?

3. How much do the Millers spend each month on house payments, heat, and electricity?

4. How much do the Millers spend each month in income taxes, Social Security, and pensions?

5. What is the total amount of the Millers' monthly budget?

For questions 6-10, choose the correct answer.

6. What percent of the Millers' budget goes for food? Round your answer off to the nearest percent.

   (1) 84% (2) 50% (3) 32% (4) 16% (5) 10%

7. The Millers spend about what fraction of their monthly budget on cars and gasoline?

   (1) $\frac{1}{10}$ (2) $\frac{1}{8}$ (3) $\frac{1}{6}$ (4) $\frac{1}{5}$ (5) $\frac{1}{4}$

8. Which statement is false?

   (1) House payments, heat, electricity, income taxes, Social Security, and pensions use up more than half the Millers' budget.
   (2) The Millers spend almost as much on food, cars, and gasoline as they spend on income taxes, Social Security, and pensions.
   (3) The Millers spend about twice as much on cars and gasoline as they spend on food.
   (4) The amount the Millers have left for "everything else" is a little more than the amount they pay for food.
   (5) The Millers spend less on food than they spend on house payments, heat, and electricity.

9. What percent of the Millers' budget goes for income taxes, Social Security, and pensions? Round off your answer to the nearest percent.

   (1) 54% (2) 27% (3) 20% (4) 18% (5) 14%

10. The amount the Millers spend on cars and gasoline is about what percent of the amount they spend on food?

   (1) 5% (2) 10% (3) 20% (4) 50% (5) 60%

Check your answers on page 184.

Bar Graphs

A **bar graph** is a good way to compare several numbers. The length of each bar corresponds to the size of a number. Bars may run up and down or from side to side. To find the number each bar stands for, use the **scale** along the side or across the bottom of the graph.

The bar graph below shows the population of the U.S. from 1900 to 1970. The scale at the left shows numbers in millions.

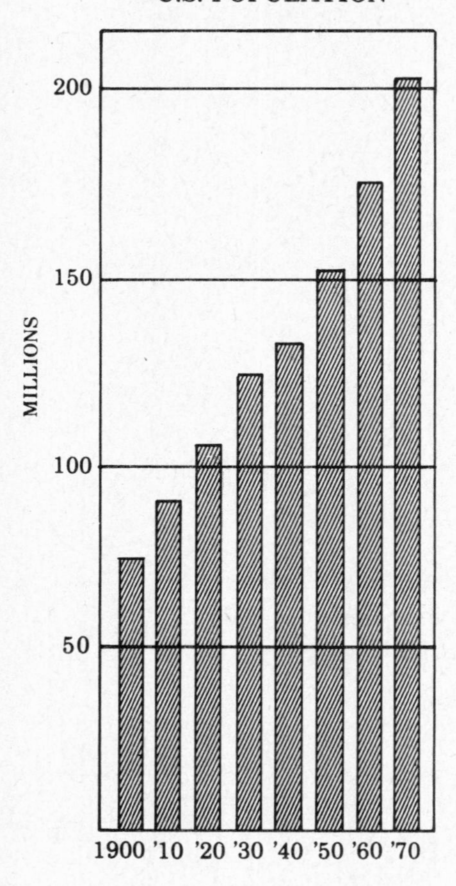

EXAMPLE: What was the population of the U.S. in 1900?

Step 1: Find 1900 on the list across the bottom of the graph.

Step 2: Follow the bar above 1900 to the top and read the population on the scale at the left.

Step 3: The bar ends about halfway between 50 and 100. Halfway between 50 and 100 is 75.

The population of the U.S. in 1900 was about 75 million.

EXERCISE 7

DIRECTIONS: Use the graph of U.S. population to answer the next questions.

1. What was the population of the U.S. in 1920?

   (1) 95 million (2) 107 million (3) 120 million
   (4) 130 million (5) 140 million

2. What was the population of the U.S. in 1940?

   (1) 100 million (2) 107 million (3) 133 million
   (4) 147 million (5) 160 million

3. What was the population of the U.S. in 1960?

   (1) 140 million (2) 153 million (3) 160 million
   (4) 163 million (5) 177 million

4. Which was the first year on the graph that the population went over 100 million?

   (1) 1910 (2) 1920 (3) 1930 (4) 1940 (5) 1950

5. The U.S. population in 1900 was about what fraction of the population in 1950?

   (1) $\frac{1}{4}$ (2) $\frac{1}{3}$ (3) $\frac{1}{2}$ (4) $\frac{2}{3}$ (5) $\frac{3}{4}$

6. By about how much did the U.S. population grow from 1950 to 1960?

   (1) 50 million (2) 40 million (3) 25 million
   (4) 10 million (5) 5 million

7. During what 10-year period did the U.S. population grow the least?

   (1) 1900–1910 (2) 1920–1930 (3) 1930–1940
   (4) 1940–1950 (5) 1960–1970

8. The increase in population from 1950 to 1960 was about the same as the increase in what other 10-year period?

   (1) 1960–1970 (2) 1930–1940 (3) 1920–1930
   (4) 1910–1920 (5) 1900–1910

9. The population of the U.S. in 1970 was about how many times as big as the population in 1900?

   (1) $\frac{3}{8}$ (2) $\frac{3}{4}$ (3) $1\frac{1}{2}$ (4) 2 (5) $2\frac{2}{3}$

10. Which statement best describes the trend of U.S. population shown on the graph?

    (1) From 1900, the population has steadily gone down.
    (2) From 1900, the population has stayed about the same.
    (3) From 1900, the population has steadily gone up.
    (4) From 1900, the population increased steadily for about 30 years; then it began to decrease.
    (5) From 1900, the population decreased steadily for about 30 years; then it began to increase.

Check your answers on page 184.

Divided Bar Graphs

A **divided bar graph** splits whole numbers into parts. The next graph is a divided bar graph that shows the number of people living in the four major areas of the U.S. for the years 1972, 1976, and 1979. The parts of each bar show the number of people living in the West, the North Central states, the South, and the Northeast.

U.S. POPULATION in millions

	West	North Central states	South	Northeast
1972	36	57	65	50
1976	39	58	69	49
1979	41	58	72	49

EXAMPLE: Find the total population of the four major areas of the U.S. in 1972.

$$\begin{array}{r} 36 \text{ million} \\ 57 \\ 65 \\ +50 \\ \hline \mathbf{208 \text{ million}} \end{array}$$

Add the population for each area in 1972.

EXERCISE 8

DIRECTIONS: Use the divided bar graph of U.S. population to answer the next questions.

1. What was the population of the South in 1972?
2. What was the population of the Northeast in 1976?
3. What was the population of the West in 1979?
4. What the the total population of the four major areas in 1976?
5. What was the total population of the four major areas in 1979?
6. What area had the biggest population increase from 1972 to 1976?
7. What major area lost population from 1972 to 1976?
8. What two major areas stayed the same from 1976 to 1979?
9. By how much did the population of the South increase from 1972 to 1979?
10. Which statement best describes the information on the divided bar graph?
    (1) From 1972 to 1979, the four major areas have grown about the same amount.
    (2) From 1972 to 1979, the North Central states and the Northeast have steadily grown, while the West and the South have stayed about the same.
    (3) From 1972 to 1979, the four major areas have lost population by about the same amount.
    (4) From 1972 to 1979, the North Central states and the Northeast have stayed about the same, while the West and South have steadily grown.

Check your answers on page 185.

Line Graphs

A **line graph** is a good way to measure a changing amount. A line that goes up from left to right shows an *upward trend.* A line that goes down from left to right shows a *downward trend.* To read a line graph, be sure you understand the vertical scale going up one side and the horizontal scale going across the graph.

This line graph shows the budget for the Safety and Health Administration from 1971 to 1980.

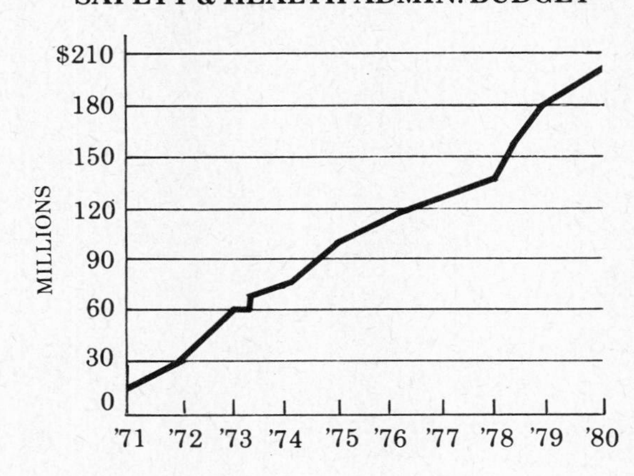

EXAMPLE: At the beginning of 1975, what was the budget of the Safety and Health Administration?

$90 million

Step 1: Find '75 on the horizontal scale at the bottom of the graph.

Step 2: Follow the line marked '75 straight up until you reach the changing line.

Step 3: Look straight across to the vertical scale. The '75 budget was about $\$90 \times 1{,}000{,}000 = \$90{,}000{,}000$.

EXERCISE 9

DIRECTIONS: Use the graph of the Safety and Health Administration's budget to answer the next questions.

1. What was the budget at the beginning of '73?

2. In which year did the budget first reach $30 million?

3. In which year did the budget first reach $120 million?

4. What was the budget at the beginning of '78?

For questions 5-10, choose the correct answer.

5. About how many times bigger did the budget get from 1971 to 1980?

   (1) 2 times (2) 3 times (3) 4 times
   (4) 5 times (5) 10 times

6. During which one-year period did the budget grow the most?

   (1) '71 to '72 (2) '73 to '74 (3) '77 to '78
   (4) '78 to '79 (5) '79 to '80

7. The budget at the beginning of '73 was about what fraction of the budget at the beginning of '75?

   (1) $\frac{3}{4}$ (2) $\frac{2}{3}$ (3) $\frac{1}{2}$ (4) $\frac{1}{3}$ (5) $\frac{1}{4}$

8. The budget at the beginning of '72 was about what fraction of the budget at the beginning of '79?

   (1) $\frac{1}{6}$ (2) $\frac{1}{5}$ (3) $\frac{1}{4}$ (4) $\frac{1}{3}$ (5) $\frac{1}{2}$

9. By how much did the budget increase from the beginning of '79 to the beginning of '80?

   (1) $10 million (2) $20 million (3) $30 million
   (4) $180 million (5) $200 million

10. If the budget grows from '80 to '81 as it grew from '79 to '80, what will be the amount of the budget at the beginning of '81?

    (1) $15 million (2) $30 million (3) $210 million
    (4) $220 million (5) $240 million

Check your answers on page 185.

The next line graph compares the travel in the U.S. by rail and by air. One line measures rail travel as a percent of total travel. The other line measures air travel as a percent of total travel.

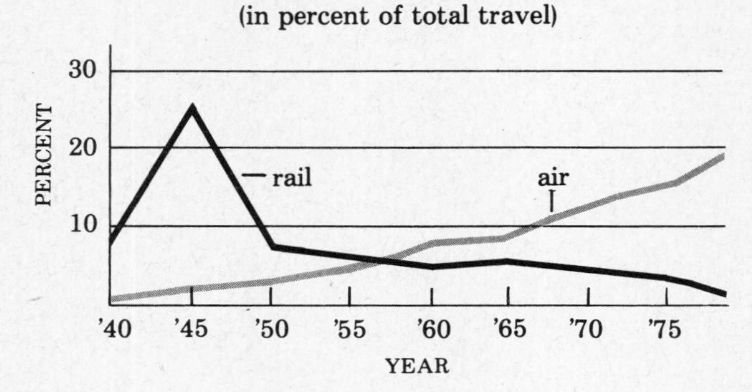

EXERCISE 10

DIRECTIONS: Use the graph about rail and air travel to answer the next questions.

1. Rail travel was what percent of total travel in 1940?

   (1) 40% (2) 25% (3) 20% (4) 15% (5) 7%

2. Rail travel was what percent of total travel in 1945?

   (1) 30% (2) 26% (3) 20% (4) 15% (5) 12%

3. Air travel was what percent of total travel in 1970?

   (1) 6% (2) 8% (3) 10% (4) 12% (5) 15%

4. Air travel was what percent of total travel in 1945?

   (1) 0% (2) 2% (3) 5% (4) 8% (5) 12%

5. During which year did rail travel and air travel represent the same percent of total travel?

   (1) 1950 (2) 1955 (3) 1957 (4) 1961 (5) 1963

6. Together, rail travel and air travel made up what percent of total travel in 1945?

   (1) 12% (2) 18% (3) 20% (4) 28% (5) 35%

7. Together, rail travel and air travel made up what percent of total travel in 1957?

   (1) 5% (2) 10% (3) 15% (4) 20% (5) 25%

8. During which year did air travel first represent 15% of total travel?

   (1) 1960 (2) 1965 (3) 1970 (4) 1975 (5) 1980

9. In which five-year period did rail travel decline the most?

   (1) '40–'45 (2) '45–'50 (3) '50–'55
   (4) '60–'65 (5) '70–'75

10. Which statement best describes the travel trend shown on the graph?

    (1) From 1940 to 1975, both rail travel and air travel increased steadily.
    (2) From 1940 to 1975, both rail travel and air travel declined steadily.
    (3) From 1940 to 1975, rail travel increased steadily, while air travel declined steadily.
    (4) From 1940 to 1975, rail travel experienced a brief increase followed by a steady decline, while air travel increased steadily.
    (5) From 1940 to 1945, rail travel and air travel represented about the same percent of total travel.

Check your answers on page 186.

ANSWERS & SOLUTIONS

EXERCISE 1

1. **3.4¢** 2. **6.7¢** 3. **1.2¢** 4. **21¢**

5. **12¢**

2.9¢
×4
11.6¢ to the nearest cent = **12¢**

6. **29¢**

1.0¢
×4
4.0
×5
20.0¢

0.6¢
×3
1.8
×5
9.0¢

20¢
+9
29¢

7. **2.4¢**

23.4¢
−21.0
2.4¢

8. **13¢**

0.3¢
×6
1.8
×7
12.6¢ to the nearest cent = **13¢**

9. **$7.50**

$.25
×30
$7.50

10. **2¢**

$$\begin{array}{r} 3.9¢ \\ \times 3.5 \\ \hline 195 \\ 11\,7 \\ \hline 13.65¢ \end{array} \qquad \begin{array}{r} 3.4¢ \\ \times 3.5 \\ \hline 170 \\ 10\,2 \\ \hline 11.90¢ \end{array} \qquad \begin{array}{r} 13.65¢ \\ -11.90 \\ \hline \end{array}$$

1.75¢ to the nearest cent = **2¢**

EXERCISE 2

1. **$1.86** 2. **$1.36** 3. **$1.35** 4. **$1.65** 5. **$3.12**

$$\begin{array}{r} \$1.56 \\ \times\ 2 \\ \hline \mathbf{\$3.12} \end{array}$$

6. **$.71**

$$\begin{array}{r} \$2.00 \\ -1.29 \\ \hline \mathbf{\$\ .71} \end{array}$$

7. **$4.83**

$$\begin{array}{r} \$1.45 \\ 1.73 \\ +1.65 \\ \hline \mathbf{\$4.83} \end{array}$$

8. **$1.24**

$$\begin{array}{r} \$1.45 \\ \times\ 2 \\ \hline \$2.90 \end{array} \qquad \begin{array}{r} \$2.90 \\ -1.66 \\ \hline \mathbf{\$1.24} \end{array}$$

9. **$.14**

$$\begin{array}{r} \$1.86 \\ -1.72 \\ \hline \mathbf{\$\ .14} \end{array}$$

10. $\mathbf{33\frac{1}{3}\%}$

$$\begin{array}{r} \$2.00 \\ -1.50 \\ \hline \$\ .50 \end{array} \qquad \frac{.50}{1.50} = \frac{1}{3} = \mathbf{33\frac{1}{3}\%}$$

EXERCISE 3

1. **4 billion** 2. $\mathbf{8\frac{1}{2}}$ **billion** 3. $\mathbf{5\frac{1}{2}}$ **billion**

4. $\mathbf{21\frac{1}{2}}$ **billion**

$$\begin{array}{l} 4 \text{ billion} \\ 8\frac{1}{2} \\ 5\frac{1}{2} \\ +3\frac{1}{2} \\ \hline 20\frac{3}{2} = \mathbf{21\frac{1}{2}\ billion} \end{array}$$

5. **3 billion**

$$\begin{array}{l} 8\frac{1}{2} \text{ billion} \\ -5\frac{1}{2} \\ \hline \mathbf{3\ billion} \end{array}$$

6. **1 billion**

$$\begin{array}{l} 4 \text{ billion} \\ +3\frac{1}{2} \\ \hline 7\frac{1}{2} \text{ billion} \end{array} \qquad \begin{array}{l} 8\frac{1}{2} \text{ billion} \\ -7\frac{1}{2} \\ \hline \mathbf{1\ billion} \end{array}$$

7. **3 eggs** $\frac{3{,}081{,}000{,}000}{1{,}000{,}000{,}000}$ = 3.081 to the nearest half egg = **3 eggs**

8. $\mathbf{\frac{1}{3}}$ $21\frac{1}{2}$ billion is close to $\frac{1}{3}$ of 67 billion.

9. **50%** 2,104,000,000 = about 2 billion $\frac{\text{2 billion}}{\text{4 billion}} = \frac{1}{2} = \mathbf{50\%}$

10. **$87\frac{1}{2}\%$** $\frac{3\frac{1}{2}}{4} = 3\frac{1}{2} \div 4 = \frac{7}{2} \times \frac{1}{4} = \frac{7}{8} = \mathbf{87\frac{1}{2}\%}$

EXERCISE 4

1. (4) **500,000**
2. (4) **9,500,000** $19 \times 500{,}000 = \mathbf{9{,}500{,}000}$

3. (1) $\mathbf{\frac{1}{8}}$ $\frac{500{,}000}{4{,}000{,}000} = \frac{1}{8}$

4. (3) **$285,000,000**

$$\begin{array}{r} 9{,}500{,}000 \\ \times \quad \$30 \\ \hline \mathbf{\$285{,}000{,}000} \end{array}$$

5. (1) **Iran, Iraq, and Kuwait**

$$\begin{array}{rr} \text{Iran} = & 4{,}000{,}000 \\ \text{Iraq} = & 3{,}250{,}000 \\ \text{Kuwait} = & +2{,}250{,}000 \\ \hline & \mathbf{9{,}500{,}000} = \text{Saudi Arabia} \end{array}$$

6. (2) **21 million gals.**

$$\begin{array}{r} 500{,}000 \\ \times \quad 42 \\ \hline 1\ 000\ 000 \\ 20\ 000\ 00 \\ \hline \mathbf{21{,}000{,}000} \end{array}$$

EXERCISE 5

1. **8%**
2. **21%**
3. **31%**

4. $\mathbf{\frac{3}{20}}$

$\frac{15}{100} = \mathbf{\frac{3}{20}}$

5. **$496.00**

31% = .31

$$\begin{array}{r} \$1{,}600 \\ \times \quad .31 \\ \hline 16\ 00 \\ 480\ 0 \\ \hline \mathbf{\$496.00} \end{array}$$

6. $\mathbf{\frac{2}{25}}$

$\frac{8}{100} = \mathbf{\frac{2}{25}}$

7. **$4,032**

$$\begin{array}{r} \$1{,}600 \\ \times \quad 12 \\ \hline 3\ 200 \\ 16\ 00 \\ \hline \$19{,}200 \end{array}$$

21% = .21

$$\begin{array}{r} \$19{,}200 \\ \times \quad .21 \\ \hline 192\ 00 \\ 3\ 840\ 0 \\ \hline \mathbf{\$4{,}032.00} \end{array}$$

8. **15¢**

$15\% = \frac{15}{100} = \mathbf{15¢}$

9. **$400**

$25\% = \frac{1}{4}$

$\frac{1}{\cancel{4}_1} \times \frac{\cancel{1,600}^{400}}{1} = \mathbf{\$400}$

10. **10%**

$\frac{\$160}{\$1,600} = \frac{1}{10} = \mathbf{10\%}$

EXERCISE 6

1. **$323** 2. **$194** 3. **$562** 4. **$536** 5. **$1,960**

$$\begin{array}{r} \$562 \\ 536 \\ 194 \\ 323 \\ +345 \\ \hline \mathbf{\$1,960} \end{array}$$

6. (4) **16%** $\frac{323}{1,960} =$ $1,960\overline{)323.000}$ = .164 to the nearest hundredth = .16 = **16%**

$$\begin{array}{r} .164 \\ 1,960\overline{)323.000} \\ \underline{196\ 0} \\ 127\ 00 \\ \underline{117\ 60} \\ 9\ 400 \\ \underline{7\ 840} \end{array}$$

7. (1) $\mathbf{\frac{1}{10}}$ $\frac{194}{1,960} = \text{about } \frac{1}{10}$

8. (3) **Number 3 is false.** The Millers spend much less on cars and gasoline than on food.

9. (2) **27%** $\frac{536}{1,960} =$ .273 to the nearest hundredth = .27 = **27%**

$$\begin{array}{r} .273 \\ 1,960\overline{)536.000} \\ \underline{392\ 0} \\ 144\ 00 \\ \underline{137\ 20} \\ 6\ 800 \\ \underline{5\ 880} \end{array}$$

10. (5) **60%** $\frac{194}{323} =$ $323\overline{)194.000}$ = .600 = **60%**

EXERCISE 7

1. (2) **107 million** 2. (3) **133 million** 3. (5) **177 million**

4. (2) **1920** 5. (3) $\frac{1}{2}$ $\frac{75 \text{ million}}{150 \text{ million}} = \frac{1}{2}$ 6. (3) **25 million**

$$\begin{array}{r} 177 \text{ million} \\ -\underline{152} \\ \mathbf{25 \text{ million}} \end{array}$$

7. (3) **1930–1940** The population increased by only about 10 million from 1930 to 1940.

8. (1) **1960–1970** Both increases were about 25 million.

9. (5) $2\frac{2}{3}$

$$\begin{array}{r} 2\frac{50}{75} = 2\frac{2}{3} \\ 75\overline{)200} \\ \underline{150} \\ 50 \end{array}$$

10. (3) **From 1900, the population has steadily gone up.**

EXERCISE 8

1. **65 million** 2. **49 million** 3. **41 million**

4. **215 million**

$$\begin{array}{r} 39 \text{ million} \\ 58 \\ 69 \\ \underline{+49} \\ \mathbf{215 \text{ million}} \end{array}$$

5. **220 million**

$$\begin{array}{r} 41 \text{ million} \\ 58 \\ 72 \\ \underline{+49} \\ \mathbf{220 \text{ million}} \end{array}$$

6. **South** It increased by 4 million.

7. **Northeast** It dropped from 50 million to 49 million.

8. **North Central states and Northeast** 9. **7 million**

$$\begin{array}{r} 72 \text{ million} \\ \underline{-65} \\ \mathbf{7 \text{ million}} \end{array}$$

10. (4) **From 1972 to 1979, the North Central states and the Northeast have stayed about the same, while the West and South have steadily grown.**

EXERCISE 9

1. **$60,000,000** 2. **1972** 3. **1976** 4. **$135 million** (Halfway between 120 and 150 is 135.)

5. (5) **10 times** The budget grew from about $20 million to about $200 million. $20\overline{)200}$ = **10**

6. (4) **'78 to '79** 7. (2) $\frac{2}{3}$ $\frac{60}{90} = \frac{2}{3}$ 8. (1) $\frac{1}{6}$ $\frac{30}{180} = \frac{1}{6}$

9. (2) **$20 million** $200 million − 180 = **$ 20 million** 10. (4) **$220 million** $200 million + 20 = **$220 million**

EXERCISE 10

1. (5) 7% 2. (2) 26% 3. (4) 12% 4. (2) 2%

5. (3) **1957** 6. (4) **28%** 26% + 2% = **28%** 7. (2) **10%** 5% + 5% = **10%**

8. (4) **1975** 9. (2) **'45–'50**

10. (4) **From 1940 to 1975, rail travel experienced a brief increase followed by a steady decline, while air travel increased steadily.**

Review Test

The next test will give you a chance to review the skills that you have studied so far in this book. The problems are similar to the types of problems that you may find on the GED Test.

After you take the test, check your answers. Circle the numbers of the problems that you missed. After you check your answers, look at the chart that follows. The chart will show you which sections of the book explain the skills that appear on the review test.

DIRECTIONS: Choose the correct answer for each problem.

1. To drive 100 miles, Steve used 6.5 gallons of gas. To the nearest whole mile, what was Steve's average number of miles per gallon?

   (1) 6 (2) 10 (3) 12 (4) 15 (5) 18

2. In 1970, the population of Buffalo was 1,349,000. In 1980, the population was 1,240,000. By how much did the population drop in 10 years?

   (1) 149,000 (2) 109,000 (3) 101,000
   (4) 59,000 (5) 49,000

3. Find the interest on \$2,000 at $5\frac{1}{4}\%$ annual interest for 8 months.

   (1) \$70 (2) \$84 (3) \$105 (4) \$108 (5) \$840

4. Ellen bought a turkey weighing 9 pounds 11 ounces and pork chops weighing 2 pounds 8 ounces. Find the total weight of the meat she bought.

   (1) 12 lbs. 9 oz. (2) 12 lbs. 6 oz. (3) 12 lbs. 3 oz.
   (4) 10 lbs. 3 oz. (5) 7 lbs. 3 oz.

5. Find the cost of $3\frac{3}{4}$ yards of lumber at \$8 a yard.

   (1) \$24 (2) \$27 (3) \$27.80 (4) \$30 (5) \$30.75

6. Marcus Adler died in 1962 after his 84th birthday. In what year was Marcus Adler born?

   (1) 1898 (2) 1888 (3) 1884 (4) 1878 (5) 1874

7. Jim's gross wages for 1980 were $13,297. His total deductions for the year were $2,678. What was Jim's net income?

   (1) $15,975 (2) $11,619 (3) $10,629
   (4) $10,619 (5) $10,519

8. How many $5\frac{1}{2}$-inch-long sticks can Phil cut from a piece of wood 77 inches long?

   (1) 42 (2) 16 (3) 15 (4) 14 (5) 4.2

Questions 9–11 refer to the table below.

STATE TAX RATE SCHEDULE

If amount on page 1, line 5 is:

over	but not over	your tax is
$ 0	$ 1,000	2% of amount on line 5
1,000	3,000	$ 20 plus 3% of excess over $ 1,000
3,000	5,000	80 plus 4% of excess over 3,000
5,000	7,000	160 plus 5% of excess over 5,000
7,000	9,000	260 plus 6% of excess over 7,000
9,000	11,000	380 plus 7% of excess over 9,000
11,000	13,000	520 plus 8% of excess over 11,000

9. Joe Samuels got $9,400 on page 1, line 5 of his state tax form. How much is his tax?

   (1) $552 (2) $408 (3) $288 (4) $284 (5) $185

10. If the amount on page 1, line 5 is exactly $5,000, what is the tax?

    (1) $160 (2) $165 (3) $250 (4) $256 (5) $266

11. Sue Patterson got $7,500 on page 1, line 5 of her state tax form. Find her tax on the tax table.

    (1) $300 (2) $295 (3) $290 (4) $280 (5) $260

12. Derrick bought 7 pounds of fish on sale. He wanted to divide the fish into three packages of equal amounts to freeze it. How much would each package weigh?

    (1) 7 (2) 4 (3) $3\frac{1}{3}$ (4) $2\frac{1}{3}$ (5) $1\frac{2}{3}$

13. The Ajax Welding Company is shipping a load weighing 20 tons 800 pounds on 6 trucks. If the weight is divided equally, how much will each truck carry?

    (1) 3 t. 100 lbs. (2) 3 t. 800 lbs. (3) 4 t. 500 lbs.
    (4) 4 t. 800 lbs. (5) 4 t. 900 lbs.

14. Bill cut a piece of copper tubing that was 2.25 meters long into 3 equal pieces. How long was each piece?

    (1) .75 m. (2) 1.75 m. (3) 2 m.
    (4) 2.75 m. (5) 6.75 m.

15. Marilyn made $680 last month. She put $85 into her savings account. What fraction of her monthly earnings did she save?

    (1) $\frac{1}{8}$ (2) $\frac{1}{6}$ (3) $\frac{1}{4}$ (4) $\frac{1}{2}$ (5) $\frac{7}{8}$

16. Frank works on commission at a shoe store. The first week in June, his commission was $165; the second week, $213; the third week, $302; and the fourth week $188. Find his average weekly commission for the month.

    (1) $168 (2) $217 (3) $302 (4) $317 (5) $868

17. 320 people work at the Midtown Paper Company. 208 of them belong to a union. What percent of the employees belong to a union?

    (1) 75% (2) 65% (3) 50% (4) 35% (5) 25%

18. A batting average times a number of times at bat tells how many hits a player gets. Manny was at bat 71 times last season. His batting average was .294. How many hits did he get?

    (1) 19 (2) 20 (3) 21 (4) 24 (5) 42

Questions 19–20 refer to the graph below.

HOW U.S. FREIGHT IS MOVED

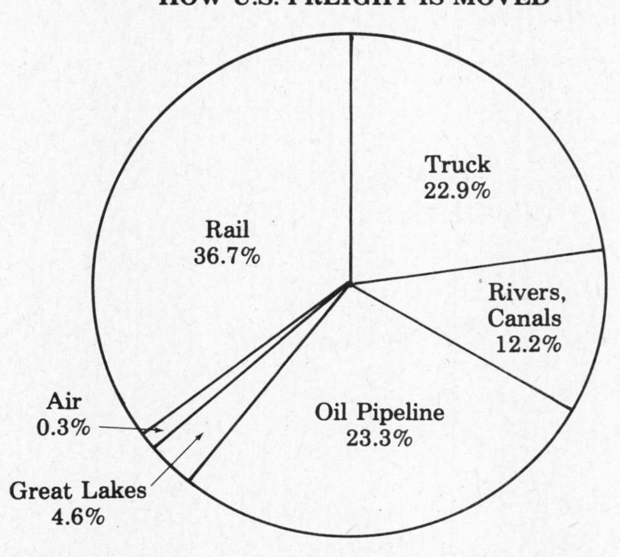

19. Which method carries the greatest amount of freight?

    (1) air (2) rail (3) truck
    (4) oil pipeline (5) rivers, canals

20. Together, rivers, canals, and rail represent about what fraction of freight hauling?

(1) $\frac{1}{4}$ (2) $\frac{1}{3}$ (3) $\frac{1}{2}$ (4) $\frac{2}{3}$ (5) $\frac{3}{4}$

21. Last year, a pound of coffee cost \$1.88. This year, a pound of coffee costs \$2.35. By what percent did the price of coffee go up in a year?

(1) $12\frac{1}{2}\%$ (2) 25% (3) $37\frac{1}{2}\%$ (4) 47% (5) 50%

22. Rent accounts for $\frac{1}{3}$ of Jose's total expenses. If his rent bills for a year total \$3,420, what are Jose's total expenses for a year?

(1) \$1,140 (2) \$2,280 (3) \$6,840
(4) \$10,260 (5) \$13,680

23. Mr. and Mrs. Allen want to buy a retirement home for \$38,000. They have to make a down payment of 15%. How much is the down payment?

(1) \$15,000 (2) \$5,700 (3) \$4,900
(4) \$4,600 (5) \$3,900

24. 1 pound = .453 kilograms. Sam weighs 160 pounds. Find his weight to the nearest kilogram.

(1) 353 kg. (2) 293 kg. (3) 160 kg.
(4) 72 kg. (5) 49 kg.

25. When Pablo left his house Friday morning, the mileage on his car read 14,285.6 miles. When he returned Sunday night, the reading was exactly 15,000 miles. How far did he drive that weekend?

(1) 185.6 mi. (2) 714.4 mi. (3) 825.6 mi.
(4) 1,285.6 mi. (5) 1,714.4 mi.

26. Find the seventh term in the series 1, 6, 4, 9, 7 . . . .

(1) 12 (2) 10 (3) 9 (4) 8 (5) 5

27. On Wednesday, a share of Allied Aluminum stock sold for $12\frac{1}{2}$. Thursday it was up $\frac{5}{8}$, and Friday it was down $\frac{1}{8}$. At what price did the stock close on Friday?

(1) $13\frac{7}{8}$ (2) 13 (3) $12\frac{7}{8}$ (4) $12\frac{1}{2}$ (5) $11\frac{7}{8}$

28. A recipe calls for 1 lb. 9 oz. of chopped beef. Kate wants to make twice as much as the recipe calls for. How much beef does she need?

(1) 1 lb. 11 oz. (2) 1 lb. 18 oz. (3) 2 lbs. 4 oz.
(4) 3 lbs. 2 oz. (5) 3 lbs. 15 oz.

29. Don drove 132 miles in $2\frac{3}{4}$ hours. Find his average speed in miles per hour.

(1) 24 mph (2) 48 mph (3) 58 mph
(4) 66 mph (5) 71 mph

30. The sales tax in Deborah's state is 6%. Find the price she paid for a $19.80 shirt, including tax.

(1) $18.61 (2) $19.86 (3) $19.94
(4) $20.06 (5) $20.99

ANSWERS & SOLUTIONS

1. (4) **15**

```
         1 5.3  to the nearest
6.5 )100.0 0   mile = 15
      65
      35 0
      32 5
       2 5 0
       1 9 5
```

2. (2) **109,000**

$$\begin{array}{r} 1{,}349{,}000 \\ -\ 1{,}240{,}000 \\ \hline \mathbf{109{,}000} \end{array}$$

3. (1) **$70**

$8 \text{ mos.} = \frac{8}{12} = \frac{2}{3} \text{ yr.}$

$\$2{,}000 \times 5\frac{1}{4}\% \times \frac{2}{3} =$

$\frac{\cancel{2{,}000}}{\cancel{100}} \times \frac{\cancel{21}}{\cancel{4}} \times \frac{2}{\cancel{3}} = \mathbf{\$70}$ (2,000/100 → 20 → 5 over 1; 21/4 → 7 over 1; 2/3 → 2 over 1)

4. (3) **12 lbs. 3 oz.**

$$\begin{array}{r} 9 \text{ lbs. } 11 \text{ oz.} \\ +\ 2 \text{ lbs. } \ 8 \text{ oz.} \\ \hline 11 \text{ lbs. } 19 \text{ oz.} \end{array} = \mathbf{12 \text{ lbs. } 3 \text{ oz.}}$$

5. (4) **$30**

$3\frac{3}{4} \times \$8 =$

$\frac{15}{\cancel{4}_1} \times \frac{\cancel{8}^2}{1} = \mathbf{\$30}$

6. (4) **1878**

$$\begin{array}{r} 1962 \\ -\ 84 \\ \hline \mathbf{1878} \end{array}$$

7. (4) **$10,619**

$$\begin{array}{r} \$13{,}297 \\ -\ 2{,}678 \\ \hline \mathbf{\$10{,}619} \end{array}$$

8. **(4) 14**

$77 \div 5\frac{1}{2} =$

$\frac{77}{1} \div \frac{11}{2} =$

$\frac{\overset{7}{\cancel{77}}}{1} \times \frac{2}{\underset{1}{\cancel{11}}} = \mathbf{14}$

9. **(2) $408**

$$\begin{array}{r} \$9{,}400 \\ -9{,}000 \\ \hline 400 \\ \times .07 \\ \hline 28.00 \\ +380 \\ \hline \mathbf{\$408} \end{array}$$

10. **(1) $160**

11. **(3) $290**

$$\begin{array}{r} \$7{,}500 \\ -\ 7{,}000 \\ \hline 500 \\ \times\ .06 \\ \hline 30.00 \\ +\ 260 \\ \hline \mathbf{\$290} \end{array}$$

12. **(4) $2\frac{1}{3}$**

$\frac{7}{1} \div \frac{3}{1} =$

$\frac{7}{1} \times \frac{1}{3} = \frac{7}{3} = \mathbf{2\frac{1}{3}}$

13. **(2) 3t. 800 lbs.**

$$\begin{array}{rl} \mathbf{3t.} & \mathbf{800\ lbs.} \\ 6\overline{)20t.} & 800\ lbs. \\ \underline{18} & \\ 2t. = & \underline{4000\ lbs.} \\ & 4800 \\ & \underline{48} \\ & 000 \end{array}$$

14. **(1) .75 m.**

$$\begin{array}{r} \mathbf{.75\ m.} \\ 3\overline{)2.25\ m.} \end{array}$$

15. **(1) $\frac{1}{8}$**

$\frac{85}{680} = \mathbf{\frac{1}{8}}$

16. **(2) $217**

$$\begin{array}{r} \$165 \\ 213 \\ 302 \\ +\ 188 \\ \hline \$868 \end{array}$$

$$\begin{array}{r} \mathbf{\$217} \\ 4\overline{)\$868} \end{array}$$

17. **(2) 65%**

$\frac{208}{320} = \frac{13}{20}$

$\frac{13}{\underset{1}{20}} \times \frac{\overset{5}{100}}{1} = \mathbf{65\%}$

18. **(3) 21**

$$\begin{array}{r} .294 \\ \times\ 71 \\ \hline 294 \\ 20\ 58\ \ \\ \hline 20.874 \end{array}$$

20.874 to the nearest whole = **21**

19. (2) **rail**

20. (3) $\frac{1}{2}$

$$\begin{array}{lll} \text{rivers, canals} & = & 12.2\% \\ \text{rail} & = & \underline{36.7\%} \\ & & 48.9\% \end{array}$$

48.9% almost = 50%

$50\% = \frac{1}{2}$

21. (2) **25%**

$$\begin{array}{r} \$2.35 \\ \underline{-\ 1.88} \\ 47 \end{array} \qquad \frac{47}{188} = \frac{1}{4} = \mathbf{25\%}$$

22. (4) **$10,260**

$\$3{,}420 \div \frac{1}{3} =$

$\$3{,}420 \times \frac{3}{1} = \mathbf{\$10{,}260}$

23. (2) **$5,700**

$$\begin{array}{r} \$38{,}000 \\ \underline{\times\ .15} \\ 1900\ 00 \\ \underline{3800} \\ \mathbf{\$5{,}700.00} \end{array}$$

24. (4) **72 kg.**

$$\begin{array}{r} .452 \\ \underline{\times\ 160} \\ 27\ 180 \\ \underline{45\ 3} \\ 72.480 \end{array}$$

72.480 to the nearest whole = **72**

25. (2) **714.4 mi.**

$$\begin{array}{r} 15{,}000.0 \\ \underline{-\ 14{,}285.6} \\ \mathbf{714.4} \end{array}$$

26. (2) **10**

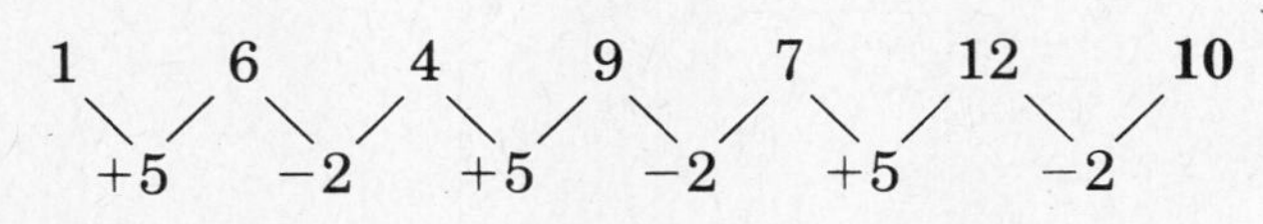

27. (2) **13**

$$\begin{array}{rcr} 12\frac{1}{2} & = & 12\frac{4}{8} \\ + \quad \frac{5}{8} & = & \frac{5}{8} \\ \hline & & 12\frac{9}{8} = 13\frac{1}{8} \end{array} \qquad \begin{array}{r} 13\frac{1}{8} \\ - \quad \frac{1}{8} \\ \hline \mathbf{13} \end{array}$$

28. (4) **3 lbs. 2 oz.**

$$\begin{array}{rr} 1 \text{ lb.} & 9 \text{ oz.} \\ \times\ 2 & \times\ 2 \\ \hline 2 \text{ lbs.} & 18 \text{ oz.} \end{array} = \mathbf{3\ lbs.\ 2\ oz.}$$

29. (2) **48 mph**

$132 \div 2\frac{3}{4} =$

$\frac{132}{1} \div \frac{11}{4} =$

$\frac{\overset{12}{\cancel{132}}}{1} \times \frac{4}{\underset{1}{\cancel{11}}} = \mathbf{48}$

30. (5) **$20.99**

6% = .06

$$\begin{array}{r} \$19.80 \\ \underline{\times\ .06} \\ \$1.18\ 80 \end{array}$$

$1.18 80 to the nearest cent = $1.19

$$\begin{array}{r} \$19.80 \\ \underline{+\ 1.19} \\ \mathbf{\$20.99} \end{array}$$

After you check your answers, look at the chart below. On the chart, circle the number of each problem you missed. Then review the section in which the skills for that problem are explained.

Section	Problem Number						
Whole Numbers	2	6	7	16	26		
Fractions	5	8	12	15	22	27	29
Decimals	1	14	18	25			
Percents	3	17	21	23	30		
Measurement	4	13	24	28			
Tables and Graphs	9	10	11	19	20		

Predictor Test: Algebra and Geometry

Before you continue in this book, take the next test. The test will show you how much you already know about algebra and geometry.

After you take the test, check your answers. Circle the numbers of the problems that you missed. After you check your answers, look at the chart that follows. The chart will show you which sections of the book explain the skills that appeared in the different parts of the test.

Don't worry if you miss problems on the test. The test is meant only to show you which areas you should study the most. All of the skills that are tested are explained in the next part of the book.

DIRECTIONS: Solve each problem.

PART A

1. Find the value of $5^3 - 2^4$.

2. Find the value of $\sqrt{6{,}889}$.

3. What lettered point on the number line stands for $-1\frac{1}{2}$?

   −4 −3 −2 −1 0 +1 +2 +3 +4

   A B C D E

4. $-9 + 14 - 12 =$

5. $(-23) - (-15) =$

6. $(-6)(-5)(+4) =$

7. $\frac{+64}{-16} =$

8. $a = 15$ and $c = 5$. Find the value of ac.

9. $m = 20$ and $n = 4$. Find the value of $\frac{1}{2}(m - n)$.

10. $W = \frac{P - 2L}{2}$. Find W when $P = 90$ and $L = 26$.

PART B

1. Simplify the expression $7a - 9 + 6a + 4$.

2. Simplify the expression $5m + 4(m - 7)$.

3. Solve the equation $18e = 12$.

4. Solve the equation $8f - 17 = 79$.

5. Solve the equation $7y + 6 = 78 - 2y$.

6. Solve the equation $6(s - 4) + 7 = 43$.

7. 120 men work at the Acme Steel Company. 35 women work there. What is the ratio of women to men working at the company?

8. Solve the proportion $\frac{9}{14} = \frac{x}{4}$.

9. The ratio of men to women in Ellen's night school class is 2:3. There are 12 women in the class. How many men are in the class?

10. Write an expression for six less than eight times a number. Use x for the number.

11. Seven times a number increased by 4 equals 88. Find the number.

12. Morris and Tony work at a car wash. One day, Morris washed 10 more than twice as many cars as Tony. Together, they washed 58 cars. How many cars did Tony wash?

Use the graph at the right to answer questions 13 and 14.

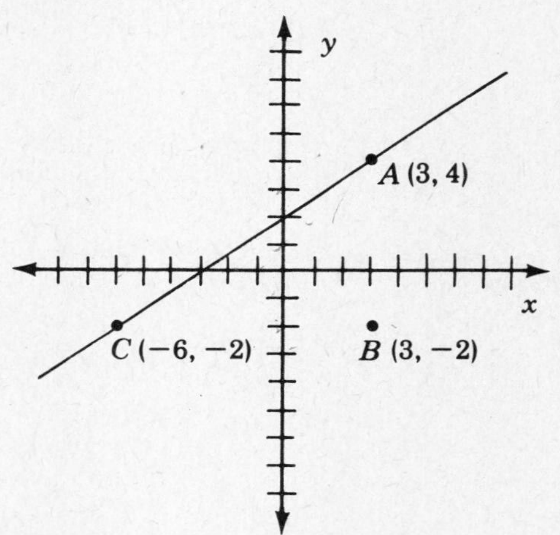

13. What is the distance from point A to point B?

14. What are the coordinates of the point where line AC crosses the y-axis?

15. Is $x = 4$ in the solution set of $x \leq 9$?

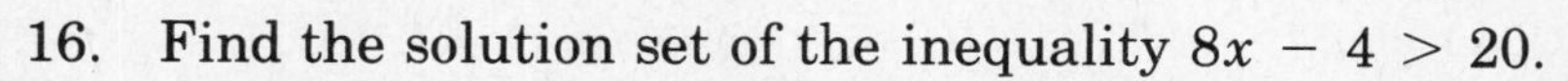

16. Find the solution set of the inequality $8x - 4 > 20$.

PART C

1. The name for an angle measuring 125° is ____________.

2. Find the supplement of an angle measuring 38° 15′ 40″.

3. A triangle with two equal angles is called ____________.

4. A figure with four equal sides and four right angles is called a ____________.

5. The distance around a circle is called the ____________.

6. George has a photograph 3 inches wide and 5 inches long. He wants to enlarge the photograph to make it 30 inches long. How wide will the enlarged picture be?

7. One leg of a right triangle measures 24 feet. The other leg measures 32 feet. Find the measurement of the hypotenuse.

8. Find the perimeter of a rectangle that is 14 yards long and $6\frac{1}{2}$ yards wide.

9. Find the circumference of a circle that has a diameter of 10 inches. Use $\pi = 3.14$.

10. A triangle has a base of 12 feet and a height of 14 feet. Find the area of the triangle.

11. Find the area of a circle that has a radius of 14 inches. Use $\pi = \frac{22}{7}$.

12. Find the volume of a rectangular boxcar 7 feet wide, 18 feet long, and 6 feet high.

ANSWERS

PART A

1. **109** 2. **83** 3. **B** 4. **−7**

5. **−8** 6. **+120** 7. **−4** 8. **75**

9. **8** 10. **19**

PART B

1. **$13a - 5$** 2. **$9m - 28$** 3. **$e = \frac{2}{3}$** 4. **$f = 12$**

5. **$y = 8$** 6. **$s = 10$** 7. **7:24** 8. **$x = 2\frac{4}{7}$**

9. **8 men** 10. **$8x - 6$** 11. **$x = 12$** 12. **16 cars**

13. **6** 14. **(0, +2)** 15. **yes** 16. **$x > 3$**

PART C

1. **obtuse** 2. **141° 44′ 20″** 3. **isosceles** 4. **square**

5. **circumference** 6. **18 in.** 7. **40 ft.** 8. **41 yds.**

9. **31.4 in.** 10. **84 sq. ft.** 11. **616 sq. in.**

12. **756 cu. ft.**

Math Skills Chart

After you check your answers, look at the chart that follows. The chart shows which parts of the book explain the skills that were tested on the different parts of the test. Use the chart to find out which parts of the book you should study the most.

Questions in:	**are explained on pages:**
Part A	199–226 (The "Language" of Algebra)
Part B	227–273 (The Uses of Algebra)
Part C	274–314 (Geometry)

The "Language" of Algebra

In the first part of this book, you worked with basic math. Whole numbers, fractions, decimals, and percents are all basic math skills.

In this part of the book, you will study algebra. Algebra is a special math "language." The operations that you used in basic math (addition, subtraction, multiplication, and division) also are used in algebra. The main difference between the algebra "language" and the basic math "language" is that, in algebra, you use both numbers and letters. In basic math, you work only with numbers.

At first, using letters in math problems may seem strange to you. However, you should keep in mind when you work with algebra that the letters always represent a missing numerical value. The letters take the place of numbers.

The algebra section of this book contains all of the basic rules of algebra that you need to know for the GED Test. The first part of the algebra section will show you the rules for reading the algebra "language." The second part of the algebra section will explain the types of algebra problems that appear on the GED Test. By the end of the algebra section, you will have studied the basic algebra skills that you need for the GED Test.

Powers

In basic math, you solve problems by using addition, subtraction, multiplication, or division. There are two more operations that you will use in algebra: **powers** and **roots.**

Here is an example of a power:

$$\mathbf{7^2}$$

The number 7 in this power is the base. The smaller number 2 that is raised above the 7 is the power, or the **exponent.** The power shows how many times the number in the base is multiplied by itself. 7^2 is read "seven to the second power." It means the same as 7×7.

EXAMPLE: What is the value of 7^2?

7 × 7	7 × 7 = **49**
Step 1	Step 2

Step 1: The power is 2. Write the base (7) 2 times.
Step 2: Multiply 7 by 7.

EXAMPLE: What is the value of 5^3?

$5 \times 5 \times 5$	$5 \times 5 = 25$	$25 \times 5 = 125$
Step 1	Step 2	Step 3

Step 1: The power is 3. Write the base (5) 3 times.
Step 2: Multiply the first 5 by the second 5.
Step 3: Multiply the product (25) by the third 5.

EXAMPLE: What is the value of $3^4 - 6^2 + 2^3$?

Step 1	Step 2
$3^4 = 3 \times 3 \times 3 \times 3 = 81$	$81 - 36 + 8$
$6^2 = 6 \times 6 = 36$	$81 - 36 = 45$
$2^3 = 2 \times 2 \times 2 = 8$	$45 + 8 = 53$

Step 1: Find the values of the powers.
Step 2: Use the values to solve the problem. Work from left to right. First, subtract 36 from 81. Second, add 8 to the difference. The answer is 53.

EXAMPLE: What is the value of $\left(\frac{3}{4}\right)^2$?

$\frac{3}{4} \times \frac{3}{4}$	$\frac{3}{4} \times \frac{3}{4} = \frac{9}{16}$
Step 1	Step 2

Step 1: Write the base 2 times.
Step 2: Multiply $\frac{3}{4}$ by $\frac{3}{4}$.

EXAMPLE: What is the value of $(.03)^2$?

$.03 \times .03$	$.03 \times .03 = .0009$
Step 1	Step 2

Step 1: Write the base 2 times.
Step 2: Multiply .03 by .03.

EXERCISE 1

DIRECTIONS: Find the value of each problem.

1. $2^4 =$ $\quad$ $9^2 =$ $\quad$ $3^3 =$
2. $13^2 =$ $\quad$ $50^2 =$ $\quad$ $6^3 =$
3. $2^5 =$ $\quad$ $1^5 =$ $\quad$ $16^2 =$
4. $10^3 =$ $\quad$ $25^2 =$ $\quad$ $4^4 =$

5. $5^2 - 2^3 =$ $\qquad$ $8^2 + 3^3 =$ $\qquad$ $10^2 - 4^2 + 5^2 =$

6. $4^3 + 6^2 - 2^4 =$ $\qquad$ $12^2 - 5^2 - 3^2 =$ $\qquad$ $10^3 - 10^2 =$

7. $\left(\frac{1}{2}\right)^2 =$ $\qquad$ $\left(\frac{1}{9}\right)^2 =$ $\qquad$ $\left(\frac{2}{3}\right)^3 =$

8. $\left(\frac{3}{10}\right)^3 =$ $\qquad$ $\left(\frac{5}{6}\right)^2 =$ $\qquad$ $\left(\frac{7}{12}\right)^2 =$

9. $(.3)^2 =$ $\qquad$ $(.08)^2 =$ $\qquad$ $(.009)^2 =$

10. $(.15)^2 =$ $\qquad$ $(.2)^3 =$ $\qquad$ $(.01)^2 =$

Check your answers on page 220.

Roots

Roots are the opposite of powers. On the GED Test, you will be expected to solve problems with **square roots.** A square root is the opposite of a number to the second power. The sign for square root is $\sqrt{\ }$. When you find a square root, you find the number that is multiplied by itself to give the number inside the $\sqrt{\ }$ sign.

EXAMPLE: Find the value of $\sqrt{36}$.

$$\sqrt{36} = 6$$

Ask yourself, "What number times itself is 36?" 6 times 6 is 36. 6 is the square root of 36.

Here is a list of some square roots:

$\sqrt{1} = 1$	$\sqrt{49} = 7$	$\sqrt{169} = 13$	$\sqrt{3,600} = 60$
$\sqrt{4} = 2$	$\sqrt{64} = 8$	$\sqrt{225} = 15$	$\sqrt{4,900} = 70$
$\sqrt{9} = 3$	$\sqrt{81} = 9$	$\sqrt{400} = 20$	$\sqrt{6,400} = 80$
$\sqrt{16} = 4$	$\sqrt{100} = 10$	$\sqrt{900} = 30$	$\sqrt{8,100} = 90$
$\sqrt{25} = 5$	$\sqrt{121} = 11$	$\sqrt{1,600} = 40$	$\sqrt{10,000} = 100$
$\sqrt{36} = 6$	$\sqrt{144} = 12$	$\sqrt{2,500} = 50$	

The square roots of most numbers are decimals. For example, the square root of the number 2 is 1.1414. On the GED Test, you will not be expected to find the decimal square root of a number. However, you may be asked to *estimate* what the square root of a number is. Here is the type of square root question that you may find on the GED Test:

EXAMPLE: The square root of 12 is

(1) 2.74 $\qquad$ (2) 3.46 $\qquad$ (3) 4.06
(4) 4.11 $\qquad$ (5) 4.83

If you know the common square roots, you can solve this problem. Look at the number 12. The number 12 falls between 9 and 16. The square root of 9 is 3. The square root of 16 is 4. Therefore, the square root of 12 must fall somewhere between the square roots of 9 and 16. The square root of 12 is between 3 and 4.

Now, look at the answer choices. Only one of the choices (answer choice 2) falls between 3 and 4. You can make a good guess that answer choice 2 is correct.

Sometimes, you may have to find the square root of a large number. Here is a good shortcut to finding the square root of a large number:

Step 1: Guess an answer.
Step 2: Divide the guess into the large number.
Step 3: Average the guess and the answer to the division problem.
Step 4: Check.

Study the next examples carefully.

EXAMPLE: Find the value of $\sqrt{1{,}024}$.

Step 1	Step 2	Step 3	Step 4
$\sqrt{900} = 30$	$30\overline{)1{,}024}$ = **34**; -90; 124; -120	$30 + 34 = $ **64**; $64 \div 2 = $ **32**	$32 \times 32 = $ **1,024**; $\sqrt{1{,}024} = 32$

Step 1: Guess. In the list of square roots, $\sqrt{900} = 30$. 30 is a good guess. It is too small, but it is easy to divide by.
Step 2: Divide 1,024 by 30. Drop the remainder.
Step 3: Average 30 and 34.
Step 4: Check. Multiply 32 by itself. 32 is the square root of 1,024.

Always guess a number that ends in zero. It's easier and faster to divide by a number that ends in zero.

To find the square root of a fraction, find the square roots of both the top number and the bottom number.

EXAMPLE: Find the value of $\sqrt{\frac{4}{9}}$.

Step 1	Step 2	Step 3
$\sqrt{4} = 2$	$\sqrt{9} = 3$	$\frac{2}{3} \times \frac{2}{3} = \frac{4}{9}$ $\sqrt{\frac{4}{9}} = \frac{2}{3}$

Step 1: Find the square root of the top number (4).
Step 2: Find the square root of the bottom number (9).
Step 3: Check. Multiply $\frac{2}{3}$ by itself. $\frac{2}{3}$ is the square root of $\frac{4}{9}$.

The square root of a decimal has half as many decimal places as the number inside the $\sqrt{}$ sign.

EXAMPLE: Find the value of $\sqrt{.0049}$.

Step 1	Step 2	Step 3	
$\sqrt{49} = \mathbf{7}$	**.07**	$\begin{array}{r} .07 \\ \times\ .07 \\ \hline \mathbf{.0049} \end{array}$	$\sqrt{\mathbf{.0049}} = \mathbf{.07}$

Step 1: Find the square root of 49. The square root of 49 is 7.
Step 2: .0049 has four decimal places. Give the answer (7) half as many, or 2, places.
Step 3: Check. Multiply .07 by itself. .07 is the square root of .0049.

EXERCISE 2

DIRECTIONS: Find the value of each square root. Use the list of square roots to make your first guess.

1. $\sqrt{289} =$ $\quad \sqrt{784} =$ $\quad \sqrt{1{,}444} =$ $\quad \sqrt{484} =$
2. $\sqrt{1{,}521} =$ $\quad \sqrt{1{,}849} =$ $\quad \sqrt{529} =$ $\quad \sqrt{2{,}704} =$
3. $\sqrt{4{,}624} =$ $\quad \sqrt{8{,}836} =$ $\quad \sqrt{4{,}761} =$ $\quad \sqrt{7{,}056} =$
4. $\sqrt{\frac{25}{36}} =$ $\quad \sqrt{\frac{4}{49}} =$ $\quad \sqrt{\frac{1}{81}} =$ $\quad \sqrt{\frac{9}{100}} =$
5. $\sqrt{\frac{64}{81}} =$ $\quad \sqrt{\frac{1}{144}} =$ $\quad \sqrt{\frac{9}{16}} =$ $\quad \sqrt{\frac{36}{49}} =$
6. $\sqrt{.04} =$ $\quad \sqrt{.09} =$ $\quad \sqrt{.0016} =$ $\quad \sqrt{.01} =$
7. $\sqrt{.0009} =$ $\quad \sqrt{.000025} =$ $\quad \sqrt{.0036} =$ $\quad \sqrt{.000064} =$

Check your answers on page 221.

SIGNED NUMBERS

The whole numbers, fractions, and decimals you have used in arithmetic are all **positive numbers.** A positive number has a value of more than zero. All positive numbers are shown on this number line:

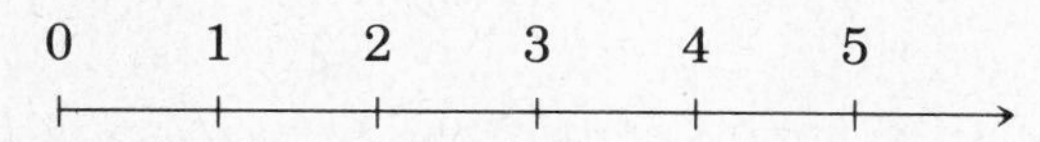

(The arrow at the end means that the numbers go on and on.)

In algebra, you use both positive numbers and **negative numbers.** A negative number has a value of less than zero. The entire set of positive and negative numbers are shown on this number line:

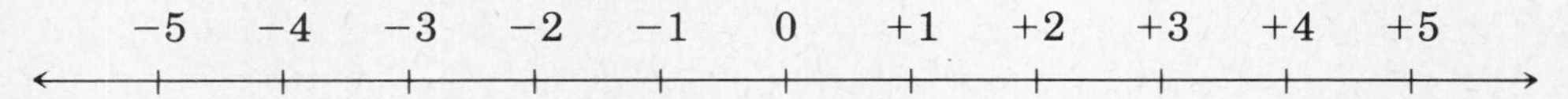

(The arrows mean that the numbers go on and on in both directions.)

In arithmetic, the + sign means to add. In algebra, the + sign also means a positive (or *plus*) number. The − sign means to subtract in arithmetic. In algebra, the − sign also means a negative (or *minus*) number. The zero separates the number line in two parts. Every number at the right of zero is positive. Every number at the left of zero is negative. All the numbers on the number line, including the positive numbers, the negative numbers, and zero, make up the set of **signed numbers.**

Use the number line below to answer the next question.

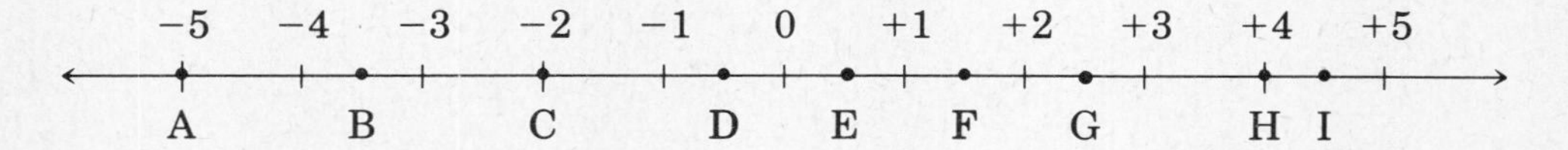

EXAMPLE: What lettered point on the number line stands for the number $-\frac{1}{2}$?

$-\frac{1}{2}$ is at the left of zero and halfway between zero and -1. The letter **D** stands for $-\frac{1}{2}$.

Adding Signed Numbers

Before you read about adding signed numbers, try to answer this question:

At 7 A.M., the temperature in Buffalo was three degrees below zero. By 11 A.M., the temperature had risen seven degrees. What was the temperature in Buffalo at 11 A.M.?

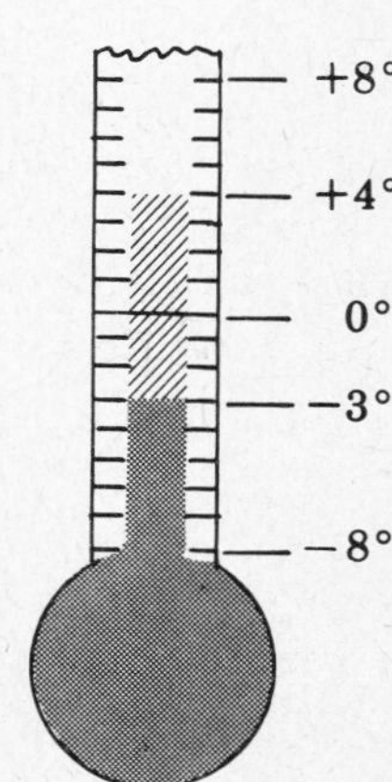

To find the answer to this question, you could use a thermometer.

Count seven degrees up from three degrees below zero (−3). Seven places up from −3 is four degrees (+4). The temperature in Buffalo at 11 A.M. was four degrees above zero, or +4 degrees.

In the last problem, you worked with adding signed numbers. Think of the thermometer as a number line. You started counting up at −3. You counted up seven degrees, or +7. Your answer was four degrees, or +4.

$$(-3) + (+7) = +4$$

When you added whole numbers in the first part of this book, you worked only with positive (+) numbers. In algebra, you work with both positive and negative (−) numbers. Adding signed numbers is a little more complicated than adding regular whole numbers. To add signed numbers, follow these two rules:

Rule 1: Add numbers with the same signs and give the total of that sign.

Rule 2: Subtract numbers with different signs. Give the answer the sign of the bigger number.

Study the next examples carefully.

EXAMPLE: Add −3, −8, and −2.

$$\begin{array}{r} (-3) \\ (-8) \\ \underline{+\ (-2)} \\ -13 \end{array}$$

Step 1: Add the numbers.
Step 2: All of the numbers have negative (−) signs. Give the answer a negative sign.

If you have trouble figuring out why the negative sign is added, remember the number line. You start at the point marked −3. You "add" −8 and −2. To add a negative number, you move to the left on the number line. Adding a negative number is the same as subtracting a positive number.

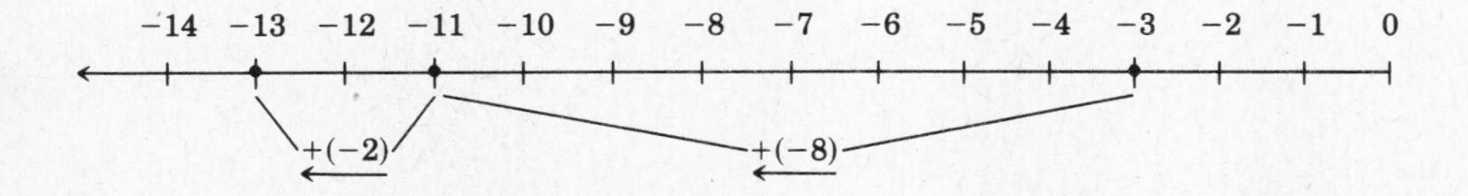

EXAMPLE: $-8 + 14 =$

$$\begin{array}{rl} 14 & \text{(larger number—a } \textbf{positive} \text{ value)} \\ -\ 8 & \text{(smaller number—a } \textbf{negative} \text{ value)} \\ \hline 6 & \text{(difference—a } \textbf{positive} \text{ value: +6)} \end{array}$$

Step 1: The signs are different, so you have to subtract. The number 14 is larger than 8; put the 14 on top and subtract.
Step 2: Give the answer a positive (+) sign (+6) because the number 14 is bigger.

EXAMPLE: $(+9) + (-17) =$

$$\begin{array}{rl} 17 & \text{(larger number—a } \textbf{negative} \text{ value)} \\ -\ 9 & \text{(smaller number—a } \textbf{positive} \text{ value)} \\ \hline 8 & \text{(difference—a } \textbf{negative} \text{ value: −8)} \end{array}$$

Step 1: The signs are different, so you have to subtract. The number 17 is larger than 9; put the 17 on top and subtract.
Step 2: Give the answer a negative (−) sign (−8) because the number 17 is bigger.

Notice that when you subtract in Step 1, you subtract as if both numbers are regular whole numbers. The bigger number goes on top. After you subtract, then you decide which sign to use.

Sometimes, you will need both rules for adding signed numbers in a problem.

EXAMPLE: $-4 + 7 - 10 + 6 =$

7 (+) + 6 (+) **13** (+)	10 (−) + 4 (−) **14** (−)	**14** (−) **− 13** (+) **1** (−)
Step 1	Step 2	Steps 3-4

Step 1: Add the positive numbers together.
Step 2: Add the negative numbers together.
Step 3: Use the two sums to subtract. Put the bigger number (14) on top.
Step 4: Give the answer a negative sign (−1) because the number 14 is bigger.

EXERCISE 3

DIRECTIONS: Add each problem.

1. $-12 - 3 - 6 =$ $\qquad$ $(8) + (21) + (+13) =$
2. Add +11 and −17. $\qquad$ Add −36 and 4.

3. Add -3 and -13. Add 42 and -18.

4. $(-6) + (+4) + (+7) =$ $(-10) + (-5) + (8) =$

5. Add -14, -13, and $+16$. Add 23, -18, and $+14$.

6. $-35 + 18 - 11 =$ $29 + 32 - 15 =$

7. $(-17) + (-23) + (-19) =$ $(+15) + (-29) + (+3) =$

8. $-9 + 8 + 18 - 16 =$ $+13 + 27 - 18 - 22 =$

9. $(-8) + (35) + (-16) + (-13) =$ $(15) + (-19) + (+9) + (-6) =$

Check your answers on page 222.

Subtracting Signed Numbers

Subtracting signed numbers is almost the same as adding signed numbers except for one thing. You must change the sign of the number that is being subtracted. After you change the sign, you follow the same rules for adding signed numbers. Subtraction of signed numbers problems usually have a minus (−) sign between numbers in parentheses.

EXAMPLE: $(-7) - (-4) =$

Step 1	Steps 2-3
$-(-4) = +4$	$\begin{array}{rl} 7 & (-) \\ \underline{-\ 4} & (+) \\ 3 & (-) \end{array}$

Step 1: You are subtracting -4. Change the sign of -4 to $+4$.
Step 2: Put the bigger number (7) on top and subtract.
Step 3: Give the answer the sign of the bigger number: -3.

Again, if you have trouble figuring out why you have to change the sign, remember the number line. In the first part of this book, you learned that subtraction is the opposite of addition. When you add a negative (−) number, you move to the *left* on the number line. When you subtract a negative number, you move to the *right* on the number line.

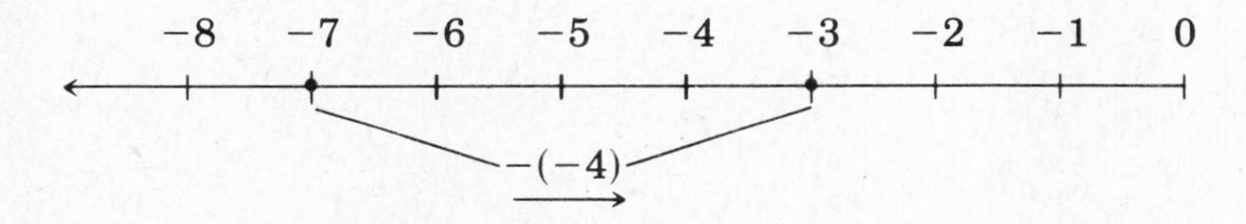

EXAMPLE: $(-5) - (+9) =$

Step 1	Steps 2-3
$-(+9) = -9$	5 (−) + 9 (−) 14 (−)

Step 1: Change the sign of the number that you are subtracting (+9).
Step 2: Both numbers now have the same sign. Therefore, you add them together.
Step 3: Give the answer the same sign as the numbers that you added: −14.

EXAMPLE: $(-8) + (-6) - (-2) =$

Step 1	Step 2	Steps 3-4
$-(-2) = +2$	8 (−) + 6 (−) 14 (−)	14 (−) − 2 (+) 12 (−)

Step 1: Change the sign of −2 and drop the parentheses. −2 is the only number being subtracted.
Step 2: Add the negative numbers together.
Step 3: Put the larger number (14) on top and subtract.
Step 4: Give the answer a negative (−) sign (−12) because the number 14 is bigger.

EXERCISE 4

DIRECTIONS: Subtract each problem.

1. $(+12) - (-7) =$ $(-3) - (+24) =$
2. $(+18) - (+9) =$ $(-14) - (-19) =$
3. $(-21) - (+6) =$ $(+16) - (+16) =$
4. $(+5) - (-17) =$ $(-17) - (-3) =$
5. $(-26) - (+9) =$ $(+4) - (+17) =$
6. $(+11) - (-11) =$ $(-7) - (-13) =$
7. $(+8) + (+12) - (+7) =$ $(+9) + (+13) - (-6) =$
8. $(-14) - (+3) - (+10) =$ $(-20) + (-7) - (+7) =$
9. $(-13) - (-5) + (-8) =$ $(+4) - (-22) - (+26) =$

Check your answers on page 223.

Multiplying Signed Numbers

The rules for multiplying signed numbers are different from the rules for adding or subtracting signed numbers. Here are the rules for multiplying two signed numbers:

Rule 1: If the signs are the same, multiply and make the answer positive.
Rule 2: If the signs are different, multiply and make the answer negative.

One way to write a multiplication of signed numbers problem is to use parentheses. No sign between the parentheses means to multiply.

EXAMPLE: $(-6)(+8) =$

$$\mathbf{(-6)(+8) = -48}$$

Multiply and make the answer negative because the signs are different.

A dot between numbers also means to multiply.

EXAMPLE: $-9 \cdot 12 =$

$$\mathbf{-9 \cdot 12 = -108}$$

Multiply and make the answer negative because the signs are different.

EXAMPLE: $\left(-\frac{3}{4}\right)\left(-\frac{8}{9}\right) =$

$$\left(-\frac{\overset{1}{\cancel{3}}}{\underset{1}{\cancel{4}}}\right)\left(-\frac{\overset{2}{\cancel{8}}}{\underset{3}{\cancel{9}}}\right) = +\frac{2}{3}$$

Cancel, multiply, and make the answer positive because the signs are the same.

Here are the rules for multiplying more than two signed numbers:

Rule 1: If there is an even number (2, 4, 6, etc.) of negative (−) signs, make the answer positive (+).
Rule 2: If there is an odd number (1, 3, 5, etc.) of negative signs, make the answer negative.

EXAMPLE: $(-2)(+5)(-3)(-4) =$

$$\mathbf{(-2)(+5)(-3)(-4) = -120}$$

Multiply: $2 \times 5 \times 3 \times 4 = 120$. Make the answer negative because there are three negative (−) signs.

If you multiply two negative numbers together, you get a positive number. Therefore, if you multiply a negative number by itself, you get a positive number.

$$(-7) \times (-7) = (+49) \text{ or}$$
$$-7^2 = (+49)$$

Since the product of -7×-7 is $+49$, you can say that -7 is the *square root* of 49. Therefore, 49 has two square roots: $+7$ and -7. This is true for all square roots. Each number has both a *positive* and a *negative* square root.

$$(+5) \times (+5) = 25$$
$$(-5) \times (-5) = 25$$
$$\sqrt{25} = +5 \text{ or } -5$$

EXERCISE 5

DIRECTIONS: Multiply each problem.

1. $(+12)(-7) =$ $\qquad$ $(-6)(+15) =$
2. $\left(+\frac{1}{2}\right)(+20) =$ $\qquad$ $(-36)\left(-\frac{3}{4}\right) =$
3. $\left(-\frac{5}{8}\right)\left(\frac{3}{10}\right) =$ $\qquad$ $(+48)\left(-\frac{2}{3}\right) =$
4. $\left(+\frac{1}{2}\right)\left(+\frac{2}{3}\right)(-4)(-15) =$ $\qquad$ $(-8)(+2)\left(+\frac{1}{4}\right)(+2) =$
5. $(-3)(+5)(-6)(+4) =$ $\qquad$ $(-1)(+9)(-3)(-2) =$

Check your answers on page 224.

Dividing Signed Numbers

The rules for dividing signed numbers are almost the same as the rules for multiplying signed numbers.

Rule 1: If the signs are the same, divide and make the answer positive.
Rule 2: If the signs are different, divide and make the answer negative.

Division of signed numbers problems are usually written like fractions.

EXAMPLE: $\frac{+16}{-8} =$

$$\frac{+16}{-8} = -2$$

Divide and make the answer negative because the signs are different.

EXAMPLE: $\frac{-45}{-9} =$

$$\frac{-45}{-9} = +5$$

Divide and make the answer positive because the signs are the same.

Sometimes, a division problem can only be reduced.

EXAMPLE: $\frac{-14}{+21} =$

$$\frac{-14}{+21} = -\frac{2}{3}$$

Reduce and make the answer negative because the signs are different.

EXERCISE 6

DIRECTIONS: Divide or reduce each problem.

1. $\frac{-48}{-4} =$ $\quad \frac{-54}{+6} =$ $\quad \frac{+19}{-19} =$ $\quad \frac{+10}{+12} =$

2. $\frac{+108}{-12} =$ $\quad \frac{-15}{+10} =$ $\quad \frac{-42}{-6} =$ $\quad \frac{-18}{+3} =$

3. $\frac{200}{-25} =$ $\quad \frac{-18}{-20} =$ $\quad \frac{+36}{+3} =$ $\quad \frac{-50}{2} =$

4. $\frac{-18}{+32} =$ $\quad \frac{+64}{+4} =$ $\quad \frac{-63}{-7} =$ $\quad \frac{24}{-20} =$

5. $\frac{+20}{-80} =$ $\quad \frac{-80}{+20} =$ $\quad \frac{-144}{12} =$ $\quad \frac{-175}{-25} =$

Check your answers on page 224.

USING LETTERS IN ALGEBRA

By now, you can perform basic operations with numbers. For example, you would not find it hard to solve this problem:

$$3 + 4 =$$

Now, you are going to study how to use *letters* in basic operations. Using letters is an important part of the language of algebra.

Here is a problem written with letters:

$$x + y =$$

What does this problem mean? What do x and y stand for?

Letters in algebra problems are called **variables.** A variable stands for an unknown value. There are two different variables in the example. The letter x stands for one unknown value. The letter y stands for another unknown value.

Now, look at this example:

Find the value of $x + y$ if $x = 3$ and $y = 4$.

In this problem, you are given the values that the letters stand for. You now know that x is equal to 3 and that y is equal to 4. You can now solve the problem.

$$x + y =$$
$$3 + 4 = 7$$

The letters stood for unknown values. Once the values were known, the numbers replaced the letters. Replacing a letter with a number is called **substitution.**

EXAMPLE: Find the value of $p - q$ if $p = 8$ and $q = -6$.

Step 1	Step 2
$p - q =$	$-(-6) = +6$
$8 - (-6) =$	$8 + 6 = 14$

Step 1: Substitute 8 for p and -6 for q.
Step 2: Subtract the signed numbers. Change the sign of -6 and add.

Many algebra problems contain values like this: ab, xy, and $5x$. These values are called **terms.** A term stands for one number. There can be two or more values in a term. For example, look at the term $5x$.

$5x$

The term $5x$ means 5 *times* x. The number 5 is a known value. The letter x stands for an unknown value. The total value of the term is 5 times whatever number x stands for. If you are given the value of x, you can find the value of the term.

EXAMPLE: Find the value of $5x$ if $x = 10$.

$5x =$ $5(10)$	$5 \times 10 = 50$
Step 1	Step 2

Step 1: Substitute. Put 10 in the place of x.
Step 2: Multiply to find the value. (When there is no sign between a number and a letter, multiply.)

EXERCISE 7

DIRECTIONS: Find the value of each term.

1. Find the value of st if $s = -12$ and $t = -6$.
2. $a = 7$. Find a^2.
3. $x = 36$ and $y = 4$. Find the value of $\frac{x}{y}$.
4. $c = 15$ and $f = 9$. Find $c - f$.
5. Find the value of $\sqrt{w}$ if $w = 100$.
6. $k = 24$. Find $\frac{3}{8}k$.
7. $x = -8$ and $y = -7$. Find the value of $(x + y)$.
8. $n = \frac{2}{3}$. Find the value of n^2.
9. $b = .06$ and $d = .3$. Find the value of bd.
10. Find the value of $\frac{l}{w}$ if $l = 15$ and $w = 21$.

Check your answers on page 224.

Order of Operations

In most algebra problems, you will be asked to do more than one operation. There is a certain order in which the operations in a problem should be done. Here is the **order of operations:**

1. Do operations inside parentheses or above and below fraction bars first.
2. Do powers and roots second.
3. Multiply and divide third.
4. Add and subtract last.

It is important to memorize the order of operations. In the following examples, you will see how the order of operations is used in algebra.

EXAMPLE: $c = 15$, $d = 2$, and $e = 3$. Find the value of $c - de$.

$c - de =$ $\mathbf{15 - 2 \times 3 =}$	$15 - 2 \times 3 =$ $15 - \mathbf{6}$	$\mathbf{15 - 6 = 9}$
Step 1	Step 2	Step 3

Step 1: Substitute the numbers for the letters.
Step 2: Look at the operations in the problem. The problem has subtraction and multiplication. Find subtraction and multiplication in the order of operations. Multiplication comes first.
Step 3: Subtract.

In the last problem, look at what happens if the order of operations is not followed.

$$15 - 2 \times 3 =$$
$$15 - 2 = 13$$
$$13 \times 3 = \mathbf{39}$$

The answer looks right, *but it is wrong*. The answer can only be right if the order of operations is followed.

EXAMPLE: $x = 4$, $y = 5$, and $t = 2$. Find the value of $x(y + t)$.

$x(y + t) =$ $\mathbf{4(5 + 2) =}$	$4(5 + 2) = 4(\mathbf{7})$	$4(7) = \mathbf{4 \times 7 = 28}$
Step 1	Step 2	Step 3

Step 1: Substitute the numbers for the letters.

Step 2: Look at the operations in the problem. The problem has multiplication. It also has addition inside parentheses. According to the order of operations, any operation inside parentheses comes first.

Step 3: Multiply.

EXERCISE 8

DIRECTIONS: Find the value of each statement.

1. $m = 8$. Find the value of $2m^2$.
2. $u = 12$ and $v = 4$. Find the value of $uv + \frac{u}{v}$.
3. $p = 9$ and $s = 5$. Find the value of $3p - 2s$.
4. $k = 6$ and $h = 7$. Find the value of $k(k + h)$.
5. $a = 12$, $b = 6$, and $c = 2$. Find the value of $a - bc$.
6. $r = 15$ and $t = 9$. Find the value of $\frac{r + t}{2}$.
7. $e = 8$ and $f = 4$. Find the value of $(e + f)^2$.
8. $e = 8$ and $f = 4$. Find the value of $e + f^2$.
9. $e = 8$ and $f = 4$. Find the value of $e^2 + f$.
10. $w = 5$ and $y = 4$. Find the value of $(w + y)(w - y)$.
11. $g = 3$ and $h = 2$. Find the value of $g^2 + gh$.
12. $j = 5$, $k = 6$, and $m = 4$. Find the value of $jk^2 - m$.
13. $p = 10$ and $q = 5$. Find the value of $\frac{p + q}{p - q}$.
14. $b = 7$ and $c = 4$. Find the value of $(b - c)^2$.
15. $m = 9$ and $n = 5$. Find the value of $\frac{1}{2}(m - n)$.

Check your answers on page 225.

Expressions and Equations

Earlier in this section, you learned the definition of a term. $2y$, ab and $7t^2$ are examples of terms. Now, look at this statement:

$$2a + 7b - 6$$

This statement is an example of an **expression.** An expression is a group of terms. An expression equals the value of the terms in it. When you work with regular whole numbers, you also work with expressions. The statement "3 + 4" is an expression. The value of the expression is the value of 3 + 4, which is 7.

Here is another statement that you've worked with:

$$3 + 4 = 7$$

This statement is an example of an **equation.** An equation shows that two things are equal. In an equation, whatever is on one side of the equal sign (=) has the same value as whatever is on the other side of the sign. An expression makes up each side of an equation. In the example, 3 + 4 has the same value as 7.

Equations in algebra are the same as equations written with whole numbers. Both sides of an equation in algebra are equal, just as 3 + 4 is equal to 7. The main difference between equations in algebra and equations written with whole numbers is that, in algebra, you usually have to work with variables.

Although you may not realize it, equations written with terms are used often in everyday life. One everyday use of equations is in a **formula.** A formula is a rule that is written in the language of algebra. You have already used a formula in this book when you worked with interest problems. Here is the formula for finding interest:

I = **P** **R** **T**
(Interest) = (Principal) × (Rate) × (Time)

or

$$I = PRT$$

In order to find the interest on something, you use the formula. You substitute the amount of money that is being borrowed for P. You substitute the percent of interest for R. You substitute the length of time on the loan for T.

Here is an example of how a formula is used to solve problems:

EXAMPLE: The formula $D = RT$ stands for "distance equals rate multiplied by time." Using the formula, find the distance that Joe traveled in 4 hours if he drove at a steady speed of 55 miles per hour.

$D = rt$	$55 \times 4 = 220$
$D = (55)(4)$	$D = \mathbf{220}$
Step 1	Step 2

Step 1: Substitute 55 for r and 4 for t.
Step 2: Multiply. D is equal to the answer.

EXERCISE 9

DIRECTIONS: Find the values of each formula with the numbers given in the problem.

1. $A = bh$. Find A when $b = 16$ and $h = 10$.
2. $P = a + b + c$. Find P when $a = 8.5$, $b = 6$, and $c = 4.75$.
3. $F = \frac{9}{5}C + 32$. Find F when $C = 25$.
4. $A = s^2$. Find A when $s = 15$.
5. $I = PRT$. Find I when $P = 2{,}000$, $R = \frac{8}{100}$, and $T = 3$.
6. $V = e^3$. Find V when $e = 5$.
7. $W = \frac{P - 2L}{2}$. Find W when $P = 60$ and $L = 12$.
8. $A = \frac{1}{2}bh$. Find A when $b = 30$ and $h = 8$.
9. $I = \frac{bd^3}{3}$. Find I when $b = 2$ and $d = 4$.
10. $C = \frac{5}{9}(F - 32)$. Find C when $F = 95$.

Check your answers on page 226.

Inequalities

You learned that an equation is a statement that two expressions are equal. An **inequality** is a statement that two expressions are *not* equal. Four symbols are used to write inequalities. Memorize them before you go on.

Symbol	Meaning
$<$	less than
$\leq$	less than or equal to
$>$	more than
$\geq$	more than or equal to

The inequality $x < 9$ means "x is less than 9." The solution for this inequality is any number left of 9 on the number line. 5 is a solution. -12 is a solution. $8\frac{3}{4}$ is a solution. 10 is *not* a solution because 10 is greater than 9. All the numbers that make an inequality true are called the **solution set.**

The inequality $x \geq 5$ means "x is more than or equal to 5." The solution set for this inequality is every number to the right of 5 and including 5 on the number line. 20 is a solution. 5 is a solution. 4 is *not* a solution. Notice that $x = 5$ is a solution because the inequality gives a choice. 5 is not more than 5, but 5 is equal to 5.

EXAMPLE: Is $x = 7$ in the solution set of $x > 10$?

Substitute 7 for x in the inequality $x > 10$. $7 > 10$ is not true. 7 does *not* belong to the solution set.

EXAMPLE: Is $x = -3$ in the solution set of $x \geq -3$?

$-3 \geq -3$ is true. -3 belongs to the solution set.

EXERCISE 10

DIRECTIONS: Answer each question about inequalities.

1. Is $a = -3$ in the solution set of $a \leq -1$?
2. Is $y = 12$ in the solution set of $y < 9$?
3. Is $c = 5\frac{1}{2}$ in the solution set of $c > 5$?
4. Is $w = -6$ in the solution set of $w \geq -7$?
5. Is $r = -1\frac{1}{4}$ in the solution set of $r > -1$?
6. Is $d = 17$ in the solution set of $d \leq 17$?
7. Is $e = \frac{3}{8}$ in the solution set of $e > \frac{1}{4}$?
8. Is $t = -10$ in the solution set of $t < -10$?
9. Is $s = 1$ in the solution set of $s \leq 1$?
10. Is $g = 0$ in the solution set of $g \geq \frac{1}{2}$?

Check your answers on page 226.

WORD REVIEW

Before you go on to study other problems in algebra, take time to review the key words of the algebra language. The key words that you need to know for the next part of this book are listed below. Make sure that you are familiar with the meanings of all of the key words.

Key Word	Meaning	Example
power	the smaller number that shows how many times a number is multiplied by itself	$2^4 =$ $2 \times 2 \times 2 \times 2$ (4 is the power)
square root	the number that is multiplied by itself to give the number inside the $\sqrt{}$ sign	$\sqrt{4} = 2$ or -2 $2 \times 2 = 4$ $(-2) \times (-2) = 4$
signed number	a number that has a positive (+) or negative (−) sign added to it	$+4, -7$
variable	a letter that stands for an unknown value	$x + 7$ (x is the variable)
substitution	replacing a letter with a number to solve a problem	$x + 7 =$ $(x = 3)$ $3 + 7 = 10$
term	one or more values that stand for one number	$5x, 3y$
expression	a group of terms	$7 - 4$ $3x + 2$
equation	a mathematical statement that shows that two things are equal	$7 - 4 = 3$ $3x + 2 = 8$
formula	a rule that is written in the language of algebra	$I = PRT$ (the formula for interest)
inequality	a mathematical statement that shows that two things are *not* equal	$x > 3$

ANSWERS & SOLUTIONS

EXERCISE 1

1. **16**

$2^4 = 2 \times 2 \times 2 \times 2 = \mathbf{16}$

81

$9^2 = 9 \times 9 = \mathbf{81}$

27

$3^3 = 3 \times 3 \times 3 = \mathbf{27}$

2. **169**

$13^2 = 13 \times 13 = \mathbf{169}$

2,500

$50^2 = 50 \times 50 = \mathbf{2{,}500}$

216

$6^3 = 6 \times 6 \times 6 = \mathbf{216}$

3. **32**

$2^5 = 2 \times 2 \times 2 \times 2 \times 2 = \mathbf{32}$

1

$1^5 = 1 \times 1 \times 1 \times 1 \times 1 = \mathbf{1}$

256

$16^2 = 16 \times 16 = \mathbf{256}$

4. **1,000**

$10^3 = 10 \times 10 \times 10 = \mathbf{1{,}000}$

625

$25^2 = 25 \times 25 = \mathbf{625}$

256

$4^4 = 4 \times 4 \times 4 \times 4 = \mathbf{256}$

5. **17**

$5^2 - 2^3 =$
$5 \times 5 - 2 \times 2 \times 2 =$
$25 - 8 = \mathbf{17}$

91

$8^2 + 3^3 =$
$8 \times 8 + 3 \times 3 \times 3 =$
$64 + 27 = \mathbf{91}$

109

$10^2 - 4^2 + 5^2 =$
$10 \times 10 - 4 \times 4 + 5 \times 5 =$
$100 - 16 + 25 = \mathbf{109}$

6. **84**

$4^3 + 6^2 - 2^4 =$
$4 \times 4 \times 4 + 6 \times 6 - 2 \times 2 \times 2 \times 2 =$
$64 + 36 - 16 = \mathbf{84}$

110

$12^2 - 5^2 - 3^2 =$
$12 \times 12 - 5 \times 5 - 3 \times 3 =$
$144 - 25 - 9 = \mathbf{110}$

900

$10^3 - 10^2$
$10 \times 10 \times 10 - 10 \times 10 =$
$1000 - 100 = \mathbf{900}$

7. $\mathbf{\frac{1}{4}}$

$\left(\frac{1}{2}\right)^2 =$

$\frac{1}{2} \times \frac{1}{2} = \mathbf{\frac{1}{4}}$

$\mathbf{\frac{1}{81}}$

$\left(\frac{1}{9}\right)^2 =$

$\frac{1}{9} \times \frac{1}{9} = \mathbf{\frac{1}{81}}$

$\mathbf{\frac{8}{27}}$

$\left(\frac{2}{3}\right)^3 =$

$\frac{2}{3} \times \frac{2}{3} \times \frac{2}{3} = \mathbf{\frac{8}{27}}$

8. $\mathbf{\frac{27}{1,000}}$ $\quad$ $\mathbf{\frac{25}{36}}$ $\quad$ $\mathbf{\frac{49}{144}}$

$\left(\frac{3}{10}\right)^3 =$ $\quad$ $\left(\frac{5}{6}\right)^2$ $\quad$ $\left(\frac{7}{12}\right)^2$

$\frac{3}{10} \times \frac{3}{10} \times \frac{3}{10} = \mathbf{\frac{27}{1,000}}$ $\quad$ $\frac{5}{6} \times \frac{5}{6} = \mathbf{\frac{25}{36}}$ $\quad$ $\frac{7}{12} \times \frac{7}{12} = \mathbf{\frac{49}{144}}$

9. **.09** $\quad$ **.0064** $\quad$ **.000081**

$(.3)^2 = \begin{array}{r} .3 \\ \times\ .3 \\ \hline \mathbf{.09} \end{array}$ $\quad$ $(.08)^2 = \begin{array}{r} .08 \\ \times\ .08 \\ \hline \mathbf{.0064} \end{array}$ $\quad$ $(.009)^2 = \begin{array}{r} .009 \\ \times\ .009 \\ \hline \mathbf{.000081} \end{array}$

10. **.0225** $\quad$ **.008** $\quad$ **.0001**

$(.15)^2 = \begin{array}{r} .15 \\ \times\ .15 \\ \hline 75 \\ \underline{15\ } \\ \mathbf{.0225} \end{array}$ $\quad$ $(.2)^3 = \begin{array}{r} .2 \\ \times\ .2 \\ \hline .04 \\ \times\ .2 \\ \hline \mathbf{.008} \end{array}$ $\quad$ $(.01)^2 = \begin{array}{r} .01 \\ \times\ .01 \\ \hline \mathbf{.0001} \end{array}$

EXERCISE 2

1. $\sqrt{289} = \mathbf{17}$

Guess 20.

$\begin{array}{r} 14 \\ 20\overline{)289} \\ \underline{20\ } \\ 89 \\ \underline{80} \\ 9 \end{array}$ $\quad$ $\begin{array}{r} 14 \\ +\ 20 \\ \hline 34 \end{array}$ $\quad$ $\begin{array}{r} 17 \\ 2\overline{)34} \end{array}$

$\sqrt{784} = \mathbf{28}$

Guess 30.

$\begin{array}{r} 26 \\ 30\overline{)784} \\ \underline{60\ } \\ 184 \\ \underline{180} \\ 4 \end{array}$ $\quad$ $\begin{array}{r} 26 \\ +\ 30 \\ \hline 56 \end{array}$ $\quad$ $\begin{array}{r} 28 \\ 2\overline{)56} \end{array}$

$\sqrt{1,444} = \mathbf{38}$

Guess 40.

$\begin{array}{r} 36 \\ 40\overline{)1,444} \\ \underline{1\ 20\ } \\ 244 \\ \underline{240} \end{array}$ $\quad$ $\begin{array}{r} 36 \\ +\ 40 \\ \hline 76 \end{array}$ $\quad$ $\begin{array}{r} 38 \\ 2\overline{)76} \end{array}$

$\sqrt{484} = \mathbf{22}$

Guess 20.

$\begin{array}{r} 24 \\ 20\overline{)484} \\ \underline{40\ } \\ 84 \\ \underline{80} \end{array}$ $\quad$ $\begin{array}{r} 24 \\ +\ 20 \\ \hline 44 \end{array}$ $\quad$ $\begin{array}{r} 22 \\ 2\overline{)44} \end{array}$

2. $\sqrt{1,521} = \mathbf{39}$

Guess 40.

$\begin{array}{r} 38 \\ 40\overline{)1,521} \\ \underline{1\ 20\ } \\ 321 \\ \underline{320} \end{array}$ $\quad$ $\begin{array}{r} 38 \\ +\ 40 \\ \hline 78 \end{array}$ $\quad$ $\begin{array}{r} 39 \\ 2\overline{)78} \end{array}$

$\sqrt{1,849} = \mathbf{43}$

Guess 40.

$\begin{array}{r} 46 \\ 40\overline{)1,849} \\ \underline{1\ 60\ } \\ 249 \\ \underline{240} \end{array}$ $\quad$ $\begin{array}{r} 46 \\ +\ 40 \\ \hline 86 \end{array}$ $\quad$ $\begin{array}{r} 43 \\ 2\overline{)86} \end{array}$

$\sqrt{529} = \mathbf{23}$

Guess 20.

$$\begin{array}{r} 26 \\ 20\overline{)529} \\ \underline{40} \\ 129 \\ \underline{120} \end{array} \qquad \begin{array}{r} 26 \\ \underline{+\ 20} \\ 46 \end{array} \qquad \begin{array}{r} 23 \\ 2\overline{)46} \end{array}$$

$\sqrt{2{,}704} = \mathbf{52}$

Guess 50.

$$\begin{array}{r} 54 \\ 50\overline{)2{,}704} \\ \underline{2\ 50} \\ 204 \\ \underline{200} \end{array} \qquad \begin{array}{r} 54 \\ \underline{+\ 50} \\ 104 \end{array} \qquad \begin{array}{r} 52 \\ 2\overline{)104} \end{array}$$

3. $\sqrt{4{,}624} = \mathbf{68}$

Guess 70.

$$\begin{array}{r} 66 \\ 70\overline{)4{,}624} \\ \underline{4\ 20} \\ 424 \\ \underline{420} \end{array} \qquad \begin{array}{r} 66 \\ \underline{+\ 70} \\ 136 \end{array} \qquad \begin{array}{r} 68 \\ 2\overline{)136} \end{array}$$

$\sqrt{8{,}836} = \mathbf{94}$

Guess 90.

$$\begin{array}{r} 98 \\ 90\overline{)8{,}836} \\ \underline{8\ 10} \\ 736 \\ \underline{720} \end{array} \qquad \begin{array}{r} 98 \\ \underline{+\ 90} \\ 188 \end{array} \qquad \begin{array}{r} 98 \\ 2\overline{)188} \end{array}$$

$\sqrt{4{,}761} = \mathbf{69}$

Guess 70.

$$\begin{array}{r} 68 \\ 70\overline{)4{,}761} \\ \underline{4\ 20} \\ 561 \\ \underline{560} \end{array} \qquad \begin{array}{r} 68 \\ \underline{+\ 70} \\ 138 \end{array} \qquad \begin{array}{r} 69 \\ 2\overline{)138} \end{array}$$

$\sqrt{7{,}056} = \mathbf{84}$

Guess 80.

$$\begin{array}{r} 88 \\ 80\overline{)7{,}056} \\ \underline{6\ 40} \\ 656 \\ \underline{640} \end{array} \qquad \begin{array}{r} 88 \\ \underline{+\ 80} \\ 168 \end{array} \qquad \begin{array}{r} 84 \\ 2\overline{)168} \end{array}$$

4. $\sqrt{\frac{25}{36}} = \mathbf{\frac{5}{6}}$ $\quad \sqrt{\frac{4}{49}} = \mathbf{\frac{2}{7}}$ $\quad \sqrt{\frac{1}{81}} = \mathbf{\frac{1}{9}}$ $\quad \sqrt{\frac{9}{100}} = \mathbf{\frac{3}{10}}$

5. $\sqrt{\frac{64}{81}} = \mathbf{\frac{8}{9}}$ $\quad \sqrt{\frac{1}{144}} = \mathbf{\frac{1}{12}}$ $\quad \sqrt{\frac{9}{16}} = \mathbf{\frac{3}{4}}$ $\quad \sqrt{\frac{36}{49}} = \mathbf{\frac{6}{7}}$

6. $\sqrt{.04} = \mathbf{.2}$ $\quad \sqrt{.09} = \mathbf{.3}$ $\quad \sqrt{.0016} = \mathbf{.04}$ $\quad \sqrt{.01} = \mathbf{.1}$

7. $\sqrt{.0009} = \mathbf{.03}$ $\quad \sqrt{.000025} = \mathbf{.005}$ $\quad \sqrt{.0036} = \mathbf{.06}$ $\quad \sqrt{.000064} = \mathbf{.008}$

EXERCISE 3

1. **−21** **+42**

$$\begin{array}{r} -12 \\ -\ 3 \\ \underline{-\ 6} \\ \mathbf{-21} \end{array} \qquad \begin{array}{r} +\ 8 \\ +21 \\ \underline{+13} \\ \mathbf{+42} \end{array}$$

2. **−6** **−32**

$$\begin{array}{r} -17 \\ \underline{+11} \\ \mathbf{-\ 6} \end{array} \qquad \begin{array}{r} -36 \\ \underline{+\ 4} \\ \mathbf{-32} \end{array}$$

3. **−16** **+24**

$$\begin{array}{r} -\ 3 \\ \underline{-13} \\ \mathbf{-16} \end{array} \qquad \begin{array}{r} +42 \\ \underline{-18} \\ \mathbf{+24} \end{array}$$

4. **+5** **−7**

$$\begin{array}{r} +\ 4 \\ \underline{+\ 7} \\ +11 \end{array} \qquad \begin{array}{r} +11 \\ \underline{-\ 6} \\ \mathbf{+\ 5} \end{array} \qquad \begin{array}{r} -10 \\ \underline{-\ 5} \\ -15 \end{array} \qquad \begin{array}{r} -15 \\ \underline{+\ 8} \\ \mathbf{-\ 7} \end{array}$$

5. **−11** **+19**

−14	−27	+23	+37
−13	+16	+14	−18
−27	**−11**	+37	**+19**

6. **−28** **+46**

−35	−46	+29	+61
−11	+18	+32	−15
−46	**−28**	+61	**+46**

7. **−59** **−11**

−17	+15	−29
−23	+ 3	+18
−19	+18	**−11**
−59		

8. **+1** **0**

+ 8	− 9	+26	+13	−18	+40
+18	−16	−25	+27	−22	−40
+26	−25	**+ 1**	+40	−40	**0**

9. **−2** **−1**

− 8	−37	+15	−19	−25
−16	+35	+ 9	− 6	+24
−13	**− 2**	+24	−25	**− 1**
−37				

EXERCISE 4

1. **+19** **−27**

   +12 + 7 = **+19** −3 − 24 = **−27**

2. **+9** **+5**

   +18 − 9 = **+9** −14 + 19 = **+5**

3. **−27** **0**

   −21 − 6 = **−27** +16 − 16 = **0**

4. **+22** **−14**

   +5 + 17 = **+22** −17 + 3 = **−14**

5. **−35** **−13**

   −26 − 9 = **−35** +4 − 17 = **−13**

6. **+22** **+6**

   +11 + 11 = **+22** −7 + 13 = **+6**

7. **+13** **+28**

   +8 + 12 − 7 =
   +20 − 7 = **+13** +9 + 13 + 6 = **+28**

8. **−27** **−34**

   −14 − 3 − 10 = **−27** −20 − 7 − 7 = **−34**

9. **−16** **0**

$-13 + 5 - 8 =$ $+4 + 22 - 26 =$

$-21 + 5 = \mathbf{-16}$ $+26 - 26 = \mathbf{0}$

EXERCISE 5

1. **−84** **−90** 2. **+10** **+27** 3. $\mathbf{-\frac{3}{16}}$ **−32**

4. **+20** **−8** 5. **+360** **−54**

EXERCISE 6

1. **+12** **−9** **−1** $\mathbf{+\frac{5}{6}}$

2. **−9** $\mathbf{-\frac{3}{2}}$ or $\mathbf{-1\frac{1}{2}}$ **+7** **−6**

3. **−8** $\mathbf{+\frac{9}{10}}$ **+12** **−25**

4. $\mathbf{-\frac{9}{16}}$ **+16** **+9** $\mathbf{-\frac{6}{5}}$ or $\mathbf{-1\frac{1}{5}}$

5. $\mathbf{-\frac{1}{4}}$ **−4** **−12** **+7**

EXERCISE 7

1. **+72**

$st = (-12)(-6) = \mathbf{+72}$

2. **49**

$a^2 = 7^2 = \mathbf{49}$

3. **9**

$\frac{x}{y} = \frac{36}{4} = \mathbf{9}$

4. **6**

$c - f = 15 - 9 = \mathbf{6}$

5. **10**

$\sqrt{w} = \sqrt{100} = \mathbf{10}$

6. **9**

$\frac{3}{8}k = \frac{3}{8} \times 24 = \mathbf{9}$

7. **−15**

$x + y = (-8) + (-7) = \mathbf{-15}$

8. $\mathbf{\frac{4}{9}}$

$n^2 = \left(\frac{2}{3}\right)^2 = \frac{2}{3} \cdot \frac{2}{3} = \mathbf{\frac{4}{9}}$

9. **.018**

$bd = (.06)(.3) = \mathbf{.018}$

10. $\mathbf{\frac{5}{7}}$

$\frac{l}{w} = \frac{15}{21} = \mathbf{\frac{5}{7}}$

EXERCISE 8

1. **128**

$2m^2 =$
$2 \cdot 8^2 =$
$2 \cdot 64 =$
128

2. **51**

$uv + \frac{u}{v} =$
$12 \cdot 4 + \frac{12}{4} =$
$48 + 3 =$
51

3. **17**

$3p - 2s =$
$3 \cdot 9 - 2 \cdot 5 =$
$27 - 10 =$
17

4. **78**

$k(k + h) =$
$6(6 + 7) =$
$6(13) =$
78

5. **0**

$a - bc =$
$12 - 6 \cdot 2 =$
$12 - 12 =$
0

6. **12**

$\frac{r + t}{2} =$
$\frac{15 + 9}{2} =$
$\frac{24}{2} =$
12

7. **144**

$(e + f)^2 =$
$(8 + 4)^2 =$
$12^2 =$
144

8. **24**

$e + f^2 =$
$8 + 4^2 =$
$8 + 16 =$
24

9. **68**

$e^2 + f =$
$8^2 + 4 =$
$64 + 4 =$
68

10. **9**

$(w + y)(w - y) =$
$(5 + 4)(5 - 4)$
$9 \cdot 1 =$
9

11. **15**

$g^2 + gh =$
$3^2 + 3 \cdot 2 =$
$9 + 6 =$
15

12. **176**

$jk^2 - m =$
$5 \cdot 6^2 - 4 =$
$5 \cdot 36 - 4 =$
$180 - 4 =$
176

13. **3**

$\frac{p + q}{p - q} =$
$\frac{10 + 5}{10 - 5} =$
$\frac{15}{5} =$
3

14. **9**

$(b - c)^2 =$
$(7 - 4)^2 =$
$3^2 =$
9

15. **2**

$\frac{1}{2}(m - n) =$
$\frac{1}{2}(9 - 5) =$
$\frac{1}{2} \cdot 4 =$
2

EXERCISE 9

1. **160**

$A = bh$
$= 16 \cdot 10$
$= \mathbf{160}$

2. **19.25**

$P = a + b + c$
$= 8.5 + 6 + 4.75$
$= \mathbf{19.25}$

3. **77**

$F = \frac{9}{5}C + 32$
$= \frac{9}{5} \cdot 25 + 32$
$= 45 + 32$
$= \mathbf{77}$

4. **225**

$A = s^2$
$= 15^2$
$= \mathbf{225}$

5. **480**

$I = PRT$
$= 2000 \cdot \frac{8}{100} \cdot 3$
$= \mathbf{480}$

6. **125**

$V = e^3$
$= 5^3$
$= \mathbf{125}$

7. **18**

$W = \frac{P - 2L}{2}$
$= \frac{60 - 2 \cdot 12}{2}$
$= \frac{60 - 24}{2}$
$= \frac{36}{2}$
$= \mathbf{18}$

8. **120**

$A = \frac{1}{2}bh$
$= \frac{1}{2} \cdot 30 \cdot 8$
$= \mathbf{120}$

9. $\mathbf{42\frac{2}{3}}$

$I = \frac{bd^3}{3}$
$= \frac{2 \cdot 4^3}{3}$
$= \frac{2 \cdot 64}{3}$
$= \frac{128}{3}$
$= \mathbf{42\frac{2}{3}}$

10. **35**

$C = \frac{5}{9}(F - 32)$
$= \frac{5}{9}(95 - 32)$
$= \frac{5}{9}(63)$
$= \mathbf{35}$

EXERCISE 10

1. **Yes** 2. **No** 3. **Yes** 4. **Yes** 5. **No**

6. **Yes** 7. **Yes** 8. **No** 9. **Yes** 10. **No**

The Uses of Algebra

So far, you have studied the key words and ideas that make up the algebra language. You have seen the ways in which algebra is similar to basic math. You also have seen the ways in which algebra is different from basic math.

In the next section, you will study how to solve problems with algebra. You also will find out how to write algebra problems.

Study this section slowly and carefully. The section contains the algebra skills that you will need for the GED Test.

WRITING AN EXPRESSION

Before you study how to write an expression in algebra, look at the following problem.

> On Monday, Jeff had $14. He spent $7 for food on Tuesday. He spent $5 for gas on Wednesday. How much money did Jeff have left after Wednesday?

In the basic math part of this book, you worked with many word problems like this. To solve the problem, you have to decide what is being asked for, decide which facts to use, and decide which math operation to use. Then you have to take the important facts and perform the math operation.

In the example above, you are asked to find out how much money Jeff has. To solve the problem, you have to subtract the money that he spent from the money that he started with. Here is how you could set the problem up:

$$\$14 - \$7 - \$5 =$$

Whenever you set up a problem, you write an expression. You "translate" the everyday language of the word problem into math language. The expression contains the facts and the operations that are needed to solve the problem.

Study the next problem carefully.

EXAMPLE: Clyde works part-time in a shoe factory. He makes \$5 an hour. On Monday, he worked 6 hours. On Wednesday, he worked 4 hours. Which of the following expressions shows the amount of money that Clyde made for those two days?

(1) $6 \times (4 + 5)$ (2) $(5 \times 6) + 4$
(3) $5 \times (6 + 4)$ (4) $(5 \times 4) + 6$

Step 1: Read the problem to find out what is going on.

Step 2: Find out what is being asked for. The problem asks you to pick out an expression. The expression should show the amount of money that Clyde made in the 2 days he worked.

Step 3: Decide which information you need to solve the problem. You need the total number of hours that Clyde worked. You also need the amount of money that he makes per hour.

Step 4: Decide which operations to use. First, you need to add the hours. Then you need to multiply the total hours by the amount of money that he makes in an hour.

Step 5: Look at the answer choices. Decide which one expresses the amount of money that Clyde made. Clyde worked 6 hours and 4 hours $(6 + 4)$. He made \$5 an hour. The expression should show $5 \times (6 + 4)$. Answer choice 3 is correct.

When you write or choose an expression, you should first figure out which operations and which facts are needed. Then you must decide the best way to write the facts and operations in an expression. Keep in mind that the order of operations is important. In the last problem, you had to add the hours first before you multiplied the total hours by the amount of money.

$$\$5 \times (6 + 4) = 5 \times 10 = \$50$$

Another way to write the expression is: $5(6 + 4)$. Remember that, when there is no sign between two numbers, the numbers should be multiplied.

EXERCISE 1

DIRECTIONS: Choose the correct answer for each problem.

1. Ed and Phil share an apartment. The rent for the apartment is $300 a month. Ed and Phil pay an equal amount of the rent. Which expression shows the amount of money that Phil pays for rent?

   (1) 2(300) (2) $\frac{1}{2}$(300) (3) 300 ÷ $\frac{1}{2}$ (4) 300(2)

2. Fred took 3 math tests. His scores were 68, 72, and 80. Which expression shows Fred's average score?

   (1) 3(68 + 72 + 80) (2) 3(68) + 3(72) + 3(80)
   (3) (68 + 72 + 80) ÷ 3 (4) (68 + 72 + 80) ÷ $\frac{1}{3}$

3. Last year, the Clifton Bombers lost 19 softball games. Altogether, they played 31 games. Which expression shows the number of games that the Bombers won last year?

   (1) 19 + 31 (2) 19(31) (3) 31 − 19 (4) 31 ÷ 19

4. Fritz bought a used car for $700. He fixed the car and sold it for $1,300. Which expression shows the profit that Fritz made on the car?

   (1) $1,300 − $700 (2) $1,300 ÷ $700
   (3) $1,300 + $700 (4) $1,300 × $700

5. Paul went fishing. He caught 17 fish. He decided to keep 3 fish for himself and to divide the rest of the fish equally between his mother and his sister. Which expression shows the number of fish that his sister got?

   (1) 17(3 − 2) (2) 17 ÷ (3 − 2)
   (3) (17 − 3) ÷ 2 (4) 2(17 ÷ 3)

Check your answers on page 258.

Writing Expressions in Algebra

When you write an expression in basic math, you change the information given in the problem to the language of basic math. When you write an expression in algebra, you change the information to the language of algebra.

The list below gives some key words for each of the four basic operations. After each key word is an example of an expression using the key word. After each example is the expression written in the language of algebra. (Notice that any letter can be used as a variable.)

Addition

sum: The sum of a number and 6: $y + 6$
total: The total of 8 and a number: $8 + x$
increased by: A number increased by 20: $t + 20$
plus: 15 plus a number: $15 + s$
more than: 12 more than a number: $r + 12$

The order of the variable and the other number in each addition expression can be changed. For example, $12 + r$ is the same as $r + 12$. However, the order in subtraction expressions cannot be changed.

Subtraction

less than: 9 less than a number: $u - 9$
decreased by: 10 decreased by a number: $10 - p$
subtracted from: A number subtracted from 50: $50 - m$
minus: 20 minus a number: $20 - n$

In multiplication expressions, the number usually comes to the left of the unknown.

Multiplication

product: The product of 4 and a number: $4k$
times: 2 times a number: $2j$
a fraction **of:** $\frac{3}{4}$ of a number: $\frac{3}{4}h$

Division

divided by: A number divided by 13: $\frac{d}{13}$
16 divided by a number: $\frac{16}{e}$

Many expressions have more than one operation.

EXAMPLES:

A number increased by 3 times the number: $c + 3c$
6 less than a number all times 4: $4(b - 6)$
(The parentheses mean that the subtraction comes first. Then 4 is multiplied by the result.)
A number decreased by $\frac{1}{4}$ of the number: $a - \frac{1}{4}a$

EXERCISE 2

DIRECTIONS: Write these expressions in the language of algebra. Use the letter x for each unknown number.

1. The sum of a number and 23:

2. 15 less than a number:
3. The square root of a number and 12:
4. 10 more than 3 times a number:
5. A number divided by 11:
6. The total of $\frac{1}{2}$ of a number and $\frac{1}{3}$ of the number:
7. A number times itself divided by 3:
8. A number decreased by $\frac{2}{3}$ of the number:
9. 30 divided by a number:
10. $\frac{1}{2}$ of a number subtracted from 100:

Check your answers on page 259.

The next problems will give you a chance to practice writing expressions from word problems.

EXAMPLE: Let g stand for George's age now. Write an expression for his age in 6 years.

$$g + 6$$

Add 6 to George's age now (g).

EXAMPLE: Let d stand for Sally's hourly wage. Write an expression for the amount Sally earns when she works 35 hours.

$$35d$$

Multiply 35 by the amount Sally makes in an hour (d).

EXERCISE 3

DIRECTIONS: Write an expression for each problem.

1. Let x stand for the amount of money that Dorothy has. Write an expression to show that Dorothy spent $10.

2. Let p stand for the price Bill paid for a used car. He fixed the car and sold it for a profit of $400. Write an expression to show the price that Bill sold the car.

3. Let b stand for the length in feet of a board. Write an expression to show the length of the board when 2 feet have been cut from the board.

4. Let c stand for the length in feet of Sam's driveway. Write an expression to show the length of the driveway in yards.

5. Let w stand for Janet's weight in pounds. Write an expression to show the amount Janet weighed after she lost 15 pounds.

6. Let h stand for the amount Debbie makes in an hour. Debbie works overtime at $1\frac{1}{2}$ times her regular rate. Write an expression to show the amount Debbie makes for an hour of overtime work.

7. Let f stand for Fred's income. Ellen makes twice as much money as Fred. Write an expression to show Ellen's income.

8. Let k stand for Kate's age. David lacks one year in being twice as old as Kate. Write an expression to show David's age.

9. Let y stand for the amount Joe makes in an hour. Joe gets $4 more than his regular wage for an hour of overtime. Write an expression to show the amount Joe makes for 3 hours of overtime work.

10. Let e stand for the price of a dozen eggs. Write an expression to show the price of one egg.

Check your answers on page 259.

SOLVING EQUATIONS

You learned earlier that an equation is a statement that two expressions are equal. $3y + 4 = 19$ is an equation. The letter y stands for an unknown number. In everyday words, the equation means: "Three multiplied by an unknown number plus four equals nineteen." The value of y that makes the equation a true statement is called the **solution** to the equation. The solution to $3y + 4 = 19$ is $y = 5$.

It is easy to check a solution. Substitute the solution for the variable. In this example, substitute 5 for y. You should get 19 on both sides of the equation.

$$3y + 4 = 19$$

$$3(5) + 4 = 19$$

$$15 + 4 = 19$$

$$19 = 19$$

If you know the value of the unknown number, working with an equation is easy. All you have to do is substitute the number for the letter. In many algebra problems, however, you have to find the value of the unknown number. When you find the value of the unknown number, you *solve* the equation.

To solve an equation, you have to "move" numbers and terms from one side of the equation to the other. You have to change the equation so that the variable is the only thing on one side of the equal (=) sign.

Look at the equation $c + 12 = 32$. To solve the equation, you have to get c by itself. You have to "move" the $+ 12$ to the other side of the equation.

$$\begin{array}{rcr} c + 12 & = & 32 \\ - 12 & = & -12 \\ \hline c \quad\quad & = & 20 \end{array}$$

If you subtract 12 from the left side, the c will stand by itself. When you subtract 12 from the left side, you also must subtract 12 from the right side. **When you perform an operation on one side of the equation, you must perform exactly the same operation on the other side of the equation.**

To move a number from one side of the equation to the other, use the **inverse** operation. The word *inverse* means *opposite.*

Addition is the inverse of subtraction.
Subtraction is the inverse of addition.
Multiplication is the inverse of division.
Division is the inverse of multiplication.

EXAMPLE: Solve for m in $m - 19 = 15$.

$\begin{array}{rcr} m - 19 & = & 15 \\ \mathbf{+ 19} & \mathbf{=} & \mathbf{+19} \\ \hline m \quad\quad & = & \mathbf{34} \end{array}$	$\mathbf{34 - 19 = 15}$
Step 1	Step 2

Step 1: In the equation, 19 is subtracted from m. The inverse of subtraction is addition. To get m by itself, add 19 to both sides of the equation.

Step 2: Check your answer. Substitute 34 for m and work out the original problem.

In some problems, the unknown number is on the right side of the equation. The side that the unknown number is on does not matter.

EXAMPLE: Solve for r in $96 = 8r$.

$\frac{96}{8} = \frac{8r}{8}$ $\mathbf{12} = r$	$\mathbf{96 = 8 \times 12}$
Step 1	Step 2

Step 1: In the equation, r is multiplied by 8. The inverse of multiplication is division. To get r by itself, divide both sides of the equation by 8. (Notice that when you divide 8 by 8, you do not write the 1 next to r. The 1 is not needed because multiplying by 1 does not change the value of the number. $1r$ is the same as r).

Step 2: Check your answer. Substitute 12 for r in the original problem.

EXAMPLE: Solve for s in $\frac{s}{9} = 10$.

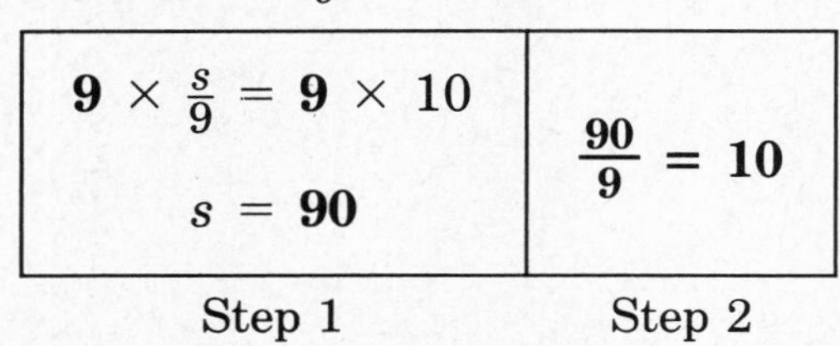

Step 1 Step 2

Step 1: In the equation, s is divided by 9. The inverse of division is multiplication. To get s by itself, multiply both sides by 9.

Step 2: Check your answer. Substitute 90 for s and work out the original problem.

EXERCISE 4

DIRECTIONS: Solve and check each equation.

1. $f - 27 = 43$ $\quad\quad$ $\frac{k}{4} = 12$ $\quad\quad$ $a + 17 = 23$
2. $9s = 63$ $\quad\quad$ $30 = b^2 + 14$ $\quad\quad$ $5 = \frac{m}{10}$
3. $100 = 4u^2$ $\quad\quad$ $c + 3\frac{1}{2} = 9\frac{1}{2}$ $\quad\quad$ $g^2 - 31 = 33$
4. $\frac{n}{8} = 16$ $\quad\quad$ $18 = h - 12$ $\quad\quad$ $120 = 12w$
5. $96 = d + 24$ $\quad\quad$ $15y = 10$ $\quad\quad$ $6 = \frac{p}{11}$

6. $.05z = 4$ $\qquad i^2 - 1\frac{1}{4} = 2\frac{3}{4}$ $\qquad 8t = 12$

7. $e + 2.5 = 6.5$ $\qquad \frac{r}{30} = 3$ $\qquad 26 = j - 19$

Check your answers on page 259.

Solving Equations with More Than One Operation

You will have to use more than one opposite operation to solve many equations. When you use more than one opposite operation, do the addition and subtraction operations first. Then do multiplication and division.

EXAMPLE: Solve for y in $3y + 4 = 19$.

Step 1	Step 2	Step 3
$3y + 4 = 19$ $\mathbf{-\,4 = -\,4}$ $3y = \mathbf{15}$	$\frac{3y}{3} = \frac{15}{3}$ $y = \mathbf{5}$	$3(\mathbf{5}) + 4 = 19$ $15 + 4 = \mathbf{19}$

Step 1: In the equation, the unknown y is multiplied by 3, and 4 is added. First, subtract 4 from both sides.
Step 2: Next, divide both sides by 3 to get y by itself.
Step 3: Check your answer. Substitute 5 for y.

EXAMPLE: Solve for c in $10 = \frac{c}{2} - 6$.

Step 1	Step 2	Step 3
$10 = \frac{c}{2} - 6$ $+\,6 = \quad +\,6$ $16 = \frac{c}{2}$	$2 \times 16 = \frac{c}{2} \times 2$ $32 = c$	$10 = \frac{32}{2} - 6$ $10 = 16 - 6$

Step 1: In the equation, c is divided by 2, and 6 is subtracted. First, add 6 to both sides.
Step 2: Next, multiply both sides by 2 to get c by itself.
Step 3: Check your answer. Substitute 32 for c.

EXERCISE 5

DIRECTIONS: Solve and check each equation.

1. $4w + 6 = 34$ $\qquad 6z - 7 = 41$ $\qquad \frac{x}{5} - 4 = 36$

2.	$28 = 12y - 8$	$51 = 20u + 11$	$43 = \frac{v}{10} + 3$
3.	$\frac{s}{4} - 13 = 2$	$\frac{2}{3}r^2 + 1 = 25$	$5q - 12 = 63$
4.	$21 = \frac{1}{2}p + 9$	$64 = 4n + 8$	$11 = \frac{m}{9} + 6$
5.	$\frac{4}{3}j - 2 = 22$	$\frac{i}{7} - 2 = 6$	$8h^2 + 3 = 75$
6.	$6f + 19 = 79$	$9e - 34 = -7$	$\frac{4}{5}d - 4 = 16$
7.	$\frac{a}{12} + 9 = 13$	$90 = 8b - 6$	$6c + 22 = 100$

Check your answers on page 260.

Special Problems with Solving Equations

In some equations, the terms are separated. First, combine the terms; then solve the equations. When the terms are on the same side of the equation, combine them according to the rules for adding signed numbers.

EXAMPLE: Solve for c in $8c + 5 - 2c = 29$.

Step 1	Step 2	Step 3
$8c + 5 - 2c = 29$ $\mathbf{6c} + 5 = 29$	$6c + 5 = 29$ $-5 = -5$ $6c = 24$	$\frac{6c}{6} = \frac{24}{6}$ $c = 4$

Step 1: Combine the terms on the left side of the equation. Subtract $2c$ from $8c$.
Step 2: Subtract 5 from both sides.
Step 3: Divide both sides of the equation by 6 to get c by itself.
Step 4: Check your answer. Substitute 4 for c in the original equation.

$$8(4) + 5 - 2(4) = 29$$
$$32 + 5 - 8 = 29$$
$$37 - 8 = 29$$

Step 4

When the terms are on both sides of the equation, you have to move the terms to one side by using opposite operations.

EXAMPLE: Solve for x in $9x - 4 = 2x + 17$.

Step 1	Step 2	Step 3
$9x - 4 = 2x + 17$ $-2x \quad = -2x$ $7x - 4 = 17$	$7x - 4 = 17$ $+4 = +4$ $7x = 21$	$\frac{7x}{7} = \frac{21}{7}$ $x = 3$

Step 1: First, combine the terms on one side of the equation. Since 2 is smaller than 9, subtract $2x$ from $9x$.
Step 2: Add 4 to both sides.
Step 3: Divide both sides by 7 to get x by itself.
Step 4: Check your answer. Substitute 3 for x in the original problem.

$$9(3) - 4 = 2(3) + 17$$
$$27 - 4 = 6 + 17$$
$$23 = 23$$

Step 4

In some problems, you will be given two separate variables. However, one of the variables is given a value with the other variable in it. Study the next example carefully.

EXAMPLE: If $a + b = 72$, and $b = 2a$, find the value of b.

Step 1	Step 2	Step 3
$a + b = 72$ $a + \mathbf{2a} = 72$	$\mathbf{3a} = 72$ $\frac{3a}{\mathbf{3}} = \frac{72}{\mathbf{3}}$ $a = \mathbf{24}$	$b = 2a$ $b = 2(\mathbf{24})$ $b = \mathbf{48}$

Step 1: Substitute $2a$ for b so that you are working with only one term.
Step 2: Combine the like terms and solve the equation for a.
Step 3: The problem asks you to find the value of b. Since $b = 2a$, solve for b by substituting 24 for a.

Some equations have numbers and terms inside parentheses. Earlier, you found that numbers or terms that are next to each other without a sign means to multiply. To solve these equations, first multiply to get rid of the parentheses. Using the number or term outside the parentheses, multiply each number or term inside the parentheses separately. Then combine the terms and solve the equation.

EXAMPLE: Solve for y in $7y = 3(y + 8)$.

Step 1	Step 2	Step 3
$7y = \mathbf{3}(y + \mathbf{8})$ $7y = \mathbf{3y} + \mathbf{24}$	$7y = 3y + 24$ $-\mathbf{3y} = -\mathbf{3y}$ $\mathbf{4y} = \mathbf{24}$	$\frac{4y}{\mathbf{4}} = \frac{24}{\mathbf{4}}$ $y = \mathbf{6}$

Step 1: Multiply to get rid of the parentheses first. Multiply 3 by y and 3 by 8. (Notice that you keep the + sign that was in the parentheses.)

Step 2: Subtract to get y on one side of the equation. Since 3 is smaller than 7, subtract $3y$ from $7y$.

Step 3: Divide each side of the equation by 4 to get y by itself.

Step 4: Check your answer. Substitute 6 for y in the original equation.

$$7(\mathbf{6}) = 3(\mathbf{6} + 8)$$
$$42 = 3(14)$$
$$\mathbf{42} = \mathbf{42}$$

Step 4

There is one more problem about solving equations that you should know about. On the GED Test, you may find a problem like this:

EXAMPLE: If $x^2 - x + 1 = 3$, then $x =$

(1) 2 only (2) 1 only (3) -1 only
(4) 2 or 1 (5) 2 or -1

There is a way to solve this equation, but it is long and complicated. The best thing to do is to *substitute* the possible values for x and then decide which answer choice is correct. Remember, on the GED Test you are given five answer choices. *One of the choices has to be right.* You can use substitution as a shortcut to "solve" the equation.

However, when you use substitution, be careful. Make sure that you try *all* of the answer choices before you choose your answer. If you don't try all the choices, you may end up choosing the wrong answer. Look at the example. If you substitute 2 for x, you "solve" the equation.

$$(\mathbf{2})^2 - (\mathbf{2}) + 1 = 3$$
$$\mathbf{4} - 2 + 1 = 3$$
$$\mathbf{2} + \mathbf{1} = \mathbf{3}$$

You may be tempted to stop after finding that the number 2 solves the equation. However, if you try all the other possible solutions, you will find that the number -1 also solves the equation.

$$(\mathbf{-1})^2 - (\mathbf{-1}) + 1 = 3$$
$$\mathbf{1} - (-1) + 1 = 3$$
$$1 + \mathbf{1} + 1 = 3$$
$$\mathbf{2 + 1 = 3}$$

Both 2 and -1 solve the equation. Therefore, the correct answer is choice 5 (2 or -1). Unless you substitute all the possible answers, you may miss the problem.

EXERCISE 6

DIRECTIONS: Solve and check each equation.

1. $7m - 3m = 32$ $\quad 40 = 2p + 6p$
2. $9r + 6 + 2r = 39$ $\quad 8s - 5s - 4 = 20$
3. $10t = 28 + 3t$ $\quad 36 - 4u = 5u$
4. $11v = 60 - v$ $\quad 20 + 2w = 7w$
5. $12x - 4 = 26 + 2x$ $\quad 9y + 3 = 29 - 4y$
6. $9a = 4(a + 20)$ $\quad 6c = 2(c + 14)$
7. $8(d + 3) = 10d$ $\quad 3(f + 6) = 5f$
8. $4(h - 2) = h + 13$ $\quad 7m - 1 = 5(m + 3)$
9. If $x + y = 15$ and $y = 2x$, then $y =$
10. If $t^2 + 7t = -12$, then $t =$

    (1) 4 only (2) -3 only (3) -4 only
    (4) 4 or -4 (5) -3 or -4

Check your answers on page 261.

Solving Inequalities

Inequalities are set up the same way as equations. The only difference is that the equal (=) sign is replaced by one of the inequality signs. You can solve inequalities the same way that you solve equations, by doing inverse operations.

EXAMPLE: Find the solution set for $x - 5 \le 6$.

$$\begin{array}{rcr} x - 5 & \le & 6 \\ +5 & & +5 \\ \hline x & \le & 11 \end{array}$$

In the inequality, 5 is subtracted from x. To get x by itself, add 5 to both sides. Carry the inequality sign down into the answer.

EXAMPLE: Find the solution set for $3x + 2 > 26$.

Step 1	Step 2
$\begin{array}{rcr} 3x + 2 & > & 26 \\ -2 & & -2 \\ \hline 3x & > & 24 \end{array}$	$\frac{3x}{3} > \frac{24}{3}$ $x > 8$

Step 1: In the inequality, 2 is added to the x term. Subtract 2 from both sides.

Step 2: Divide both sides of the inequality by 3 to get x by itself. Remember to carry the inequality sign into your answer.

EXERCISE 7

DIRECTIONS: Find the solution set for each inequality.

1. $m - 7 \le 3$
2. $9a > 30$
3. $\frac{c}{4} \ge -3$
4. $d + 8 < -6$
5. $2e - 9 \ge -1$
6. $5n + 4 < 3n + 16$
7. $8w - 7 > 2w + 23$
8. $9z + 1 \le 6z - 20$
9. $\frac{1}{2}p - 7 \ge -2$
10. $4u - 11 < 3u + 5$

Check your answers on page 263.

Writing Equations

Writing an equation is just like writing an expression. The only difference is that an equation has two sides. When you write an equation from words, watch for the words *equals* and *is*. They stand for the equal (=) sign.

EXAMPLE: In June, Victor worked twice as many plus 7 hours of overtime more than he worked in May. He worked 23 hours of overtime in June. How many hours of overtime did he work in May?

Step 1	Step 2	Step 3
$\mathbf{2x + 7 = 23}$	$2x + 7 = 23$ $\mathbf{-\ 7 = -\ 7}$ $2x = \mathbf{16}$	$\frac{2x}{2} = \frac{16}{2}$ $x = \mathbf{8}$

Step 1: Write the equation in the language of algebra. Use the equal sign for the word *equals*.
Step 2: Subtract 7 from both sides.
Step 3: Divide both sides of the equation by 2 to get x by itself. Victor worked 8 hours of overtime in May.

EXAMPLE: 5 times the sum of a number and 2 is the same as the number increased by 22. Find the number.

Steps 1–2	Step 3
$\mathbf{5(c + 2) = c + 22}$ $\mathbf{5c + 10} = c + 22$	$5c + 10 = c + 22$ $-c = -c$ $\mathbf{4c + 10} = 22$

Step 1: Write the equation in the language of algebra.
Step 2: Multiply to get rid of the parentheses.
Step 3: Combine the c terms. Subtract c from both sides of the equation.
Step 4: Subtract 10 from both sides.
Step 5: Divide both sides of the equation by 4 to get c by itself.

Step 4	Step 5
$4c + 10 = 22$ $-\ 10 = -10$ $4c = \mathbf{12}$	$\frac{4c}{4} = \frac{12}{4}$ $c = \mathbf{3}$

EXERCISE 8

DIRECTIONS: Write and solve each equation.

1. Seven times a number decreased by 6 equals 29.
2. Three more than a number divided by 2 is 9.
3. 43 is the same as 8 more than 5 times a number.
4. One more than $\frac{3}{4}$ of a number equals 13.
5. Seventy is the same as 8 times a number decreased by 2.
6. Eleven less than a number divided by 9 is 3.
7. Eight times a number decreased by 2 times the number equals 24.
8. Fourteen decreased by 3 times a number equals 4 times the number.
9. A number increased by 9 times the number is 50.
10. The product of 5 and a number is the same as 3 times the number increased by 12.

Check your answers on page 264.

The next problems will give you a chance to practice writing equations from practical situations. Read each problem carefully. Choose a letter to represent the unknown number. Write an expression for each amount described in the problem. Then write and solve an equation showing how the amounts in the problem are related.

EXAMPLE: Mary makes twice as much as Bill. Together, they make $405 in a week. How much does each person make?

Step 1	Step 2	Step 3	Step 4
x = Bill's salary $2x$ = Mary's salary	$x + 2x = 405$	$3x = 405$ $\frac{3x}{3} = \frac{405}{3}$ $x = 135$	$x = \$135$ (Bill) $2x = \$270$ (Mary)

Step 1: Write a term for each person's salary. (Mary makes twice as much as Bill. If you use x for Bill's salary, then you should use $2x$ (2 times x) for Mary's salary.)

Step 2: Write an equation in the language of algebra. Added together, the two salaries equal $405.

Step 3: Solve the equation.
Step 4: Substitute $135 for x to solve the problem.

EXERCISE 9

DIRECTIONS: Solve each problem with an equation.

1. There are twice as many women as men in the union at the Troy Paper Products Co. Altogether, 84 workers at the factory are union members. How many of the members are women?

2. Bill, Gordon, and Jeff started a business. Gordon invested twice as much as Bill. Jeff invested $2,000 more than Gordon. Altogether, they invested $22,000. How much did Bill invest?

3. Colette pays $20 more than $\frac{1}{4}$ of her salary for rent. Her monthly rent is $195. How much does she make each month?

4. Jack is five years older than his brother Frank. Six years ago, Jack was twice as old as Frank. How old are Jack and Frank now?

5. Pete has a piece of lumber 60 inches long. He wants to cut it into two pieces. One piece is to be 14 inches longer than the other. How long will each piece be?

6. Nick has a piece of wire 79 inches long. He wants to cut it into two pieces. One piece is to be 2 inches less than twice the length of the other piece. How long will each piece be?

7. Mr. Allen is now 4 years more than twice as old as his daughter. Eight years ago, he was three times as old as his daughter. How old are Mr. Allen and his daughter now?

8. Jane worked 8 hours on Friday and 5 hours on Saturday. Saturday, she got paid twice as much an hour as she did on Friday. For the two days, she made $81 altogether. How much did she get each hour on Friday?

9. Mr. Frisch makes $400 more than twice his wife's annual salary. Together, they make $20,800 in a year. How much does each make in a year?

10. Manny's net wages are 4 times the amount of the deductions from his pay. His gross pay is $325 a week. How much does Manny take home in a week?

Check your answers on page 265.

RATIO

A **ratio** is a way of comparing two numbers. The ratio of a day to a week is 1 to 7. 1 stands for a day. 7 stands for the 7 days in a week. A ratio can be written with a colon (:) or as a fraction. The ratio of a day to a week is 1:7 or $\frac{1}{7}$.

The numbers in a ratio should be in the same order as the words in the problem. Like fractions, ratios should be reduced to lowest terms.

EXAMPLE: In Charles's math class, there are 14 women and 7 men. What is the ratio of women to men in the class?

Step 1	Step 2	Step 3
$\frac{\textbf{women}}{\textbf{men}}$ or **women : men**	$\frac{\mathbf{14}}{\mathbf{7}}$ or **14 : 7**	$\frac{14 \div 7}{7 \div 7} = \frac{\mathbf{2}}{\mathbf{1}}$ or **2 : 1**

Step 1: Find the order of the words in the problem. The problem asks you to compare the number of women to the number of men.

Step 2: Write a ratio to show the number of women compared to the number of men. (You can write the ratio as a fraction or with a colon.)

Step 3: Reduce the ratio as you would reduce a fraction. (Notice that you do not change the fraction to a whole number.)

In everyday words, the last ratio means: "For every two women in the class, there is one man."

EXAMPLE: What is the ratio of the number of women in Charles's math class to the total number of students in the class?

Step 1	Step 2	Step 3
14 + 7 = 21	$\frac{\textbf{women}}{\textbf{total}} = \frac{\mathbf{14}}{\mathbf{21}}$ or **women : total =** **14 : 21**	$\frac{14 \div 7}{21 \div 7} = \frac{\mathbf{2}}{\mathbf{3}}$ or **2 : 3**

Step 1: First, find the total number of students. Add the number of women and the number of men.

Step 2: Find the order of words in the problem. The problem asks you to compare the number of women to the total number of students. Write in the numbers.

Step 3: Reduce the ratio as you would reduce a fraction. In everyday words, the ratio means that two out of every three students are women.

EXERCISE 10

DIRECTIONS: For each problem, write a ratio and reduce the ratio to lowest terms.

1. In town, Gordon gets an average of 15 miles of driving on one gallon of gas. In the country, he gets an average of 25 miles per gallon. What is the ratio of the mileage Gordon gets in town to the mileage he gets in the country?

2. At the Mohawk Power Company, 80 workers are women and 440 are men. What is the ratio of men to women at the company?

3. What is the ratio of 6 ounces to a pint? (1 pint = 16 ounces)

4. The Jacksons make $960 a month. They pay $180 for rent. What is the ratio of the Jacksons' rent to their income?

5. Peter's living room is 12 feet wide and 18 feet long. What is the ratio of the length of the room to the width?

6. Manny's gross salary is $210 a week. Manny's employer deducts $42 each week for taxes and Social Security. What is the ratio of the deductions to Manny's gross salary?

7. What is the ratio of 9 inches to a yard? (1 yard = 36 inches)

8. The Rigbys had a $30,000 mortgage on their house. So far, they have paid $12,000 toward the mortgage. What is the ratio of the amount the Rigbys have paid to the total amount of the mortgage?

9. Phil got 5 problems wrong on a test. The test had 40 problems. What was the ratio of the number of problems Phil got right to the total number of problems on the test?

10. In January, John weighed 240 pounds. By September, John lost 60 pounds. What is the ratio of John's September weight to his January weight?

Check your answers on page 266.

PROPORTION

A **proportion** is an equation that says two ratios are equal. $\frac{6}{8} = \frac{3}{4}$ is an example of a proportion. In a proportion, the "cross products" are equal. This means that the top of one side times the bottom of the other side is the same as the bottom of the first side times the top of the other. In the example, $6 \times 4 = 8 \times 3$. Both cross products equal 24.

In many proportion problems, one part of a proportion is missing. To find the missing part, write a new equation with the cross products. Then solve the new equation for the missing part.

EXAMPLE: Solve for c in $\frac{5}{6} = \frac{3}{c}$.

Step 1	Step 2	Step 3
$\frac{5}{6} \times \frac{3}{c}$		$\frac{5c}{5} = \frac{18}{5}$
$\mathbf{5} \times c = \mathbf{5}c$	$\mathbf{5}c = \mathbf{18}$	
$\mathbf{6 \times 3 = 18}$		$c = \mathbf{3\frac{3}{5}}$

Step 1: Multiply to find the cross products. Multiply 5 by c and 6 by 3.
Step 2: Write a new equation with the cross products.
Step 3: Solve the equation for c.

EXAMPLE: Solve for s in $s:9 = 4:7$.

Step 1	Step 2	Step 3
$\frac{s}{9} \times \frac{4}{7}$		$\frac{7s}{7} = \frac{36}{7}$
$s \times \mathbf{7} = \mathbf{7}s$	$\mathbf{7}s = \mathbf{36}$	
$\mathbf{9 \times 4 = 36}$		$s = \mathbf{5\frac{1}{7}}$

Step 1: Rewrite the proportion as two fractions. Multiply to find the cross products.
Step 2: Write a new equation with the cross products.
Step 3: Solve the equation for s.

EXERCISE 11

DIRECTIONS: Find the unknown number in each proportion.

1. $\frac{a}{7} = \frac{8}{5}$ $\quad \frac{6}{e} = \frac{9}{4}$ $\quad \frac{10}{11} = \frac{c}{3}$ $\quad \frac{4}{f} = \frac{7}{12}$

2. $\frac{2}{13} = \frac{5}{m}$ $\quad \frac{5}{6} = \frac{2}{r}$ $\quad \frac{4}{7} = \frac{n}{15}$ $\quad \frac{5}{24} = \frac{3}{s}$

3. $4:t = 9:10$ $\qquad 2:15 = v:3$ $\qquad 2:25 = 4:k$

4. $5:16 = h:20$ $\qquad d:9 = 3:8$ $\qquad 9:v = 7:12$

Check your answers on page 267.

Proportion Word Problems

You can use a proportion to solve many word problems. Remember to use the word labels to help you set up the right order of the proportion.

EXAMPLE: The ratio of men to women in the Surf City Disco is 1:2. 84 women are in the disco. How many men are there?

Step 1	Step 2	Step 3
$\frac{\text{men}}{\text{women}} = \frac{1}{2}$	$\frac{1}{2} = \frac{x}{84}$ $1 \times 84 = 84$ $2 \times x = 2x$	$2x = 84$ $\frac{2x}{2} = \frac{84}{2}$ $x = 42$ men

Step 1: Find the order of the words in the problem. The problem asks you to compare the number of men to the number of women. There is one man for every two women.

Step 2: Set up the proportion to solve for x. Multiply to find the cross products.

Step 3: Write a new equation with the cross products. Then solve the equation for x.

EXERCISE 12

DIRECTIONS: Solve each problem with a proportion.

1. The ratio of union members to non-union members at Tom's factory is 8:3. 96 workers at the factory do not belong to the union. How many union members work at the factory?

2. Ellen makes $23.80 for working 7 hours. How much does she make for working 12 hours?

3. The weight of 15 feet of aluminum tubing is 12 pounds. Find the weight of 25 feet of tubing.

4. Jerry can walk 9 miles in $2\frac{1}{4}$ hours. At the same speed, how far can he walk in 5 hours?

5. The scale on a map is 2 inches = 65 miles. The distance between two cities on the map is 6 inches. Find the distance between the cities in miles.

6. The ratio of the width to the length of a photograph is 5:7. The picture was enlarged to be 20 inches wide. How long is the enlarged picture?

7. In 1978, the U.S. divorce rate was 5 for every 1,000 people in the country. The population of Totowa was 36,000 in 1978. The divorce rate in Totowa was the same as the rate in the whole country. How many divorces were there in Totowa that year?

8. The price of 2 ounces of gold is $756. Find the price of 5 ounces of gold.

9. Pete mixed green paint and white paint in a ratio of 4:3. He used 12 gallons of white paint. How many gallons of green paint did he use?

10. From a 15-acre field, Mr. Douglas can get 1,350 bushels of corn. How much corn can he get from a 40-acre field?

Check your answers on page 268.

RECTANGULAR COORDINATES

In the basic math part of this book, you worked with graphs. There is another type of graph that you should know about. This type of graph is used in algebra. It is called the **rectangular coordinate** system.

A rectangular coordinate is made up of two number lines. One number line is horizontal (across). It is called the ***x*-axis.** Numbers at the right of zero are positive. Numbers at the left are negative. The other number line is vertical (up and down). It is called the ***y*-axis.** Numbers above zero are positive. Numbers below zero are negative. The two lines **intersect** at the zero points for both number lines. "Intersect" means to cross. The point where the two lines intersect is the **origin.**

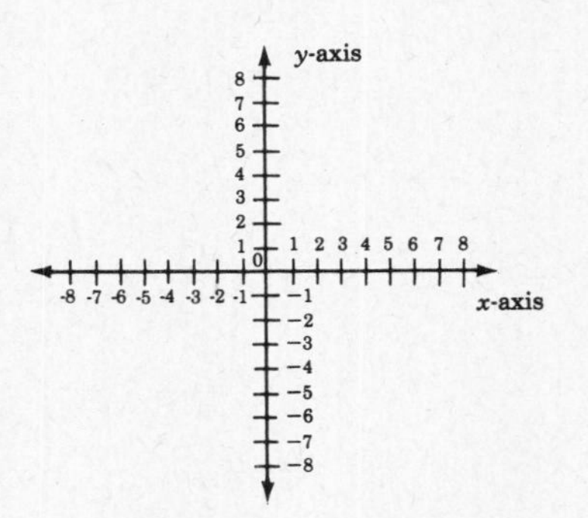

Two numbers called **coordinates** tell the exact position of any point on the system. The first number tells the distance of the point to the right or left of the y-axis. This number is called the **x-value.** A positive number means that the point is to the right of the y-axis. A negative number means that the point is to the left. The second number tells the distance of the point above or below the x-axis. This number is called the **y-value.** A positive number means that the point is above the x-axis. A negative number means that the point is below. The coordinates are usually written inside parentheses. A comma (,) separates the two numbers.

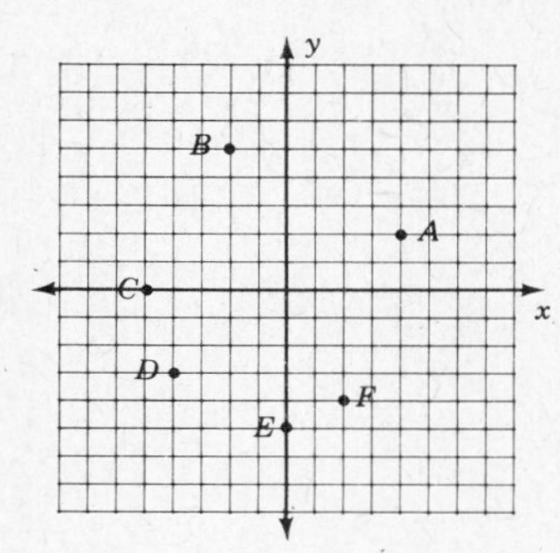

EXAMPLES: Write coordinates for each lettered point on the rectangular coordinate system at the left.

Point A is 4 spaces to the right of the y-axis and 2 spaces above the x-axis. The coordinates are $(+4, +2)$.

Point B is 2 spaces to the left of the y-axis and 5 spaces above the x-axis. Its coordinates are $(-2, +5)$.

Point C is 5 spaces to the left of the y-axis and neither above nor below the x-axis. Its coordinates are $(-5, 0)$.

Point D is 4 spaces left of the y-axis and 3 spaces below the x-axis. Its coordinates are $(-4, -3)$.

Point E is neither left nor right of the y-axis and 5 spaces below the x-axis. Its coordinates are $(0, -5)$.

Point F is 2 spaces right of the y-axis and 4 spaces below the x-axis. Its coordinates are $(+2, -4)$.

EXERCISE 13

DIRECTIONS: Write coordinates for each lettered point on the graph below.

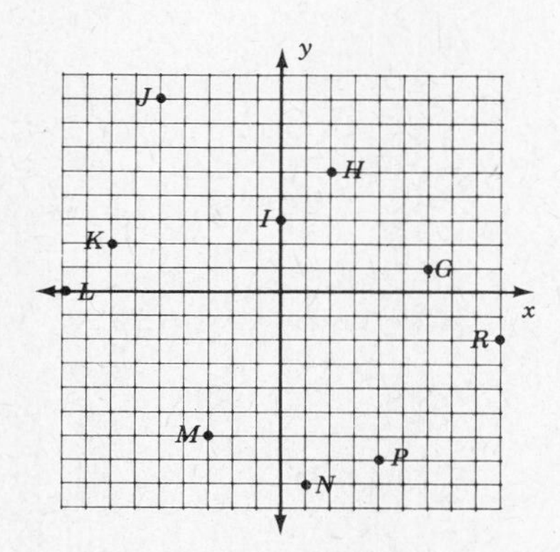

G = L =

H = M =

I = N =

J = P =

K = R =

Check your answers on page 268.

Graphing Equations

An equation can have more than one solution. Look at the equation $y = x - 2$. Try substituting different values of x into the equation. Every value of x gives a different value for y. For example, when $x = 5$, $y = 5 - 2$ (3). When $x = 2$, $y = 2 - 2$ (0). When $x = -3$, $y = -3 - 2$ (-5). These three solutions give us three pairs of coordinates: (5,3) (2,0) and ($-3,-5$). Remember, the first number, or x-value, in each pair tells the distance right or left of the y-axis. The second number, or y-value, tells the distance above or below the x-axis. The three points are shown at the left. When you connect the points, you get a straight line. The straight line is called the **graph** of the equation. Every point on the line is a solution to the equation $y = x - 2$. For example, the line goes through the point $(0, -2)$. Check to see if the values $x = 0$ and $y = -2$ give a solution to the equation. Substitute 0 for x. The value of y should be -2. $y = 0 - 2$ (-2). Any equation that makes a straight line graph is called a **linear equation.**

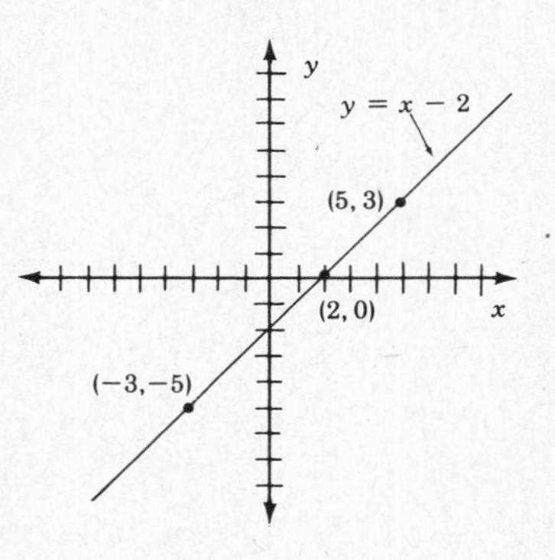

On the GED Test, you will not have to draw graphs. But you may have to answer questions about the graphs of equations. Study the next examples carefully.

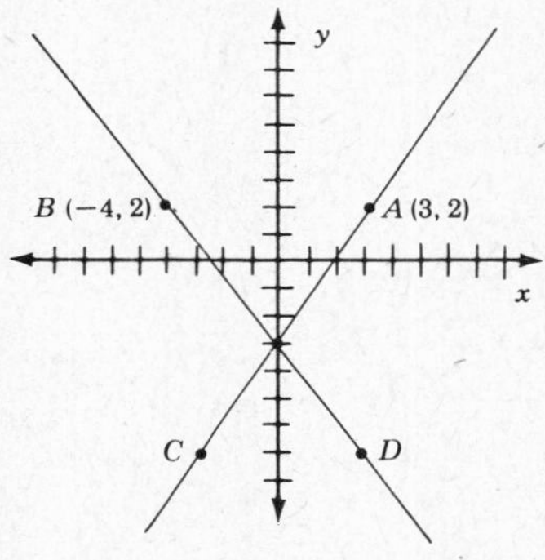

EXAMPLE: What is the distance from point A to point B on the graph?

Point A is 3 spaces to the right of the y-axis. Point B is 4 spaces to the left. The total distance between them is $3 + 4 = 7$.

EXAMPLE: At what point does line AC intersect line BD?

"Intersect" means to cross. Line AC crosses line BD on the y-axis 3 spaces below the origin. The coordinates of this point are $(0, -3)$.

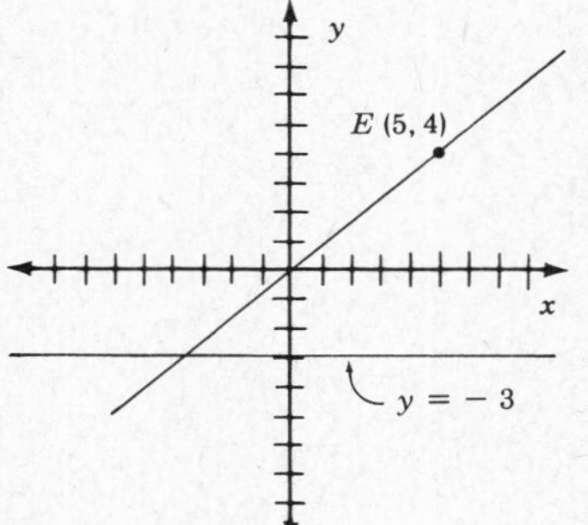

EXAMPLE: What is the perpendicular distance from point E to the graph of the equation $y = -3$?

The **perpendicular** distance means a direct drop from point E down to the graph of the equation $y = -3$. Point E is 4 spaces above the x-axis. The graph $y = -3$ is 3 spaces below the axis. The total distance from point E to the graph $y = -3$ is $4 + 3 = 7$.

Notice how the distance is measured in the examples. The distance between two points or between a point and a line is always a positive number. Distance is simply a total number of spaces.

EXERCISE 14

DIRECTIONS: Answer these questions about the graphs.

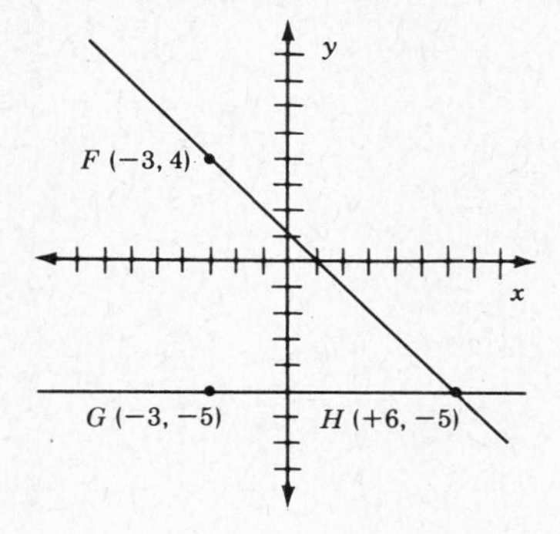

1. What is the distance from F to G?
2. What is the distance from G to H?
3. What are the coordinates of the point where line FH intersects the y-axis?

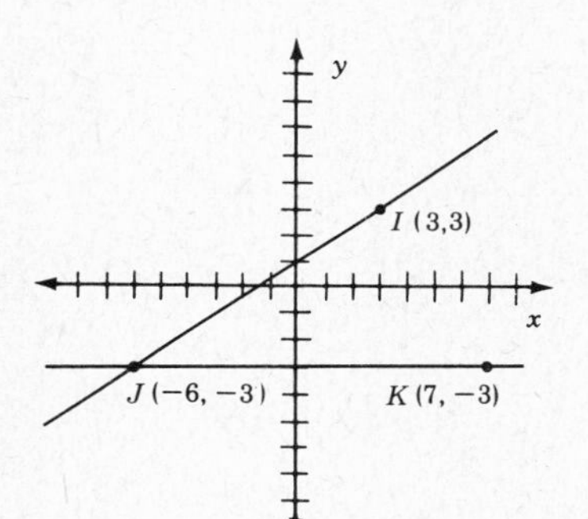

4. What is the distance from J to K?
5. What is the perpendicular distance from I to line JK?
6. Line IJ is the graph of the equation $y = \frac{2}{3}x + 1$. Is the point (6,5) on the graph of this equation?

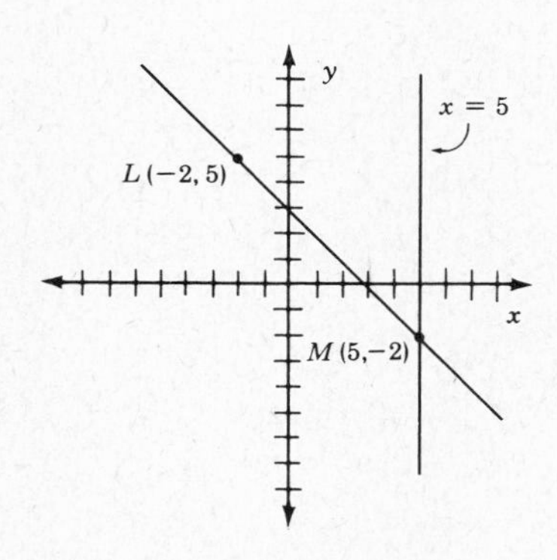

7. What is the perpendicular distance from L to the graph of the equation $x = 5$?
8. What are the coordinates of the point where line LM intersects the y-axis?
9. Line LM is the graph of the equation $y = -x + 3$. Is the point (5,8) on the graph of this equation?

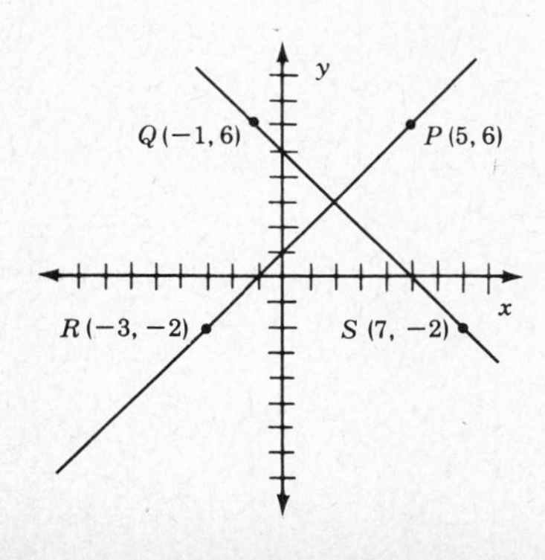

10. What is the distance from P to Q?
11. What is the distance from R to S?
12. Line PR is the graph of the equation $y = x + 1$. Is the point (2,3) on the graph of this equation?

Check your answers on page 268.

There are two other questions about graphs that appear often on the GED Test. The first type of question will ask you to find the **slope** of a graph. The second type will ask you to figure out which equation a graph stands for.

The **slope** of a graph refers to the way that the line goes up or down. For example, look at this graph:

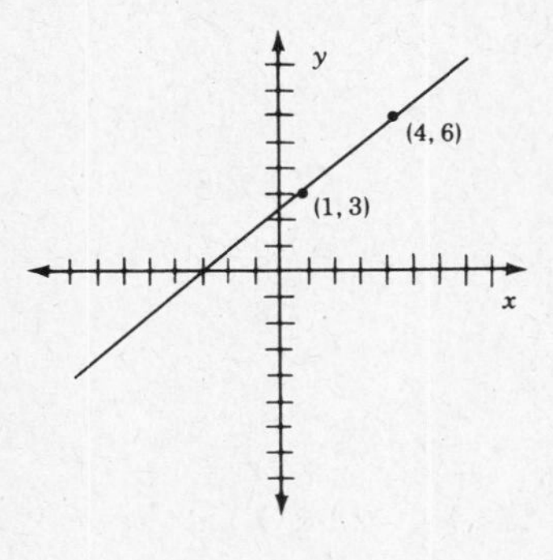

The graph is going up from left to right. When the graph is going up, it has a *positive* slope. A positive slope is always a positive number. When the graph is going down, it has a *negative* slope. A negative slope is always a negative number. A line that runs straight across has a *zero* slope (0).

On the last graph, two of the coordinates are (1,3) and (4,6). Here is the formula that you should use to find the slope:

$$s = \frac{y_2 - y_1}{x_2 - x_1}$$

The formula means that you subtract the y-value in the first coordinate from the y-value in the second coordinate. Then you subtract the x-value in the first coordinate from the x-value in the second coordinate. Finally, you divide the bottom value into the top value to get the slope.

Here is how to get the slope for the coordinates (1,3) and (4,6).

$$s = \frac{y_2 - y_1}{x_2 - x_1} = \frac{6 - 3}{4 - 1} = \frac{3}{3} = \mathbf{1}$$

The slope of the graph is 1.

When you figure out the slope of a graph, you do not have to have a picture of the graph to solve the problem. All you need are two of the graph's coordinates.

EXERCISE 15

DIRECTIONS: Use the formula for slope to answer the next questions.

1. What is the slope of a graph that has the coordinates (3,2) and (5,6)?

2. What is the slope of a graph that has the coordinates (5,3) and (2,1)?

3. What is the slope of a graph that has the coordinates $(-2,-2)$ and $(2,2)$?

4. What is the slope of a graph that has the coordinates $(0,1)$ and $(-4,6)$?

5. What is the slope of a graph that has the coordinates $(4,0)$ and $(4,8)$?

Check your answers on page 269.

You also may be asked to figure out which line represents the graph of a particular equation. An easy way to do this is to choose three values for one of the unknown numbers. Then substitute each value in the equation to get the value of the other unknown number. Study the next example carefully.

EXAMPLE: Which line represents the equation $y = 3x - 2$?

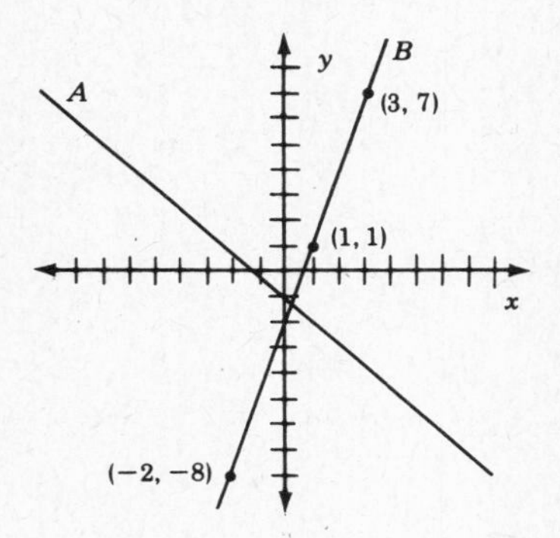

Step 1: Choose three values for x. (In this example, the chosen values are $x = 3$, $x = 1$, and $x = -2$. You could also choose other values for x.)

Step 2: Use the first chosen value. Find the value of y when $x = 3$. Substitute 3 for x in the equation: $y = 3(3) - 2 = 9 - 2 = 7$. When x is 3, y is 7. The coordinates for this point are $(3,7)$.

Step 3: Now, use the second value. Substitute 1 for x in the equation: $y = 3(1) - 2 = 1$. The coordinates for this point are $(1,1)$.

Step 4: Substitute the third value for x (-2) in the equation: $y = 3(-2) - 2 = -8$. The coordinates for this point are $(-2,-8)$.

Step 5: Look at the three sets of coordinates. Which line on the graph runs through all three points? Line B runs through the points. Line B is the line that represents the graph of the equation $y = 3x - 2$.

EXERCISE 16

DIRECTIONS: Use the equation $y = 2x - 3$ to answer questions 1 through 5.

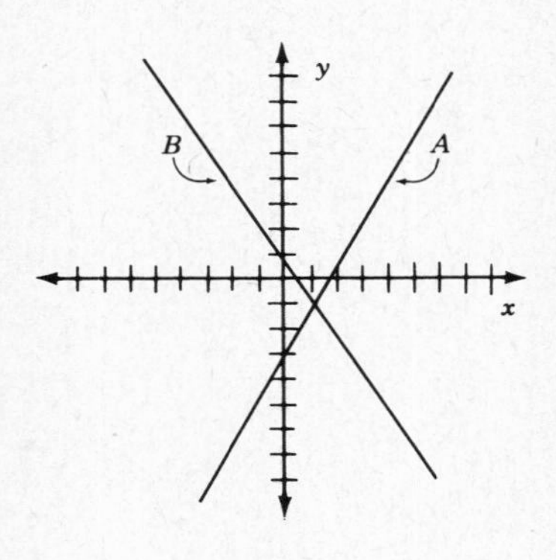

1. When $x = 5$, what is the value of y?
2. When $x = 1$, what is the value of y?
3. When $x = -2$, what is the value of y?
4. Write the coordinates for each point in problems 1 through 3.
5. Which line, A or B, represents the equation $y = 2x - 3$?

Use the equation $y = -\frac{1}{2}x + 4$ to answer questions 6 through 10.

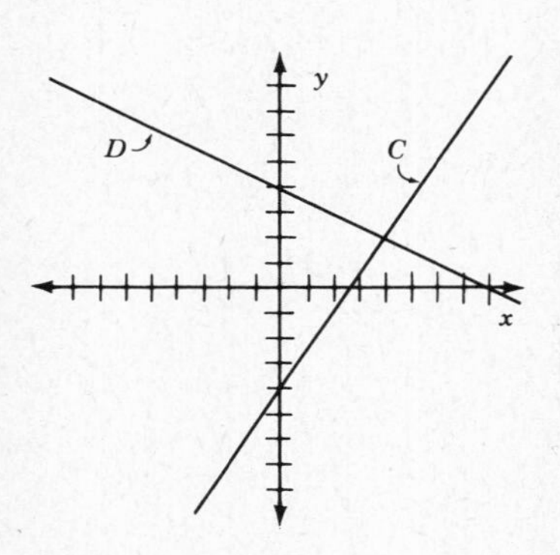

6. When $x = 8$, what is the value of y?
7. When $x = 2$, what is the value of y?
8. When $x = -4$, what is the value of y?
9. Write the coordinates for each point in problems 6 through 8.
10. Which line, C or D, represents the equation $y = -\frac{1}{2}x + 4$?

Use the equation $F = \frac{9}{5}C + 32$ to answer questions 11 through 15. This is the equation for changing centigrade temperature to Fahrenheit temperature.

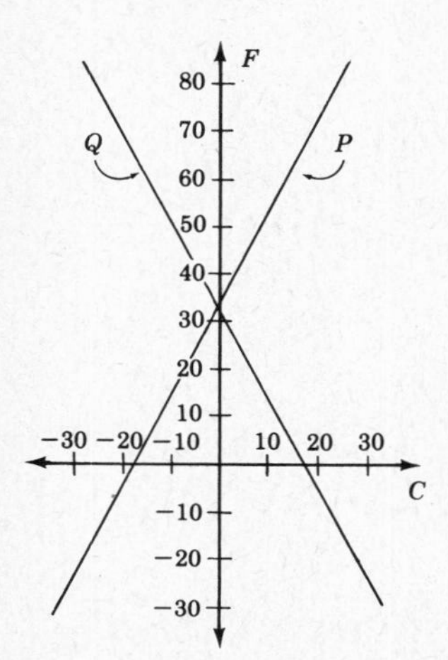

11. When $C = 20$, what is the value of F?
12. When $C = 10$, what is the value of F?
13. When $C = -20$, what is the value of F?
14. Write the coordinates for each point in problems 11 through 13.
15. Which line, P or Q, represents the equation $F = \frac{9}{5}C + 32$?

Check your answers on page 269.

Algebra Word Problems

In the next exercise, you will have a chance to work with the types of algebra word problems that you may find on the GED Test. Work each problem slowly and carefully. Use the hints and other information given in this section to solve the problems.

EXERCISE 17 (Word Problems)

DIRECTIONS: Choose the correct answer to each problem.

1. Last Wednesday, 90 planes landed at Teterboro Airport. The number of planes that landed at the airport last Wednesday was twice the number of planes that landed there last Tuesday. How many planes landed at Teterboro Airport last Tuesday?

   (1) 30 (2) 45 (3) 60 (4) 90 (5) 180

2. Phil works in the city five days a week. He drives to work each day. Each day, he pays a 75¢ toll and a $5 parking fee. Which of the following expressions show the amount of money that Phil spends on tolls and parking fees in a week?

   (1) $(75 + 5) \times 5$ (2) $(5 \times 5) + 75$ (3) $(.75 \times 5) + 5$
   (4) $5(.75 + 5)$ (5) $5(5 - .75)$

3. In a recent survey, 4 out of every 5 people said that they chewed sugarless gum. If a total of 80 people were surveyed, how many people said that they chewed sugarless gum?

   (1) 20 (2) 32 (3) 64 (4) 100 (5) 320

4. One morning in February, the temperature in Portland, Maine, was $-12°$F. At the same time, the temperature in Atlanta, Georgia, was 46°F. How many degrees colder was it in Portland than in Atlanta?

   (1) 34 (2) 46 (3) 58 (4) 60 (5) 72

5. If $x^2 = 2{,}500$, then $x =$

   (1) 25 only (2) 50 only (3) 25 or -25
   (4) 50 or -50 (5) 250 or -250

6. The formula for finding interest is $I = PRT$. If Paula paid $60 in interest on a 1-year loan at 8%, how much money did she borrow?

   (1) $480 (2) $750 (3) $960
   (4) $1,240 (5) $1,500

7. The number of employees at the Burnco Match Company is 3 times plus 70 higher than the number of employees at the Flintlock Lighter Company. Which of the following shows the difference between the numbers of employees at the two companies?

   (1) $(3x + 70) - x$ (2) $(3x + 70) + x$ (3) $(x^3 + 70) - x$
   (4) $(x^3 + 70) + x$ (5) $3(x + 70) - x$

8. If $5a = 3(a + 12)$, then a

   (1) $4\frac{1}{2}$ (2) 8 (3) 12 (4) 15 (5) 18

9. If $4e - 2 \leq 14$, then $e \leq$

   (1) 14 (2) 10 (3) 8 (4) 6 (5) 4

10. The Alvarez family pays \$30 more than $\frac{1}{3}$ of its total monthly income for rent. If the family's rent is \$370 a month, how much is its total monthly income?

    (1) \$400 (2) \$460 (3) \$920
    (4) \$1,020 (5) \$1,120

11. Marsha Watts makes \$175 a week. Her husband Sid makes \$150 a week. Which of the following show the ratio of Marsha's weekly income to the total weekly income of the two people?

    (1) 7:6 (2) 6:7 (3) 7:9
    (4) 7:13 (5) 9:16

12. If $a + b = 48$, and $b = 3a$, then $b =$

    (1) $\sqrt{48 + a}$ (2) 12 (3) 16
    (4) 24 (5) 36

13. In 1978, the ratio of wins to losses for the Philadelphia Phillies was 5:4. The Phillies lost 72 games that year. How many games did they win?

    (1) 58 (2) 72 (3) 90 (4) 102 (5) 162

14. If $y = 4$, then $y^2 + 2y - 6 =$

    (1) 10 (2) 18 (3) 26 (4) 32 (5) 40

15. Which of the following is equal to $5(r - 12)$?

    (1) $(5 + r)(5 - 12)$ (2) $5 - (r - 12)$ (3) $5 \div (r - 12)$
    (4) $5r + 60$ (5) $5r - 60$

Check your answers on page 270.

Algebra Review

These problems will give you a chance to practice the skills you learned in the algebra section of this book. With each answer is the name of the section in which you learned the skills needed for that problem. For every problem you get wrong, look over the section in which the skills for that problem are explained. Then try the problem again.

DIRECTIONS: Solve each problem.

1. Find the value of $3^2 + 4^3 - 2^3$.

2. Find the value of $\sqrt{5,329}$.

3. What lettered point on the number line stands for $+2\frac{1}{2}$?

−4 −3 −2 −1 0 1 2 3 4

A B C D E

4. $15 - 9 - 23 + 6 =$

5. $(+29) - (-14) =$

6. $(+8)(-2)(-3) =$

7. $\frac{-121}{-11} =$

8. $m = .6$ and $n = .4$. Find the value of mn.

9. $s = 7$, $t = 4$, and $u = 6$. Find the value of $s(t + u)$.

10. $F = \frac{9}{5}C + 32$. Find F when $C = 15$.

11. Simplify the expression $4p + 7 - 12 - 3p$.

12. Simplify the expression $8x - 5(x + 3)$.

13. Solve the equation $12c = 78$.

14. Solve the equation $22 = 3y - 8$.

15. Solve the equation $9d - 5 = 3d + 49$.

16. Solve the equation $8(m - 6) + 3 = 27$.

17. Fred's gross pay is \$320 a week. Fred's boss takes out \$64 each week from Fred's pay for taxes and Social Security. What is the ratio of the weekly deductions to Fred's gross pay?

18. Solve the proportion $\frac{7}{12} = \frac{8}{x}$.

19. A man 6 feet tall casts a shadow 5 feet long. At the same time, a building casts a shadow 75 feet long. How tall is the building?

20. Let p stand for Pete's age. Nora is 3 years more than twice as old as Pete. Write an expression for Nora's age.

21. Nine times a number decreased by 7 equals 101. Find the number.

22. George makes \$140 a month more than his wife Marge. Together, they make \$1,380 a month. How much does each of them make in a month?

Use the graph at the right to answer questions 23 and 24.

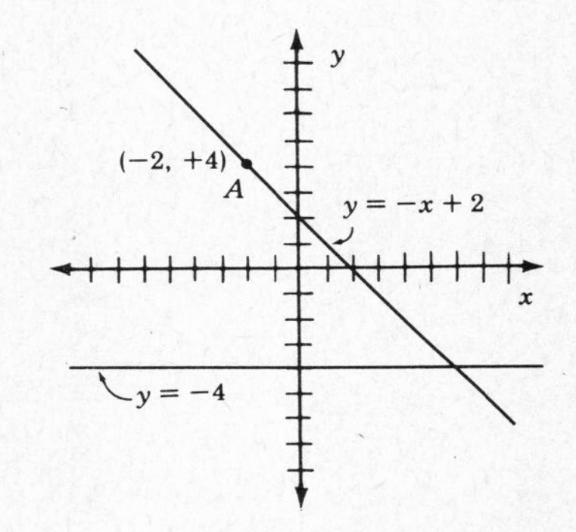

23. What is the perpendicular distance from point A to the line $y = -4$?

24. What are the coordinates of the point at which the graph $y = -x + 2$ crosses the x-axis?

25. Is $a = 5$ in the solution set of the inequality $a > 5$?

26. Find the solution set of the inequality $7m - 2 \leq 3m + 18$.

Check your answers on page 271.

ANSWERS & SOLUTIONS

EXERCISE 1

1. **(2) $\frac{1}{2}$(300)** 2. **(3) (68 + 72 + 80) ÷ 3** 3. **(3) 31 − 19**

4. **(1) \$1,300 − \$700** 5. **(3) (17 − 3) ÷ 2**

EXERCISE 2

1. $x + 23$ 2. $x - 15$ 3. $\sqrt{x} + 12$
4. $3x + 10$ 5. $\frac{x}{11}$ 6. $\frac{1}{2}x + \frac{1}{3}x$
7. $\frac{x^2}{3}$ 8. $x - \frac{2}{3}x$ 9. $\frac{30}{x}$
10. $100 - \frac{1}{2}x$

EXERCISE 3

1. $x - 10$ 2. $p + 400$ 3. $b - 2$
4. $\frac{c}{3}$ 5. $w - 15$ 6. $1\frac{1}{2}h$ or $\frac{3}{2}h$
7. $2f$ 8. $2k - 1$ 9. $3(y + 4)$
10. $\frac{e}{12}$

EXERCISE 4

1. **70**

$$f - 27 = 43$$
$$+27 \quad +27$$
$$f = \mathbf{70}$$

48

$$4 \cdot \frac{k}{4} = 12 \cdot 4$$
$$k = \mathbf{48}$$

6

$$a + 17 = 23$$
$$-17 \quad -17$$
$$a = \mathbf{6}$$

2. **7**

$$\frac{9s}{9} = \frac{63}{9}$$
$$s = \mathbf{7}$$

4 or −4

$$30 = b^2 + 14$$
$$-14 \quad -14$$
$$16 = b^2$$
$$\sqrt{16} = \sqrt{b^2}$$
$$\mathbf{4 \text{ or } -4 = b}$$

50

$$10 \cdot 5 = \frac{m}{10} \cdot 10$$
$$\mathbf{50} = m$$

3. **5 or −5**

$$\frac{100}{4} = \frac{4u^2}{4}$$
$$25 = u^2$$
$$\mathbf{5 \text{ or } -5 = u}$$

6

$$c + 3\frac{1}{2} = 9\frac{1}{2}$$
$$-3\frac{1}{2} \quad -3\frac{1}{2}$$
$$c = \mathbf{6}$$

8 or −8

$$g^2 - 31 = 33$$
$$+31 \quad +31$$
$$g^2 = 64$$
$$g = \mathbf{8 \text{ or } -8}$$

4. **128**

$$8 \cdot \frac{n}{8} = 16 \cdot 8$$
$$n = 128$$

30

$$18 = h - 12$$
$$+12 \quad +12$$
$$\mathbf{30} = h$$

10

$$\frac{120}{12} = \frac{12w}{12}$$
$$\mathbf{10} = w$$

5. **72** | $\mathbf{\frac{2}{3}}$ | **66**

$$\begin{array}{rcl} 96 &=& d + 24 \\ -24 & & \quad -24 \\ \hline \mathbf{72} &=& d \end{array} \qquad \begin{array}{c} \frac{15y}{15} = \frac{10}{15} \\[6pt] y = \mathbf{\frac{2}{3}} \end{array} \qquad \begin{array}{c} 11 \cdot 6 = \frac{p}{11} \cdot 11 \\[6pt] \mathbf{66} = p \end{array}$$

6. **80** | **2** or **−2** | $\mathbf{\frac{3}{2}}$ or $\mathbf{1\frac{1}{2}}$

$$\begin{array}{c} \frac{.05z}{.05} = \frac{4}{.05} \\[6pt] z = \mathbf{80} \end{array} \qquad \begin{array}{rcl} i^2 - 1\frac{1}{4} &=& 2\frac{3}{4} \\ +\,1\frac{1}{4} & & +1\frac{1}{4} \\ \hline i^2 &=& 4 \\ i &=& \mathbf{2} \text{ or } \mathbf{-2} \end{array} \qquad \begin{array}{c} \frac{8t}{8} = \frac{12}{8} \\[6pt] t = \mathbf{\frac{3}{2}} \text{ or } \mathbf{1\frac{1}{2}} \end{array}$$

7. **4** | **90** | **45**

$$\begin{array}{rcl} e + 2.5 &=& 6.5 \\ -\,2.5 & & -2.5 \\ \hline e &=& \mathbf{4} \end{array} \qquad \begin{array}{c} 30 \cdot \frac{r}{30} = 3 \cdot 30 \\[6pt] r = \mathbf{90} \end{array} \qquad \begin{array}{rcl} 26 &=& j - 19 \\ +19 & & \quad +19 \\ \hline \mathbf{45} &=& j \end{array}$$

EXERCISE 5

1. **7** | **8** | **200**

$$\begin{array}{rcl} 4w + 6 &=& 34 \\ -\,6 & & -\,6 \\ \hline \frac{4w}{4} &=& \frac{28}{4} \\[6pt] w &=& \mathbf{7} \end{array} \qquad \begin{array}{rcl} 6z - 7 &=& 41 \\ +\,7 & & +\,7 \\ \hline \frac{6z}{6} &=& \frac{48}{6} \\[6pt] z &=& \mathbf{8} \end{array} \qquad \begin{array}{rcl} \frac{x}{5} - 4 &=& 36 \\ +\,4 & & +\,4 \\ \hline 5 \cdot \frac{x}{5} &=& 40 \cdot 5 \\[6pt] x &=& \mathbf{200} \end{array}$$

2. **3** | **2** | **400**

$$\begin{array}{rcl} 28 &=& 12y - 8 \\ +\,8 & & \quad +\,8 \\ \hline \frac{36}{12} &=& \frac{12y}{12} \\[6pt] \mathbf{3} &=& y \end{array} \qquad \begin{array}{rcl} 51 &=& 20u + 11 \\ -11 & & \quad -\,11 \\ \hline \frac{40}{20} &=& \frac{20u}{20} \\[6pt] \mathbf{2} &=& u \end{array} \qquad \begin{array}{rcl} 43 &=& \frac{v}{10} + 3 \\ -\,3 & & \quad -\,3 \\ \hline 10 \cdot 40 &=& \frac{v}{10} \cdot 10 \\[6pt] \mathbf{400} &=& v \end{array}$$

3. **60** | **6** or **−6** | **15**

$$\begin{array}{rcl} \frac{s}{4} - 13 &=& 2 \\ +\,13 & & +13 \\ \hline 4 \cdot \frac{s}{4} &=& 15 \cdot 4 \\[6pt] s &=& \mathbf{60} \end{array} \qquad \begin{array}{rcl} \frac{2}{3}r^2 + 1 &=& 25 \\ -\,1 & & -\,1 \\ \hline \frac{3}{2} \cdot \frac{2}{3}r^2 &=& 24 \cdot \frac{3}{2} \\[6pt] r^2 &=& 36 \\ r &=& \mathbf{6} \text{ or } \mathbf{-6} \end{array} \qquad \begin{array}{rcl} 5q - 12 &=& 63 \\ +\,12 & & +12 \\ \hline \frac{5q}{5} &=& \frac{75}{5} \\[6pt] q &=& \mathbf{15} \end{array}$$

4. **24**

$$\begin{aligned} 21 &= \tfrac{1}{2}p + 9 \\ -9 & \quad\; -9 \\ 2 \cdot 12 &= \tfrac{1}{2}p \cdot 2 \\ \mathbf{24} &= p \end{aligned}$$

14

$$\begin{aligned} 64 &= 4n + 8 \\ -8 & \quad\; -8 \\ \tfrac{56}{4} &= \tfrac{4n}{4} \\ \mathbf{14} &= n \end{aligned}$$

45

$$\begin{aligned} 11 &= \tfrac{m}{9} + 6 \\ -6 & \quad\; -6 \\ 9 \cdot 5 &= \tfrac{m}{9} \cdot 9 \\ \mathbf{45} &= m \end{aligned}$$

5. **18**

$$\begin{aligned} \tfrac{4}{3}j - 2 &= 22 \\ +2 & \quad +2 \\ \tfrac{3}{4} \cdot \tfrac{4}{3}j &= 24 \cdot \tfrac{3}{4} \\ j &= \mathbf{18} \end{aligned}$$

56

$$\begin{aligned} \tfrac{i}{7} - 2 &= 6 \\ +2 & \quad +2 \\ 7 \cdot \tfrac{i}{7} &= 8 \cdot 7 \\ i &= \mathbf{56} \end{aligned}$$

3 or −3

$$\begin{aligned} 8h^2 + 3 &= 75 \\ -3 & \quad -3 \\ \tfrac{8h^2}{8} &= \tfrac{72}{8} \\ h^2 &= 9 \\ h &= \mathbf{3 \text{ or } -3} \end{aligned}$$

6. **10**

$$\begin{aligned} 6f + 19 &= 79 \\ -19 & \quad -19 \\ \tfrac{6f}{6} &= \tfrac{60}{6} \\ f &= \mathbf{10} \end{aligned}$$

3

$$\begin{aligned} 9e - 34 &= -7 \\ +34 & \quad +34 \\ \tfrac{9e}{9} &= \tfrac{+27}{9} \\ e &= \mathbf{3} \end{aligned}$$

25

$$\begin{aligned} \tfrac{4}{5}d - 4 &= 16 \\ +4 & \quad +4 \\ \tfrac{5}{4} \cdot \tfrac{4}{5}d &= 20 \cdot \tfrac{5}{4} \\ d &= \mathbf{25} \end{aligned}$$

7. **48**

$$\begin{aligned} \tfrac{a}{12} + 9 &= 13 \\ -9 & \quad -9 \\ 12 \cdot \tfrac{a}{12} &= 4 \cdot 12 \\ a &= \mathbf{48} \end{aligned}$$

12

$$\begin{aligned} 90 &= 8b - 6 \\ +6 & \quad\; +6 \\ \tfrac{96}{8} &= \tfrac{8b}{8} \\ \mathbf{12} &= b \end{aligned}$$

13

$$\begin{aligned} 6c + 22 &= 100 \\ -22 & \quad -22 \\ \tfrac{6c}{6} &= \tfrac{78}{6} \\ c &= \mathbf{13} \end{aligned}$$

EXERCISE 6

1. **8**

$$\begin{aligned} 7m - 3m &= 32 \\ \tfrac{4m}{4} &= \tfrac{32}{4} \\ m &= \mathbf{8} \end{aligned}$$

5

$$\begin{aligned} 40 &= 2p + 6p \\ \tfrac{40}{8} &= \tfrac{8p}{8} \\ \mathbf{5} &= p \end{aligned}$$

2. **3**

$$\begin{aligned} 9r + 6 + 2r &= 39 \\ 11r + 6 &= 39 \\ -6 & \quad -6 \\ \tfrac{11r}{11} &= \tfrac{33}{11} \\ r &= \mathbf{3} \end{aligned}$$

8

$$\begin{aligned} 8s - 5s - 4 &= 20 \\ 3s - 4 &= 20 \\ +4 & \quad +4 \\ \tfrac{3s}{3} &= \tfrac{24}{3} \\ s &= \mathbf{8} \end{aligned}$$

3. **4** **4**

$$\begin{array}{rcl} 10t &=& 28 + 3t \\ -\ 3t & & -\ 3t \\ \hline \frac{7t}{7} &=& \frac{28}{7} \\ t &=& \mathbf{4} \end{array} \qquad \begin{array}{rcl} 36 - 4u &=& 5u \\ +\ 4u & & +4u \\ \hline \frac{36}{9} &=& \frac{9u}{9} \\ \mathbf{4} &=& u \end{array}$$

4. **5** **4**

$$\begin{array}{rcl} 11v &=& 60 - v \\ +\ v & & +\ v \\ \hline \frac{12v}{12} &=& \frac{60}{12} \\ v &=& \mathbf{5} \end{array} \qquad \begin{array}{rcl} 20 + 2w &=& 7w \\ -\ 2w & & -2w \\ \hline \frac{20}{5} &=& \frac{5w}{5} \\ \mathbf{4} &=& w \end{array}$$

5. **3** **2**

$$\begin{array}{rcl} 12x - 4 &=& 26 + 2x \\ -\ 2x & & -\ 2x \\ \hline 10x - 4 &=& 26 \\ +\ 4 & & +\ 4 \\ \hline \frac{10x}{10} &=& \frac{30}{10} \\ x &=& \mathbf{3} \end{array} \qquad \begin{array}{rcl} 9y + 3 &=& 29 - 4y \\ +4y & & +\ 4y \\ \hline 13y + 3 &=& 29 \\ -\ 3 & & -\ 3 \\ \hline \frac{13y}{13} &=& \frac{26}{13} \\ y &=& \mathbf{2} \end{array}$$

6. **16** **7**

$$\begin{array}{rcl} 9a &=& 4(a + 20) \\ 9a &=& 4a + 80 \\ -4a & & -4a \\ \hline \frac{5a}{5} &=& \frac{80}{5} \\ a &=& \mathbf{16} \end{array} \qquad \begin{array}{rcl} 6c &=& 2(c + 14) \\ 6c &=& 2c + 28 \\ -2c & & -2c \\ \hline \frac{4c}{4} &=& \frac{28}{4} \\ c &=& \mathbf{7} \end{array}$$

7. **12** **9**

$$\begin{array}{rcl} 8(d + 3) &=& 10d \\ 8d + 24 &=& 10d \\ -8d & & -\ 8d \\ \hline \frac{24}{2} &=& \frac{2d}{2} \\ \mathbf{12} &=& d \end{array} \qquad \begin{array}{rcl} 3(f + 6) &=& 5f \\ 3f + 18 &=& 5f \\ -3f & & -3f \\ \hline \frac{18}{2} &=& \frac{2f}{2} \\ \mathbf{9} &=& f \end{array}$$

8. **7** **8**

$$\begin{array}{rcl} 4(h - 2) &=& h + 13 \\ 4h - 8 &=& h + 13 \\ -\ h & & -h \\ \hline 3h - 8 &=& 13 \\ +\ 8 & & +\ 8 \\ \hline \frac{3h}{3} &=& \frac{21}{3} \\ h &=& \mathbf{7} \end{array} \qquad \begin{array}{rcl} 7m - 1 &=& 5(m + 3) \\ 7m - 1 &=& 5m + 15 \\ -5m & & -5m \\ \hline 2m - 1 &=& 15 \\ +\ 1 & & +\ 1 \\ \hline \frac{2m}{2} &=& \frac{16}{2} \\ m &=& \mathbf{8} \end{array}$$

9. **10**

$$\begin{aligned} x + y &= 15 \\ x + 2x &= 15 \\ \frac{3x}{3} &= \frac{15}{3} \\ x &= 5 \\ y(\text{or } 2x) &= \mathbf{10} \end{aligned}$$

10. **(5) −3 or −4**

$$\begin{aligned} t^2 + 7t &= -12 \\ (-3)^2 + 7(-3) &= -12 \\ 9 + (-21) &= -12 \\ -12 &= -12 \end{aligned}$$

$$\begin{aligned} t^2 + 7t &= -12 \\ (-4)^2 + 7(-4) &= -12 \\ 16 + (-28) &= -12 \\ -12 &= -12 \end{aligned}$$

EXERCISE 7

1. **10**

$$\begin{array}{rcl} m - 7 & \leq & 3 \\ \underline{+\ 7} & & \underline{+7} \\ m & \leq & \mathbf{10} \end{array}$$

2. $\mathbf{3\frac{1}{3}}$

$$\begin{aligned} 9a &> 30 \\ \frac{9a}{9} &> \frac{30}{9} \\ a &> 3\tfrac{3}{9} \\ a &> \mathbf{3\tfrac{1}{3}} \end{aligned}$$

3. **−12**

$$\begin{aligned} \frac{c}{4} &\geq -3 \\ 4 \cdot \frac{c}{4} &\geq -3 \cdot 4 \\ c &\geq \mathbf{-12} \end{aligned}$$

4. **−14**

$$\begin{array}{rcl} d + 8 & < & -6 \\ \underline{-\ 8} & & \underline{-8} \\ d & < & \mathbf{-14} \end{array}$$

5. **4**

$$\begin{array}{rcl} 2e - 9 & \geq & -1 \\ \underline{+\ 9} & & \underline{+9} \\ \frac{2e}{2} & \geq & \frac{+8}{2} \\ e & \geq & \mathbf{4} \end{array}$$

6. **6**

$$\begin{array}{rcl} 5n + 4 & < & 3n + 16 \\ \underline{-3n\qquad} & & \underline{-3n\qquad} \\ 2n + 4 & < & +\ 16 \\ \underline{-\ 4} & & \underline{-\ \ 4} \\ \frac{2n}{2} & < & \frac{12}{2} \\ n & < & \mathbf{6} \end{array}$$

7. **5**

$$\begin{array}{rcl} 8w - 7 & > & 2w + 23 \\ \underline{-2w\qquad} & & \underline{-2w\qquad} \\ 6w - 7 & > & +\ 23 \\ \underline{+\ 7} & & \underline{+\ \ 7} \\ \frac{6w}{6} & > & \frac{30}{6} \\ w & > & \mathbf{5} \end{array}$$

8. **−7**

$$\begin{array}{rcl} 9z + 1 & \leq & 6z - 20 \\ \underline{-6z\qquad} & & \underline{-6z\qquad} \\ 3z + 1 & \leq & -\ 20 \\ \underline{-\ 1} & & \underline{-\ \ 1} \\ \frac{3z}{3} & \leq & -\frac{21}{3} \\ z & \leq & \mathbf{-7} \end{array}$$

9. **10**

$$\begin{array}{rcl} \frac{1}{2}p - 7 & \geq & -2 \\ \underline{+\ 7} & & \underline{+7} \\ 2 \cdot \frac{1}{2}p & \geq & +5 \cdot 2 \\ p & \geq & \mathbf{10} \end{array}$$

10. **16**

$$\begin{array}{rcl} 4u - 11 & < & 3u + 5 \\ \underline{-3u\qquad} & & \underline{-3u\qquad} \\ u - 11 & < & +\ 5 \\ \underline{+\ 11} & & \underline{+\ 11} \\ u & < & \mathbf{16} \end{array}$$

EXERCISE 8

1. **5**

$$\begin{aligned} 7x - 6 &= 29 \\ +6 &\quad +6 \\ \tfrac{7x}{7} &= \tfrac{35}{7} \\ x &= \mathbf{5} \end{aligned}$$

2. **12**

$$\begin{aligned} \tfrac{x}{2} + 3 &= 9 \\ -3 &\quad -3 \\ 2 \cdot \tfrac{x}{2} &= 6 \cdot 2 \\ x &= \mathbf{12} \end{aligned}$$

3. **7**

$$\begin{aligned} 43 &= 5x + 8 \\ -8 &\quad -8 \\ \tfrac{35}{5} &= \tfrac{5x}{5} \\ \mathbf{7} &= x \end{aligned}$$

4. **16**

$$\begin{aligned} \tfrac{3}{4}x + 1 &= 13 \\ -1 &\quad -1 \\ \tfrac{4}{3} \cdot \tfrac{3}{4}x &= 12 \cdot \tfrac{4}{3} \\ x &= \mathbf{16} \end{aligned}$$

5. **9**

$$\begin{aligned} 70 &= 8x - 2 \\ +2 &\quad +2 \\ \tfrac{72}{8} &= \tfrac{8x}{8} \\ \mathbf{9} &= x \end{aligned}$$

6. **126**

$$\begin{aligned} \tfrac{x}{9} - 11 &= 3 \\ +11 &\quad +11 \\ 9 \cdot \tfrac{x}{9} &= 14 \cdot 9 \\ x &= \mathbf{126} \end{aligned}$$

7. **4**

$$\begin{aligned} 8x - 2x &= 24 \\ \tfrac{6x}{6} &= \tfrac{24}{6} \\ x &= \mathbf{4} \end{aligned}$$

8. **2**

$$\begin{aligned} 14 - 3x &= 4x \\ +3x &\quad +3x \\ \tfrac{14}{7} &= \tfrac{7x}{7} \\ \mathbf{2} &= x \end{aligned}$$

9. **5**

$$\begin{aligned} x + 9x &= 50 \\ \tfrac{10x}{10} &= \tfrac{50}{10} \\ x &= \mathbf{5} \end{aligned}$$

10. **6**

$$\begin{aligned} 5x &= 3x + 12 \\ -3x &\quad -3x \\ \tfrac{2x}{2} &= \quad + \tfrac{12}{2} \\ x &= \mathbf{6} \end{aligned}$$

EXERCISE 9

1. **56 women**

x = number of men

$2x$ = number of women

$x + 2x = 84$

$\frac{3x}{3} = \frac{84}{3}$

$x = 28$

$2x = 2(28) =$ **56 women**

2. **$4,000**

x = Bill's investment

$2x$ = Gordon's investment

$2x + 2{,}000$ = Jeff's investment

$x + 2x + 2x + 2{,}000 = 22{,}000$

$5x + 2{,}000 = 22{,}000$

$-2{,}000 \quad -2{,}000$

$\frac{5x}{5} = \frac{20{,}000}{5}$

$x =$ **$4,000**

3. **$700**

x = monthly salary

$\frac{1}{4}x + 20 = 195$

$-20 \quad -20$

$\frac{4}{1} \cdot \frac{1}{4}x = 175 \cdot \frac{4}{1}$

$x =$ **$700**

4. **Frank = 11 yrs.; Jack = 16 yrs.**

x = Frank's age now

$x + 5$ = Jack's age now

$x - 6$ = Frank's age 6 yrs. ago

$x + 5 - 6 = x - 1$ = Jack's age 6 yrs. ago

$x - 1 = 2(x - 6)$

$x - 1 = 2x - 12$

$-x \quad -x$

$-1 = x - 12$

$+12 \quad +12$

$11 = x$

Frank = 11 yrs.

5. **short piece = 23 in.; long piece = 37 in.**

x = short piece

$x + 14$ = long piece

$x + x + 14 = 60$

$2x + 14 = 60$

$-14 \quad -14$

$\frac{2x}{2} = \frac{46}{2}$

$x = 23$

short piece = **23 in.**

long piece = 23 + 14 = **37 in.**

6. **short piece = 27 in.; long piece = 52 in.**

x = short piece

$2x - 2$ = long piece

$x + 2x - 2 = 79$

$3x - 2 = 79$

$+2 \quad +2$

$\frac{3x}{3} = \frac{81}{3}$

$x = 27$

short piece $ **27 in.**

long piece $ 2(27) − 2 = **52 in.**

7. **Mr. Allen = 44 yrs.; daughter = 20 yrs.**

x = daughter's age now
$2x + 4$ = man's age now
$x - 8$ = daughter's age 8 yrs. ago
$2x + 4 - 8 = 2x - 4$ = man's age 8 yrs. ago

$$2x - 4 = 3(x - 8)$$
$$2x - 4 = 3x - 24$$
$$-2x \qquad -2x$$
$$-4 = x - 24$$
$$+24 \qquad +24$$
$$20 = x$$

daughter = **20 yrs.**
man = 2(20) + 4 = **44 yrs.**

8. **$4.50**

x = hourly wage on Friday
$2x$ = hourly wage on Saturday

$$8x + 5(2x) = 81$$
$$8x + 10x = 81$$
$$\frac{18x}{18} = \frac{81}{18}$$
$$x = 4\tfrac{1}{2} = \$4.50$$

9. **Mrs. Frisch = $6,800; Mr. Frisch = $14,000**

x = Mrs. Frisch's salary
$2x + 400$ = Mr. Frisch's salary

$$x + 2x + 400 = 20{,}800$$
$$3x + 400 = 20{,}800$$
$$-400 \qquad -400$$
$$\frac{3x}{3} = \frac{20{,}400}{3}$$
$$x = 6{,}800$$

Mrs. F's salary = **$6,800**
Mr. F's salary = 2(6,800) + 400 = **$14,000**

10. **$260**

x = deductions
$4x$ = net wages

$$x + 4x = 325$$
$$\frac{5x}{5} = \frac{325}{5}$$
$$x = 65$$

net wages = 4(65) = **$260**

EXERCISE 10

1. **3:5**

town : country = 15:25 = 3:5

2. **11:2**

men : women = 440:80 = 11:2

3. **3:8**

6:16 = 3:8

4. **3:16**

rent : income = $180:$960 = 3:16

5. **3:2**

length : width = 18:12 = 3:2

6. **1:5**

deductions : gross pay = $42:$210 = 1:5

7. **1:4**

9:36 = 1:4

8. **2:5**

amt. paid : mortgage = \$12,000 : \$30,000 = 2:5

9. **7:8**

40 − 5 = 35
right : total = 35:40 = 7:8

10. **3:4**

240 − 60 = 180
Sept. wt. : Jan. wt. = 180:240 = 3:4

EXERCISE 11

1. $\mathbf{11\frac{1}{5}}$ $\quad$ $\mathbf{2\frac{2}{3}}$ $\quad$ $\mathbf{2\frac{8}{11}}$ $\quad$ $\mathbf{6\frac{6}{7}}$

$\frac{a}{7} = \frac{8}{5}$, $5a = 56$, $a = \mathbf{11\frac{1}{5}}$

$\frac{6}{e} = \frac{9}{4}$, $9e = 24$, $e = 2\frac{6}{9} = \mathbf{2\frac{2}{3}}$

$\frac{10}{11} = \frac{c}{3}$, $11c = 30$, $c = \mathbf{2\frac{8}{11}}$

$\frac{4}{f} = \frac{7}{12}$, $7f = 48$, $f = \mathbf{6\frac{6}{7}}$

2. $\mathbf{32\frac{1}{2}}$ $\quad$ $\mathbf{2\frac{2}{5}}$ $\quad$ $\mathbf{8\frac{4}{7}}$ $\quad$ $\mathbf{14\frac{2}{5}}$

$\frac{2}{13} = \frac{5}{m}$, $2m = 65$, $m = \mathbf{32\frac{1}{2}}$

$\frac{5}{6} = \frac{2}{r}$, $5r = 12$, $r = \mathbf{2\frac{2}{5}}$

$\frac{4}{7} = \frac{n}{15}$, $7n = 60$, $n = \mathbf{8\frac{4}{7}}$

$\frac{5}{24} = \frac{3}{s}$, $5s = 72$, $s = \mathbf{14\frac{2}{5}}$

3. $\mathbf{4\frac{4}{9}}$ $\quad$ $\mathbf{\frac{2}{5}}$ $\quad$ **50**

$4:t = 9:10$, $\frac{4}{t} = \frac{9}{10}$, $9t = 40$, $t = \mathbf{4\frac{4}{9}}$

$2:15 = v:3$, $\frac{2}{15} = \frac{v}{3}$, $15v = 6$, $v = \frac{6}{15} = \mathbf{\frac{2}{5}}$

$2:25 = 4:k$, $\frac{2}{25} = \frac{4}{k}$, $2k = 100$, $k = \mathbf{50}$

4. $\mathbf{6\frac{1}{4}}$ $\quad$ $\mathbf{3\frac{3}{8}}$ $\quad$ $\mathbf{15\frac{3}{7}}$

$5:16 = h:20$, $\frac{5}{16} = \frac{h}{20}$, $16h = 100$, $h = 6\frac{4}{16} = \mathbf{6\frac{1}{4}}$

$d:9 = 3:8$, $\frac{d}{9} = \frac{3}{8}$, $8d = 27$, $d = \mathbf{3\frac{3}{8}}$

$9:y = 7:12$, $\frac{9}{y} = \frac{7}{12}$, $7y = 108$, $y = \mathbf{15\frac{3}{7}}$

EXERCISE 12

1. **256 workers**

$\frac{\text{union}}{\text{non-union}}\ \frac{8}{3} = \frac{x}{96}$

$3x = 768$

$x = \mathbf{256}$

2. **$40.80**

$\frac{\$}{\text{hours}}\ \frac{23.80}{7} = \frac{x}{12}$

$7x = 285.60$

$x = \mathbf{40.80}$

3. **20 lbs.**

$\frac{\text{length}}{\text{weight}}\ \frac{15}{12} = \frac{25}{x}$

$15x = 300$

$x = \mathbf{20}$

4. **20 miles**

$\frac{\text{miles}}{\text{hours}}\ \frac{9}{2\frac{1}{4}} = \frac{x}{5}$

$2\frac{1}{4}x = 45$

$\frac{4}{9} \cdot \frac{9}{4}x = 45 \cdot \frac{4}{9}$

$x = \mathbf{20}$

5. **195 miles**

$\frac{\text{inches}}{\text{miles}}\ \frac{2}{65} = \frac{6}{x}$

$2x = 390$

$x = \mathbf{195}$

6. **28 in.**

$\frac{\text{width}}{\text{length}}\ \frac{5}{7} = \frac{20}{x}$

$5x = 140$

$x = \mathbf{28}$

7. **180 divorces**

$\frac{\text{divorces}}{\text{population}}\ \frac{5}{1{,}000} = \frac{x}{36{,}000}$

$1{,}000x = 180{,}000$

$x = \mathbf{180}$

8. **$1,890**

$\frac{\text{ounces}}{\text{price}}\ \frac{2}{756} = \frac{5}{x}$

$2x = 3{,}780$

$x = \mathbf{1{,}890}$

9. **16 gals.**

$\frac{\text{green}}{\text{white}}\ \frac{4}{3} = \frac{x}{12}$

$3x = 48$

$x = \mathbf{16}$

10. **3,600 bu.**

$\frac{\text{acres}}{\text{bushels}} = \frac{15}{1{,}350} = \frac{40}{x}$

$15x = 54{,}000$

$x = \mathbf{3{,}600}$

EXERCISE 13

G = (+6, +1)   **L = (−9, 0)**

H = (+2, +5)   **M = (−3, −6)**

I = (0, +3)   **N = (+1, −8)**

J = (−5, +8)   **P = (+4, −7)**

K = (−7, +2)   **R = (+9, −2)**

EXERCISE 14

1. **9**   2. **9**   3. **(0, +1)**

4. **13** 5. **6** 6. **Yes** Substitute 6 for x in $y = \frac{2}{3}x + 1$. If $y = 5$, then point (6,5) is on line $y = \frac{2}{3}(6) + 1 = 5$. You can guess the answer by looking at the graph. If you extend line *IJ*, you can see that it goes through point (6,5).

7. **7** 8. **(0,3)** 9. **No** Substitute 5 for x in $y = -x + 3$. If $y = 8$, then point (5,8) is on the line $y = -(5) + 3 = -5 + 3 = -2$. You can guess the answer by looking at the graph. Point (5,8) is not close to the line *LM*.

10. **6** 11. **10** 12. **Yes** Substitute 2 for x in $y = x + 1$. If $y = 3$, then (2,3) is on the line $y = (2) + 3 = 5$. You can guess the answer by looking at the graph. Point (2,3) "looks" like it is on line *PR*. In fact, (2,3) is the point at which line *PR* and *QS* intersect.

EXERCISE 15

1. **2**

$$\frac{6 - 2}{5 - 3} = s$$
$$\frac{4}{2} = s$$
$$\mathbf{2} = s$$

2. $\mathbf{\frac{2}{3}}$

$$\frac{1 - 3}{2 - 5} = s$$
$$\frac{-2}{-3} = s$$
$$\mathbf{\frac{2}{3}} = s$$

3. **1**

$$\frac{2 - (-2)}{2 - (-2)} = s$$
$$\frac{4}{4} = s$$
$$\mathbf{1} = s$$

4. $\mathbf{-1\frac{1}{4}}$

$$\frac{6 - 1}{(-4) - 0} = s$$
$$-\frac{5}{4} = s$$
$$\mathbf{-1\frac{1}{4}} = s$$

5. **0**

$$\frac{8 - 0}{4 - 4} = s$$
$$\frac{8}{0} = s$$
$$\mathbf{0} = s$$

EXERCISE 16

1. **7**

$y = 2(5) - 3$
$y = 10 - 3$
$y = \mathbf{7}$

2. **−1**

$y = 2(1) - 3$
$y = 2 - 3$
$y = \mathbf{-1}$

3. **−7**

$y = 2(-2) - 3$
$y = -4 - 3$
$y = \mathbf{-7}$

4. **(5,7) (1, −1) (−2, −7)**

5. **A** Line A passes through (5,7)(1, −1) and (−2, −7).

6. **0**

$y = -\frac{1}{2}(8) + 4$

$y = -4 + 4$

$y = \mathbf{0}$

7. **+3**

$y = -\frac{1}{2}(2) + 4$

$y = -1 + 4$

$y = \mathbf{+3}$

8. **+6**

$y = -\frac{1}{2}(-4) + 4$

$y = +2 + 4$

$y = \mathbf{+6}$

9. **(8,0) (2,3) (−4,6)**

10. **D** Line D passes through (8,0) (2,3) and (−4,6).

11. **68**

$F = \frac{9}{5}(20) + 32$

$F = 36 + 32$

$F = \mathbf{68}$

12. **50**

$F = \frac{9}{5}(10) + 32$

$F = 18 + 32$

$F = \mathbf{50}$

13. **−4**

$F = \frac{9}{5}(-20) + 32$

$F = -36 + 32$

$F = \mathbf{-4}$

14. **(20,68) (10,50) (−20, −4)**

15. **P** Line P passes through (20,68)(10,50) and (−20, −4)

EXERCISE 17 (WORD PROBLEMS)

1. **(2) 45**

$2x = 90$

$\frac{2x}{2} = \frac{90}{2}$

$x = \mathbf{45}$

2. **(4) 5(.75 + 5)**

3. **(3) 64**

$\frac{\text{people who chew gum}}{\text{total}}\ \frac{4}{5} = \frac{x}{80}$

$5x = 320$

$\frac{5x}{5} = \frac{320}{5}$

$x = \mathbf{64}$

4. **(3) 58**

$-12 + x = 46$

$+12 \quad = +12$

$x = \mathbf{58}$

5. **(4) 50 or −50**

$x^2 = 2{,}500$

$\sqrt{x^2} = \sqrt{2{,}500}$

$x = \mathbf{50}$ or $\mathbf{-50}$

6. **(2) $750**

$I = PRT$

$\$60 = P(.08)(1)$

$\frac{60}{.08} = \frac{.08P}{.08}$

$\mathbf{\$750} = P$

7. **(1)(3x + 70) − x**

8. **(5) 18**

$5a = 3(a + 12)$

$5a = 3a + 36$

$-3a = -3a$

$2a = 36$

$a = \mathbf{18}$

9. **(5) 4**

$4e - 2 \leq 14$

$+ 2 \quad + 2$

$\frac{4e}{4} \leq \frac{16}{4}$

$e \leq \mathbf{4}$

10. **(4) $1,020**

$$\frac{1}{3}x + 30 = 370$$

$$\frac{1}{3}x + 30 = 370$$
$$- 30 = -30$$
$$\frac{1}{3}x = 340$$

$$3 \times \frac{1}{3}x = 340 \times 3$$

$$x = \mathbf{\$1{,}020}$$

11. **(4) 7:13**

$$\begin{array}{r} \$175 \\ +\ 150 \\ \hline 325 \end{array}$$

$$\frac{175}{325} = \frac{\mathbf{7}}{\mathbf{13}}$$

12. **(5) 36**

$$a + b = 48$$
$$a + 3a = 48$$
$$\frac{4a}{4} = \frac{48}{4}$$
$$a = 12$$
$$3a = \mathbf{36}$$

13. **(3) 90**

$$\frac{\text{wins}}{\text{losses}}\ \frac{5}{4} = \frac{x}{72}$$
$$\frac{4x}{4} = \frac{360}{4}$$
$$x = \mathbf{90}$$

14. **(2) 18**

$$y^2 + 2y - 6 =$$
$$4^2 + 2(4) - 6 =$$
$$16 + 8 - 6 =$$
$$24 - 6 = \mathbf{18}$$

15. **(5) $5r - 60$**

ALGEBRA REVIEW

1. **65**

$3^2 + 4^3 - 2^3 =$
$3 \times 3 + 4 \times 4 \times 4 - 2 \times 2 \times 2 =$
$9 + 64 - 8 =$ **65**
(Powers)

2. $\sqrt{5{,}329} = \mathbf{73}$

Guess 70.

$$70\overline{)5329}\ \text{quotient } 76;\quad \begin{array}{r} 5329 \\ \underline{490} \\ 429 \\ \underline{420} \\ 9 \end{array} \qquad \begin{array}{r} 76 \\ +\ 70 \\ \hline 146 \end{array} \qquad 2\overline{)146}\ \text{quotient } \mathbf{73}$$

(Roots)

3. **D**
(Signed Numbers)

4. **−11**

$$\begin{array}{r} +15 \\ +\ 6 \\ \hline +21 \end{array} \qquad \begin{array}{r} -\ 9 \\ -23 \\ \hline -32 \end{array} \qquad \begin{array}{r} -32 \\ +21 \\ \hline \mathbf{-11} \end{array}$$

(Adding Signed Numbers)

5. **+43**

$(+29) - (-14) =$
$+29 + 14 =$ **+43**
(Subtracting Signed Numbers)

6. **+48**

$(+8)(-2)(-3) =$ **+48**
Two negative signs give a positive answer.
(Multiplying Signed Numbers)

7. **+11**

$\frac{-121}{-11} = \mathbf{+11}$

When the signs are the same, the answer is positive.
(Dividing Signed Numbers)

8. **.24**

$mn = (.6)(.4) = \mathbf{.24}$
(Substitution)

9. **70**

$s(t + u) =$
$7(4 + 6) =$
$7(10) = \mathbf{70}$
(Order of Operations)

10. **59**

$F = \frac{9}{5}C + 32$

$F = \left(\frac{9}{5} \times 15\right) + 32$

$F = 27 + 32$
$F = \mathbf{59}$
(Formulas)

11. **$p - 5$**

$4p + 7 - 12 - 3p =$
$p - 5$
(Special Problems with Solving Equations)

12. **$3x - 15$**

$8x - 5(x + 3) =$
$8x - 5x - 15 =$
$3x - 15$
(Special Problems with Solving Equations)

13. **$c = 6\frac{1}{2}$**

$\frac{12c}{12} = \frac{78}{12}$

$c = 6\frac{6}{12} = \mathbf{6\frac{1}{2}}$

(Solving Equations)

14. **$10 = y$**

$$\begin{array}{rcr} 22 & = & 3y - 8 \\ +\ 8 & & +\ 8 \\ \hline \frac{30}{3} & = & \frac{3y}{3} \end{array}$$

$\mathbf{10} = y$
(Solving Equations)

15. **$d = 9$**

$$\begin{array}{rcr} 9d - 5 & = & 3d + 49 \\ -3d & & -3d \\ \hline 6d - 5 & = & +\ 49 \\ +\ 5 & & +\ 5 \\ \hline \frac{6d}{6} & = & \frac{54}{6} \\ d & = & \mathbf{9} \end{array}$$

(Special Problems with Solving Equations)

16. **$m = 9$**

$$\begin{array}{rcr} 8(m - 6) + 3 & = & 27 \\ 8m - 48 + 3 & = & 27 \\ 8m - 45 & = & 27 \\ +\ 45 & & +45 \\ \hline \frac{8m}{8} & = & \frac{72}{8} \\ m & = & \mathbf{9} \end{array}$$

(Special Problems with Solving Equations)

17. **1:5**

deductions : gross pay
$64 : $320
1:5
(Ratio)

18. **$x = 13\frac{5}{7}$**

$\frac{7}{12} = \frac{8}{x}$

$7x = 96$

$x = \mathbf{13\frac{5}{7}}$

(Proportion)

19. **$x = 90$ ft.**

$$\frac{\text{height}}{\text{shadow}}\ \frac{6}{5} = \frac{x}{75}$$

$$5x = 450$$

$$x = \mathbf{90\ ft.}$$

20. **$2p + 3$**

(Writing Expressions)

21. **$x = 12$**

$$9x - 7 = 101$$

$$+7 \quad + 7$$

$$\frac{9x}{9} = \frac{108}{9}$$

$$x = 12$$

(Writing Equations)

22. Marge's salary = **\$620**
George's salary = **\$760**

x = Marge's salary
$x + 140$ = George's salary

$$x + x + 140 = 1{,}380$$

$$2x + 140 = 1{,}380$$

$$-140 \quad -140$$

$$\frac{2x}{2} = \frac{1{,}240}{2}$$

$$x = 620$$

Marge's salary = \$620
George's salary = 620 + 140 = \$760
(Equation Word Problems)

23. **8**
(Rectangular Coordinates)

24. **(+2, 0)**
(Rectangular Coordinates)

25. **No**

(Inequalities)

26. **$m \leq 5$**

$$7m - 2 \leq 3m + 18$$

$$-3m \quad -3m$$

$$4m - 2 \leq +18$$

$$+2 \quad +2$$

$$\frac{4m}{4} \leq \frac{20}{4}$$

$$m \leq 5$$

(Solving Inequalities)

Geometry

Geometry is a branch of math that deals with measuring lines, angles, surfaces, and objects. Some of the basic rules of geometry are very useful in everyday life. For example, when you buy a carpet, you need to know the *area* of the room that you will put the carpet in. Finding area is a geometry skill.

Geometry not only is useful in everyday life, but it also is a part of the GED Test. In the next part of this book, you will study the geometry skills that you will need for the GED Test. First, you will study the names of the important geometric figures. Then you will find out how to measure the different geometric figures.

ANGLES

An **angle** is made up of two lines extending from the same point. The size of an angle depends on how "open" or "closed" the lines are. The point that both lines extend from is called the **vertex.**

Angles are measured in degrees (°). The names of angles depend on the number of degrees in the angles.

Below is a list of the five angle names. Alongside, there are pictures showing examples of each angle. In the examples, the small curves show the angles that each name refers to.

A **right angle** is an angle that has a measurement of exactly 90°. A box at the vertex always means a right angle.

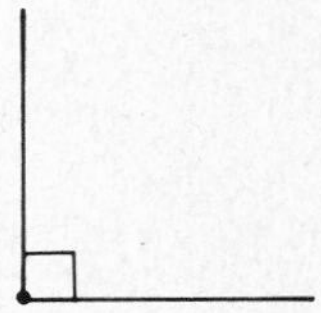

An **acute angle** is an angle that has a measurement of less than 90°. An acute angle is more closed than a right angle.

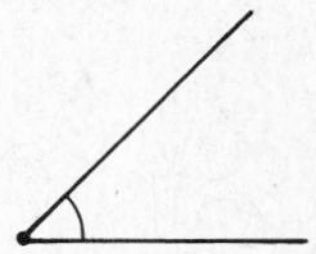

An **obtuse angle** is an angle that has a measurement of more than 90° but less than 180°. An obtuse angle is wider than a right angle.

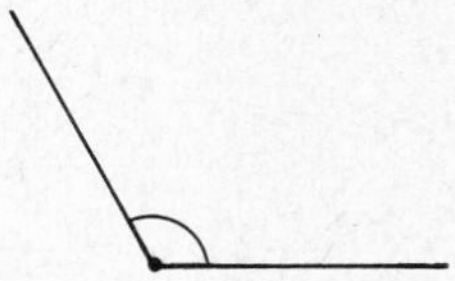

A **straight angle** is an angle that has a measurement of exactly 180°. A straight angle looks like a straight line.

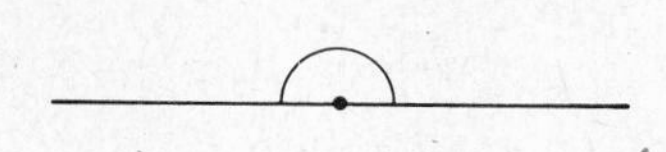

A **reflex angle** is an angle that has a measurement of more than 180° but less than 360°.

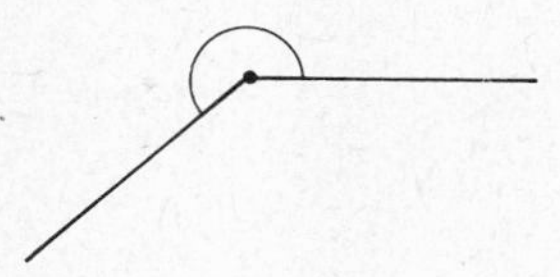

The symbol for angle is ∠. There are three common ways to refer to angles.

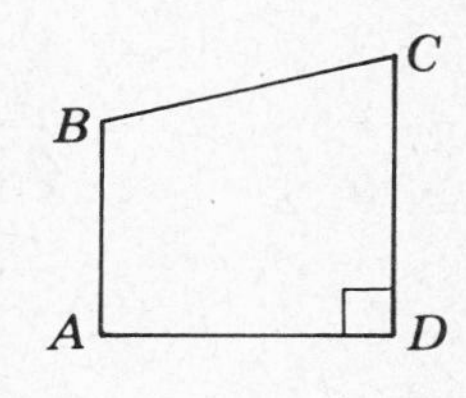

The figure at the left has 4 sides and 4 angles. Each capital letter stands for the vertex of an angle. You can refer to the four angles like this:

$\angle A$ $\angle B$ $\angle C$ $\angle D$

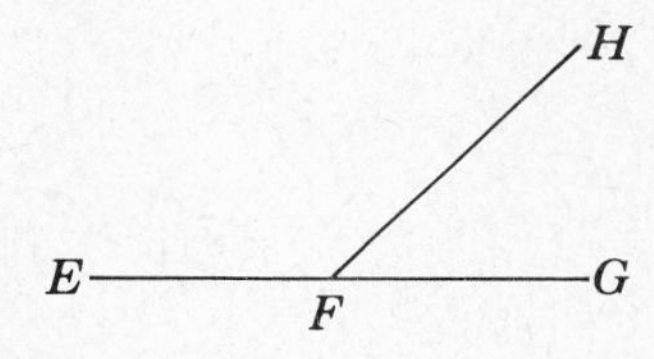

Another way to refer to an angle is to use 3 capital letters. The middle letter is always the vertex.

$\angle EFH$ $\angle GFH$ $\angle EFG$

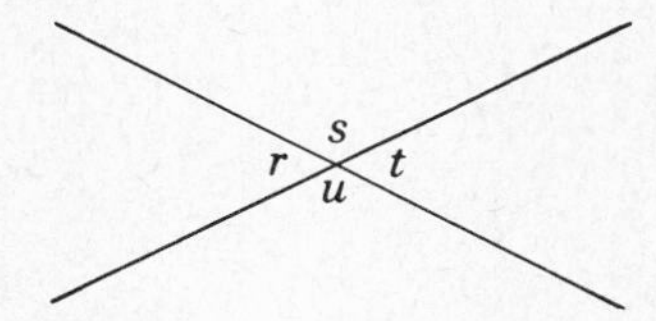

The third way to refer to an angle is to use small letters inside an angle.

$\angle r$ $\angle s$ $\angle t$ $\angle u$

EXERCISE 1

DIRECTIONS: Memorize the definition of each angle. Then identify each angle that is represented by the number of degrees below.

1. 30° 80° 110° 90°
2. 270° 180° 179° 89°

Check your answers on page 304.

Pairs of Angles

Study these three pairs of angles.

Complementary angles are two angles that add up to 90°.

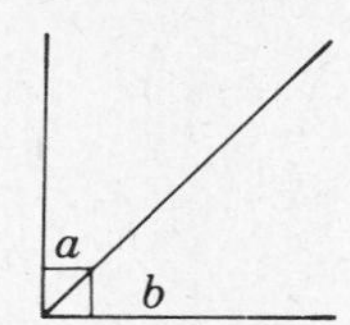

Together, $\angle a$ and $\angle b$ make a right angle. $\angle a$ and $\angle b$ are complementary angles. If $\angle a = 65°$, then $\angle b = 90° - 65° = 25°$.

Supplementary angles are two angles that add up to 180°.

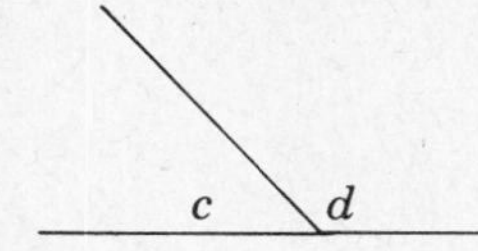

Together, $\angle c$ and $\angle d$ make a straight angle. $\angle c$ and $\angle d$ are supplementary angles. If $\angle c = 50°$, then $\angle d = 180° - 50° = 130°$.

Vertical angles are opposite pairs of angles formed when two straight lines cross. Vertical angles are equal.

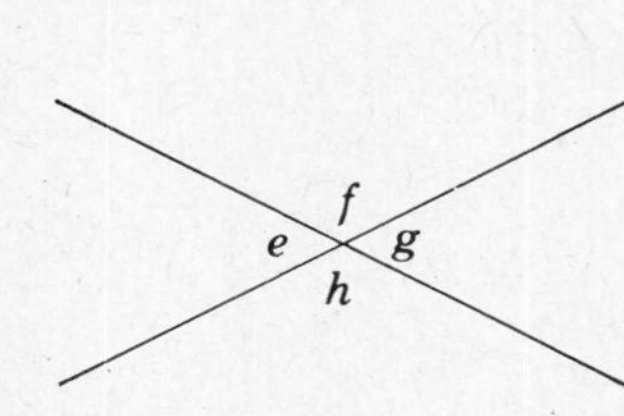

$\angle e$ and $\angle g$ are opposite (across from) each other.
$\angle e$ and $\angle g$ are vertical angles.
$\angle f$ and $\angle h$ are opposite each other.
$\angle f$ and $\angle h$ are vertical angles.
If $\angle e = 70°$, then $\angle g$ is also 70° and $\angle f = 180° - 70° = 110°$. $\angle h$ is also 110°.

A degree can be divided into the smaller units of minutes and seconds. A degree (°) equals 60 minutes (′). A minute (′) equals 60 seconds (″).

EXAMPLE: Find the complement of an angle that measures 62°35′21″.

Step 1	Step 2	Step 3
90°	**89°60′**	89°**59′60″**
− 62°35′21″	− 62°35′21″	− 62°35′21″
		27°24′39″

Step 1: Complementary angles equal 90°. To find the complement, subtract 62°35′21″ from 90°.
Step 2: Borrow 1° from 90°. Change 90° to 89° and change the degree that you borrowed to 60′.
Step 3: Borrow 1′ from 60′. Change 60′ to 59′ and change the minute that you borrowed to 60″. Subtract seconds, minutes, and degrees separately.

In the example, 27°24′39″ is the complement of 62°35′21″.

EXERCISE 2

DIRECTIONS: Answer each question about angles.

1. Find the supplement of an angle that measures $72°$.

2. Find the complement of an angle that measures $72°$.

3. Find the supplement of an angle that measures $120°28'44''$.

4. Find the complement of an angle that measures $36°52'17''$.

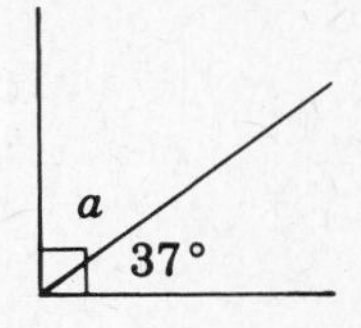

5. How many degrees are in $\angle a$?

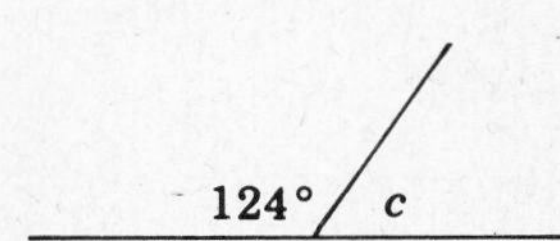

6. How many degrees are in $\angle c$?

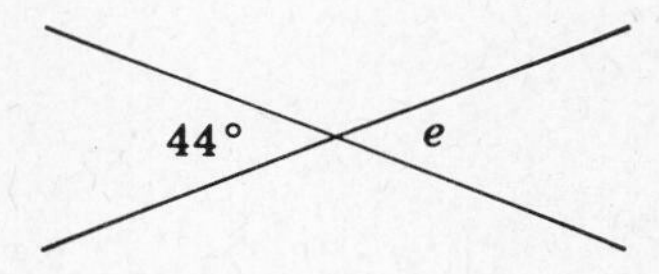

7. How many degrees are in $\angle e$?

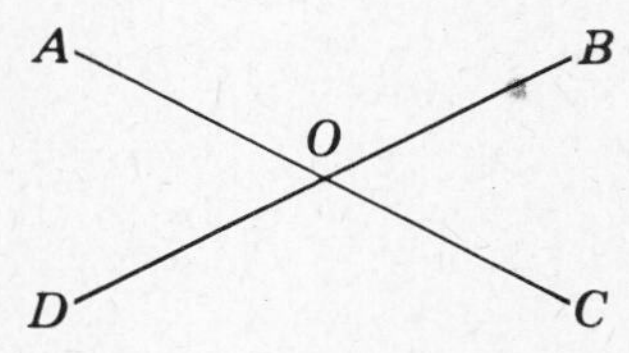

Use the picture at the left for questions 8 through 12.

8. Which angle is vertical to $\angle AOD$?

9. Which angle is vertical to $\angle AOB$?

10. If $\angle BOC = 63°$, what is $\angle AOB$?

11. If $\angle COD = 135°$, what is $\angle AOD$?

12. What is the sum of $\angle AOB$ and $\angle AOD$?

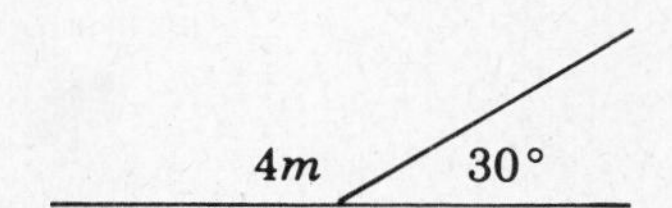

13. Solve for m.

14. One angle is four times another angle. The two angles are complementary. Find the angles.

15. One angle is 28° bigger than another. The two angles are supplementary. Find the angles.

Check your answers on page 304.

More Angle Relationships

Two lines that never meet are called **parallel lines.** The symbol for parallel lines is $\parallel$.

The statement $AB \parallel CD$ means the two lines at the left are parallel.

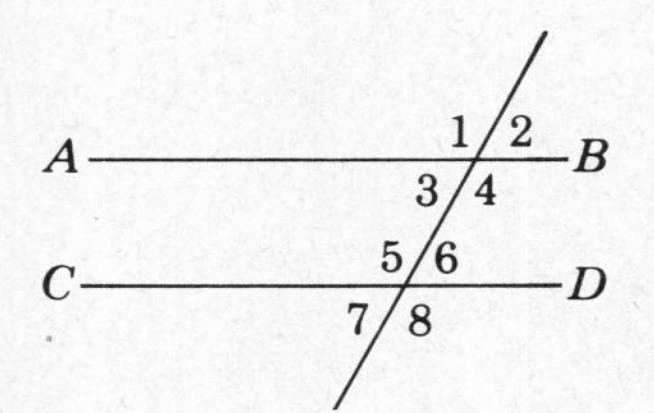

When another line, called a **transversal,** crosses two parallel lines, you get 8 angles. Below are the names of the pairs of angles shown at the left.

There are four pairs of **corresponding angles** in the diagram.

$\angle 1$ and $\angle 5$ are corresponding angles.
$\angle 2$ and $\angle 6$ are corresponding angles.
$\angle 3$ and $\angle 7$ are corresponding angles.
$\angle 4$ and $\angle 8$ are corresponding angles.

Study the figure to see how one angle "corresponds" to another. Corresponding angles are equal. In each pair, one angle is in the upper set. The other is in the lower set.

There are two pairs of **alternate interior angles.**

$\angle 3$ and $\angle 6$ are alternate interior angles.
$\angle 4$ and $\angle 5$ are alternate interior angles.

Alternate interior angles are equal. In each pair, one angle is in the upper set. The other is in the lower set. Both angles of a pair are inside the two parallel lines.

There are two pairs of **alternate exterior angles.**

$\angle 1$ and $\angle 8$ are alternate exterior angles.
$\angle 2$ and $\angle 7$ are alternate exterior angles.

Alternate exterior angles are equal. In each pair, one angle is in the upper set. The other is in the lower set. Both angles of a pair are outside the two parallel lines.

EXERCISE 3

DIRECTIONS: Use the diagram at the right to answer the next questions.

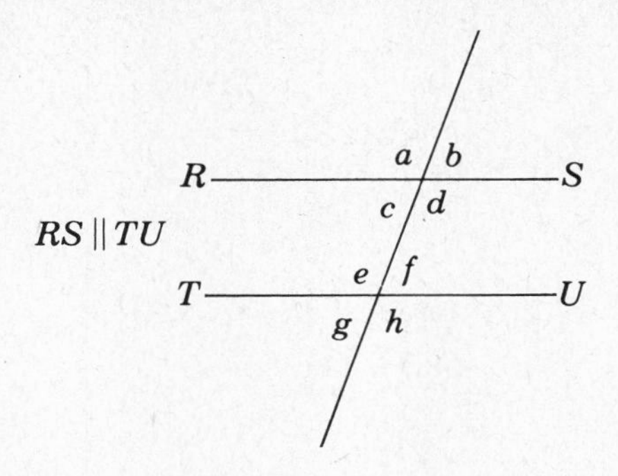

1. Which angle corresponds to $\angle f$?

2. Which angle is an alternate interior with $\angle d$?

3. Which angle is an alternate exterior with $\angle b$?

4. If $\angle a = 125°$, what does $\angle e$ equal?

5. If $\angle c$ is $60°$, what does $\angle f$ equal?

6. If $\angle b = 85°$, what does $\angle d$ equal?

7. $\angle c$ and $\angle f$ are ________________ angles.

8. $\angle d$ and $\angle h$ are ________________ angles.

9. $\angle a$ and $\angle h$ are ________________ angles.

10. $\angle b$ and $\angle c$ are ________________ angles.

Check your answers on page 305.

PLANE FIGURES

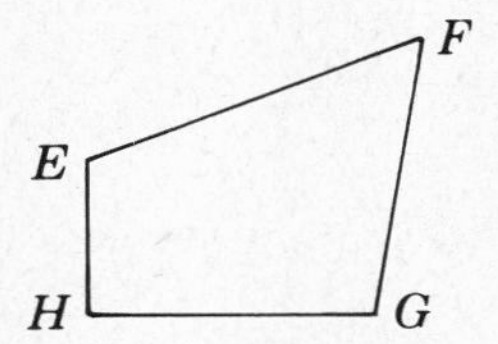

Many of the figures you will study in geometry are plane figures. Figure *EFGH* is a plane figure. *Plane* means the figure is flat. Think of the surface of this page as a plane. A figure drawn on the plane is a plane figure.

In the next sections, you will study the names of the most important plane figures. Try to memorize the names of the different figures. On the GED Test, you are expected to know the differences between figures.

Triangles

A triangle is a plane figure with three sides. The three angles inside a triangle add up to 180°. The names of triangles depend on the relationships among the sides or angles.

An **equilateral** triangle has three equal sides. The three angles are also equal. Each angle measures 60°. Triangle ABC is equilateral.

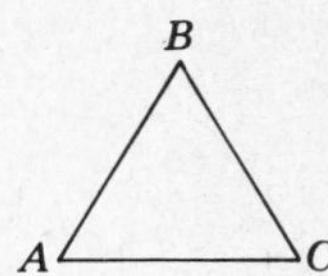

An **isosceles** triangle has two equal sides. Two angles are also equal. Triangle XYZ is an isosceles triangle. The two equal angles in an isosceles triangle are called the **base angles.** The third angle is called the **vertex angle.** In triangle XYZ, $\angle X$ and $\angle Z$ are the base angles. $\angle X$ and $\angle Z$ are equal. $\angle Y$ is the vertex angle.

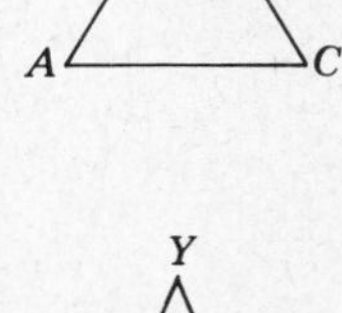

A **scalene** triangle has no equal sides and no equal angles. Triangle MNO is a scalene triangle.

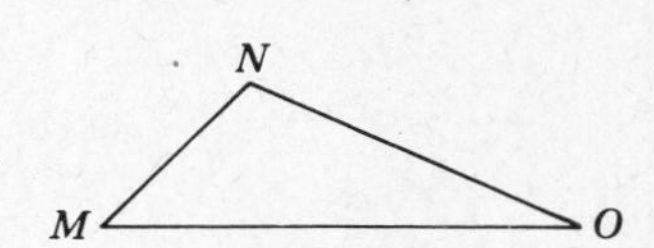

A **right** triangle has one right angle. The side across from the right angle is called the **hypotenuse.** The other two sides are called the **legs.** Triangle PQR is a right triangle. $\angle P$ is the right angle. Side QR is the hypotenuse. Side PR and side PQ are the legs.

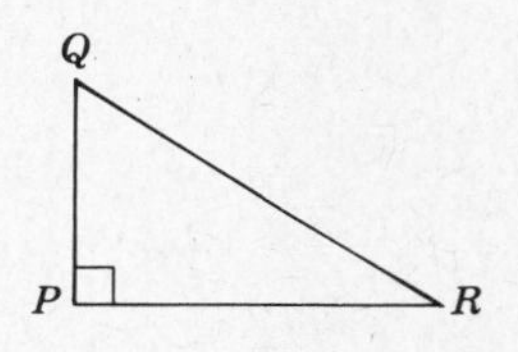

A side of a triangle is referred to by the two letters that form the ends of the side. The hypotenuse in the last triangle is QR. An angle of a triangle can be called by one letter. $\angle P$ is the right angle in the right triangle PQR. An angle can also be called by three letters. $\angle QPR$ is another name for $\angle P$. Remember that the vertex point is the middle letter.

Study the descriptions of triangles before you try the next exercise.

EXERCISE 4

DIRECTIONS: Answer each question about triangles.

1. In triangle STU, $\angle S = 65°$ and $\angle T = 45°$. Find the measure of $\angle U$.

2. In triangle ABC, $\angle B$ is twice as big as $\angle A$, and $\angle C$ is 20° less than $\angle B$. Find the measure of each angle.

3. Each base angle in an isosceles triangle is 25°. Find the measure of the vertex angle.

4. The vertex angle of an isosceles triangle is 65°. Find the measure of each base angle.

5. One acute angle in a right triangle measures 35°. Find the measure of the other acute angle.

6. In triangle DEF, $\angle E$ is 20° more than $\angle D$, and $\angle F$ is twice as big as $\angle E$. Find the measure of each angle.

7. In isosceles triangle MNO, vertex angle N is three times as big as each base angle. Find the measure of each angle.

8. In triangle PQR, $\angle P$ is 90°, and $\angle R$ is twice as big as $\angle Q$. Find the measures of $\angle Q$ and $\angle R$.

9. In triangle STU at the right, $\angle S = 90°$, $\angle T = 55°$, and $\angle U = 35°$. Which side of the triangle is the hypotenuse?

Check your answers on page 305.

Quadrilaterals

A **quadrilateral** is a plane figure with four sides. The sum of the four angles in a quadrilateral is 360°. Figure $ABCD$ is a quadrilateral. Below are five more quadrilaterals. The first two, the square and the rectangle, are the most common.

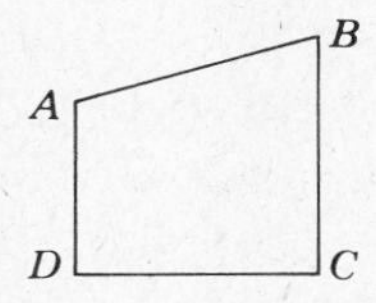

A **square** is a quadrilateral with four equal sides and four right angles. The sides across from each other are parallel. Figure $EFGH$ is a square.

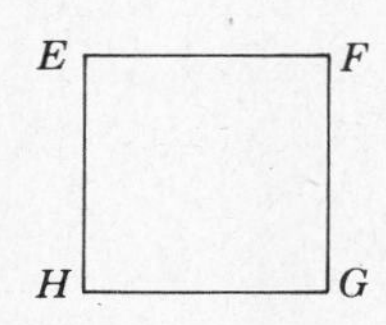

A **rectangle** is a quadrilateral with four right angles. The sides across from each other are equal and parallel. Figure $IJKL$ is a rectangle.

A **parallelogram** is a quadrilateral with two pairs of equal and parallel sides. The angles opposite each other are equal. Figure $MNOP$ is a parallelogram. $\angle M = \angle O$ and $\angle N = \angle P$.

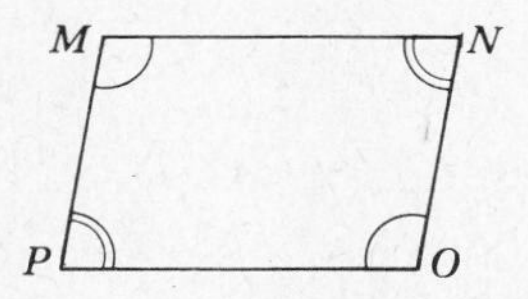

A **trapezoid** is a quadrilateral with two parallel sides. The other two sides are not parallel. The parallel sides are called the *bases*. Figure *QRST* is a trapezoid. Side *QR* is parallel to side *ST*. Side *QT* and side *RS* are not parallel. *QR* and *ST* are the bases.

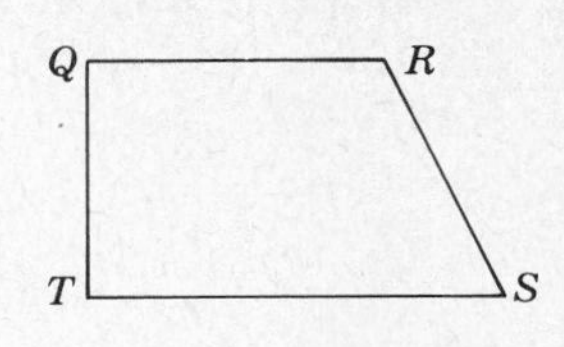

A **rhombus** is a quadrilateral with four equal sides and two pairs of parallel sides. The angles opposite each other are equal. Figure *UVWX* is a rhombus. Each side (*UV*, *VX*, *XW*, and *UW*) has the same length. $\angle U$ and $\angle X$ are equal. $\angle V$ and $\angle W$ are equal.

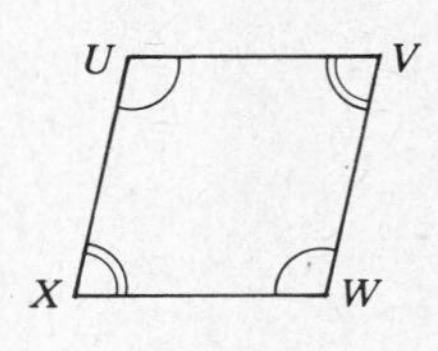

Study the descriptions of these quadrilaterals carefully before you try the next questions.

EXERCISE 5

DIRECTIONS: Answer each question about quadrilaterals.

1. In rectangle *ABCD*, side *AB* measures 12 inches, and side *BC* measures 9 inches. What is the measurement of *CD*?

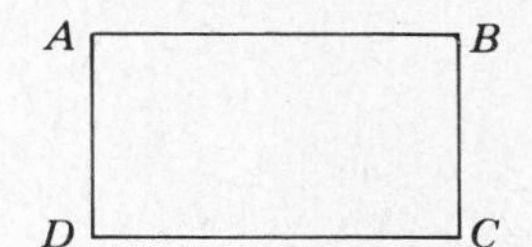

2. In rectangle *ABCD*, what is the measurement of *AD*?

3. In rectangle *ABCD*, how many degrees are there in $\angle B$?

4. The figure at the right shows two diagonal lines (*AC* and *BD*) cutting across rectangle *ABCD*. What kind of triangle is *AOB*?

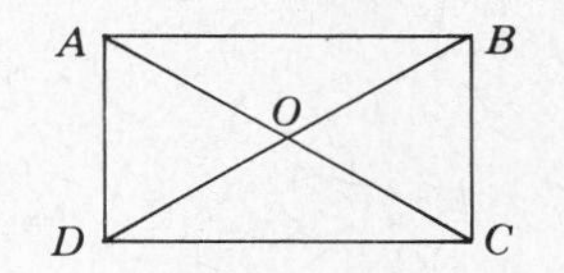

5. In figure *WXYZ*, side *WX* is parallel to side *YZ*. Sides *WZ* and *XY* are not parallel. What kind of figure is *WXYZ*?

6. Side *WX* and side *YZ* are called the ______________.

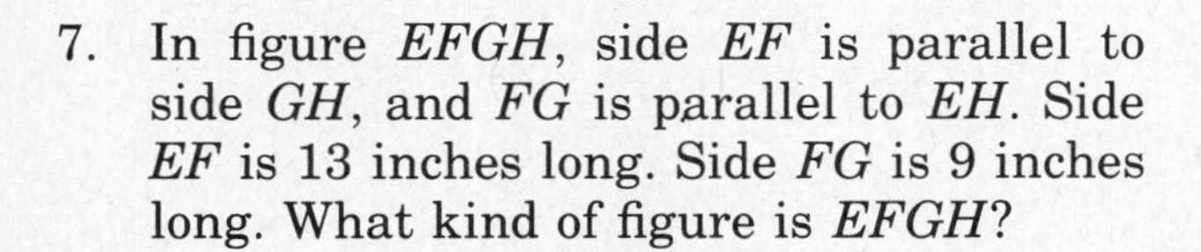

7. In figure *EFGH*, side *EF* is parallel to side *GH*, and *FG* is parallel to *EH*. Side *EF* is 13 inches long. Side *FG* is 9 inches long. What kind of figure is *EFGH*?

8. In figure *EFGH*, $\angle E = 125°$. How many degrees are there in $\angle G$?

9. In figure $STUV$, side ST is parallel to side UV, and SV is parallel to TU. Side ST is 8 inches long, and side TU is 8 inches long. What kind of figure is $STUV$?

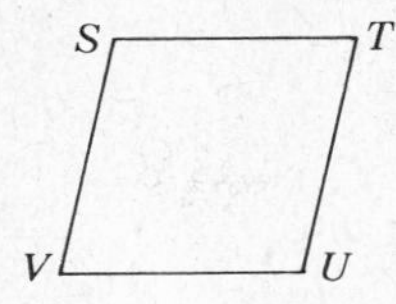

Check your answers on page 306.

Finding Perimeter

When you find the perimeter of a plane figure, you find the distance around the figure. In everyday life, you sometimes have to find the perimeter of something. For instance, if you need to know how much fencing you need for a yard or a garden, you have to find the perimeter of the yard or garden. You have to measure the length of each side.

To find the perimeter of a figure that has sides, add the lengths of each side.

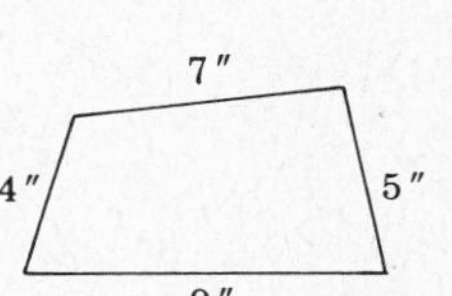

EXAMPLE: Find the perimeter of the quadrilateral at the left.

Add the lengths of each side.

$4'' + 7'' + 5'' + 9'' = 25''$

Notice that the perimeter of a figure is measured in the same units as the sides. In the last example, the units are inches.

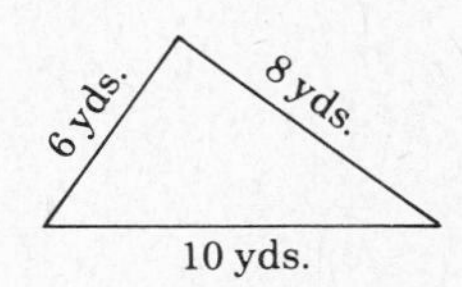

EXAMPLE: Find the perimeter of the triangle at the left.

Add the lengths of each side.

$6 + 10 + 8 = 24$ yds.

The two figures used most often in perimeter problems are rectangles and squares. In perimeter word problems, you may not be given the measurements of all sides of a rectangle or a square. However, you can still solve the problem if you remember the relationships among the lines. In a square, all four sides have the same length. In a rectangle, there are two pairs of sides that have the same length. Study the following examples carefully.

EXAMPLE: How many feet of fencing does Mrs. Davidson need to go around a rectangular garden that is 20 feet long and 12 feet wide?

The problem tells you that the garden is a rectangle. You can find the perimeter even though you have the measurements for only two sides. In a rectangle, two pairs of sides have the same length.

Therefore, two sides of the garden are 20 feet, and the other two sides are 12 feet.

$$20 + 20 + 12 + 12 = 64 \text{ feet}$$

EXAMPLE: Jill wanted to put weather stripping around a square window in her kitchen. One side of the window is 36 inches long. How many inches of weather stripping should she buy for the window?

The problem tells you that the window is square. A square has four equal sides. If you know the length of one of the sides, then you can find the perimeter of the window. Just add the length four times.

$$36 + 36 + 36 + 36 = 144 \text{ inches}$$

There are two formulas that you can use as a shortcut to finding the perimeter of a rectangle or a square. To find the perimeter of a rectangle, add 2 times the length plus 2 times the width.

$$2l + 2w = \text{perimeter of rectangle}$$

To find the perimeter of a square, multiply the length of one side times 4.

$$4s = \text{perimeter of a square}$$

EXERCISE 6

DIRECTIONS: Solve these problems about perimeter.

1. The two equal sides of an isosceles triangle measure 12 inches each. The third side measures 7 inches. Find the perimeter of the triangle.

2. A picture measures $8\frac{1}{2}$ inches wide and 11 inches long. How many inches of framing are needed to go around the picture?

3. Each side of an equilateral triangle measures 19 feet. Find the perimeter of the triangle.

4. Silvia has 90 feet of fencing to enclose a rectangular space in her yard for her dog to run. She wants the space to be 4 times as long as it is wide. Find the length and width of the space.

5. The side of a square measures 4.75 meters. Find the perimeter of the square.

6. Sheila wants to put weather stripping around four windows in her house. Each window is $30\frac{1}{2}$ inches wide and 62 inches long. How many inches of weather stripping does she need?

7. How many feet of weather stripping does Sheila, in problem 6, need?

8. The perimeter of a square is 25 yards. Find the measurement of one side of the square.

Check your answers on page 306.

AREA

Area is a measure of the amount of flat space inside a plane figure. Area is measured in square inches, square feet, square meters, etc. The area of this rectangle is the total number of one-inch squares that make up the surface of the rectangle. The area of the rectangle is 12 square inches.

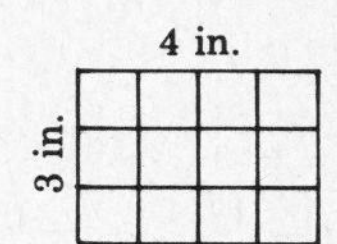

If you have studied area problems before, you probably were asked to memorize several different formulas. You can do area problems with the formulas. But there is another way to solve area problems without memorizing several formulas.

To find the area of any plane figure except a circle, you multiply two numbers. You multiply the length of the figure's bottom line (called the *base*) by the length of one of the figure's sides (called the *height*). Base and height are also called *width* and *length*. Study the next examples carefully.

EXAMPLE: The dining room in Homer's house is 12 feet long and 9 feet wide. Find the area of the room.

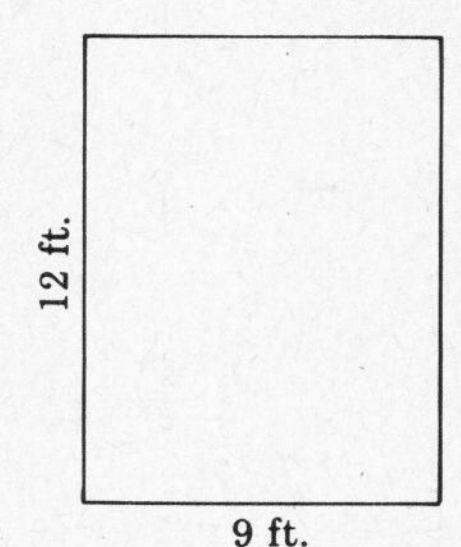

Draw a picture of the room using the measurements. The room is 9 feet wide. The base of the drawing is 9. The room is 12 feet long. The height of the drawing is 12. Multiply 9 by 12 to get the answer.

$$9 \times 12 = 108 \text{ square feet}$$

EXAMPLE: Mrs. Paul wants to put a carpet in her living room. Each side of the room measures 12 feet. How many square feet of carpet will she need if she buys a wall-to-wall carpet?

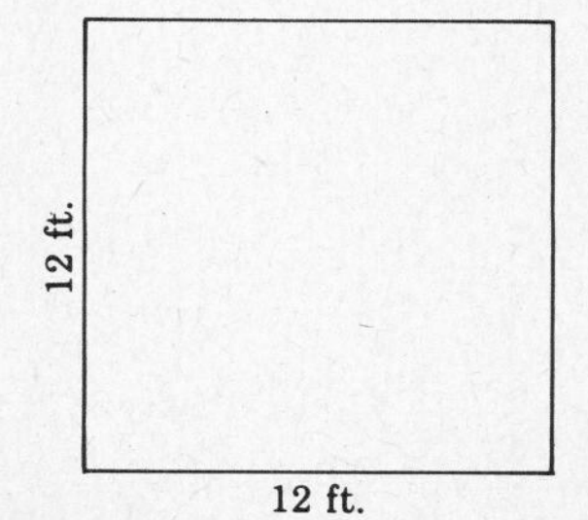

Draw a picture of the room. The base is 12 and the height is 12. Multiply 12 by 12 to get the answer.

$$12 \times 12 = 144 \text{ square feet}$$

Remember to always express area in square units.

Be careful with problems that give you a measurement in one unit and ask for the area in another unit. Study the next example carefully.

EXAMPLE: The Twelfth Street School needs a new linoleum floor in one of its classrooms. The classroom measures 30 feet long and 45 feet wide. How many square yards of linoleum are needed to cover the classroom floor?

Step 1	Step 2	Step 3
30×45	$30 \times 45 = $ **1,350** **3×3** $= 9$	$9\overline{)1,350}$ with quotient 150 $=$ **150 sq. yds.**

Step 1: Set up to multiply the base by the height.
Step 2: Multiply the base by the height. The problem asks you to convert feet into yards. Find the number of square feet in a square yard. Since there are 3 feet in 1 yard, multiply 3 by 3.
Step 3: Divide 9 into 1,350. Label the answer in square yards.

EXERCISE 7

DIRECTIONS: Find the area in each problem.

1. The hall in Linda's apartment is 3 feet wide and 16 feet long. Find the area of the floor of the hall.

2. Max wants to put tiles on the bathroom wall above the tub. Each tile is 3 inches wide and 6 inches long. The wall space Max wants to cover with tiles is 42 inches wide and 54 inches long. How many tiles does Max need to cover the space?

3. Fred built a square patio behind his garage. Each side measures 15 feet. Find the area of the patio in square yards.

4. Joan's living room is 15 feet wide. 30 square yards of carpet cover the living room floor. How long is the living room in feet?

5. Nick built a square coffee table measuring 30 inches on a side. How many square tiles each measuring 2 inches on a side does Nick need to cover the table?

6. Ellen and Gordon want to put carpet in three rooms of their apartment. Their living room is 12 feet by 15 feet. Their dining room is 9 feet by 12 feet. Their bedroom is also 9 feet by 12 feet. The carpet they want costs $13 a square yard. Find the total cost of carpet for the three rooms.

7. The walls in the Johnsons' house are 8 feet high. The total length of the walls in their house, not including windows and doors, is 206 feet. Find the total area of the walls in their house.

8. Mrs. Rose's garden is 18 feet long and 12 feet wide. A bag of fertilizer covers an area of 180 square feet. How many bags of fertilizer will Mrs. Rose need to cover her garden?

9. A large space in front of city hall in Garfield is to be paved with 4-foot by 4-foot concrete squares. The space is 120 feet wide and 200 feet long. How many squares are needed to cover the space?

10. The floor of the kitchen in Faye's apartment is 54 inches wide and 90 inches long. Faye wants to buy 9-inch by 9-inch square tiles to put on the floor. How many tiles should she buy?

Check your answers on page 307.

Area of More Plane Figures

You have found out how to find the area of a rectangle or a square. To find the area of a rectangle or a square, you multiply the base by the height. Finding the area of other plane figures is a little different. The difference is that the height of the other figures is measured with another line.

To find the area of a rhombus or a parallelogram, you also multiply the base by the height. However, the height of a rhombus or a parallelogram does *not* equal the length of its side. Instead, the height is measured by a perpendicular line. A perpendicular line is a line that runs from the base of the figure to the top of the figure. The line meets the base at a right (90°) angle.

EXAMPLE: Find the area of the parallelogram at the left.

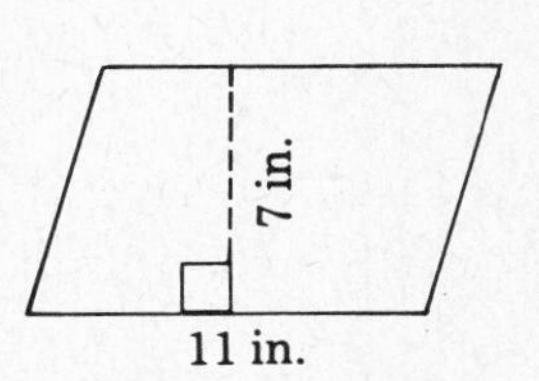

The base of the parallelogram is 11 inches. The broken line is the perpendicular line that runs from the base to the top of the figure. The perpendicular line measures the height of the figure. The height is 7 inches. To find the area, multiply 11 by 7. The answer is 77 square inches.

To find the area of a triangle, multiply the base by the height by $\frac{1}{2}$. The height of a triangle also is measured with a perpendicular line. You multiply by $\frac{1}{2}$ because a triangle is one-half of a four-sided figure. If you "flip" a triangle over, you see that two triangles put together are the same as a four-sided figure.

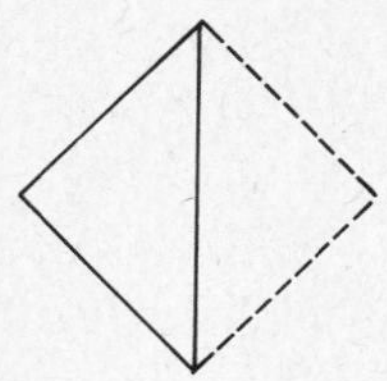

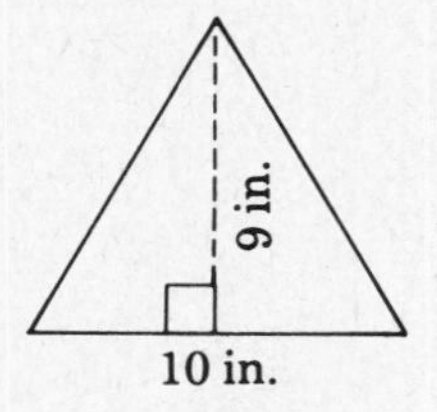

EXAMPLE: Find the area of the triangle at the left.

To find the area, multiply the base by the height and by $\frac{1}{2}$. The base of the triangle is 10 inches. The height is 9 inches.

$$10 \times 9 \times \frac{1}{2} = 45 \text{ sq. in.}$$

To find the area of a trapezoid, you first have to add the lengths of the base and the top line together. You then multiply the sum by the perpendicular height and by $\frac{1}{2}$.

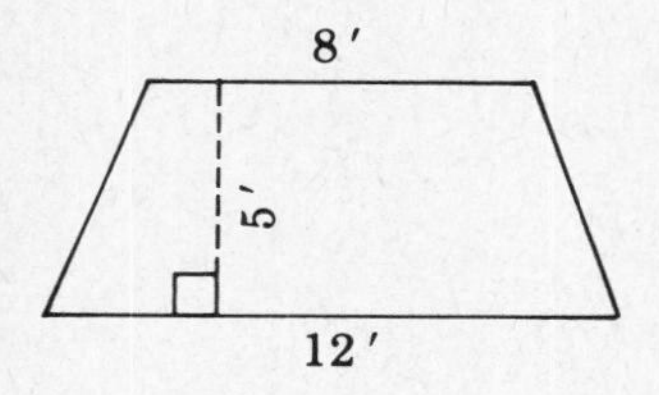

EXAMPLE: Find the area of the trapezoid at the left.

Step 1: Add the lengths of the base and the top line together.

$$12 + 8 = 20$$

Step 2: Multiply the sum by the perpendicular height and by $\frac{1}{2}$.

$$20 \times 5 \times \frac{1}{2} = 50 \text{ sq. ft.}$$

EXERCISE 8

DIRECTIONS: Find the area in each problem.

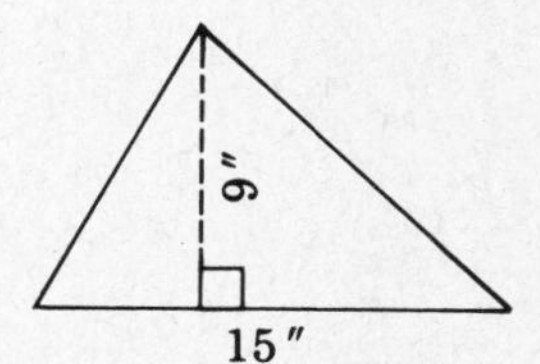

1. Find the area of the triangle at the right.

2. Find the area of a triangle with a base of 0.75 meter and a height of 0.6 meter.

3. The area of a triangle is 60 square feet. The height is 20 feet. Find the base.

4. The base of a triangle is $2\frac{1}{2}$ inches. The height is $1\frac{3}{4}$ inches. Find the area.

5. Find the area of the parallelogram at the right.

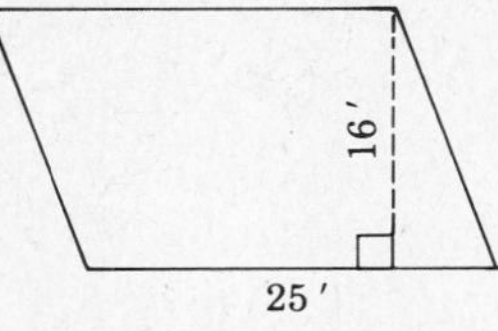

6. Find the area of the rhombus at the right.

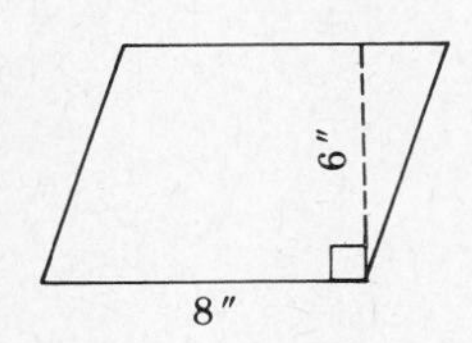

7. Find the area of the parallelogram at the right.

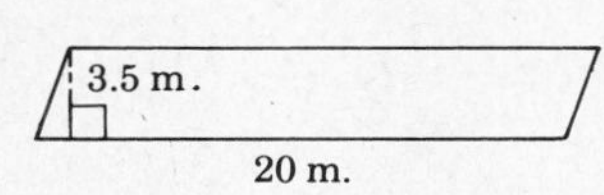

8. Find the area of the trapezoid at the right.

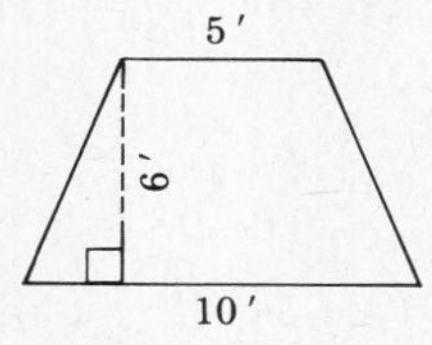

9. Find the area of the trapezoid at the right.

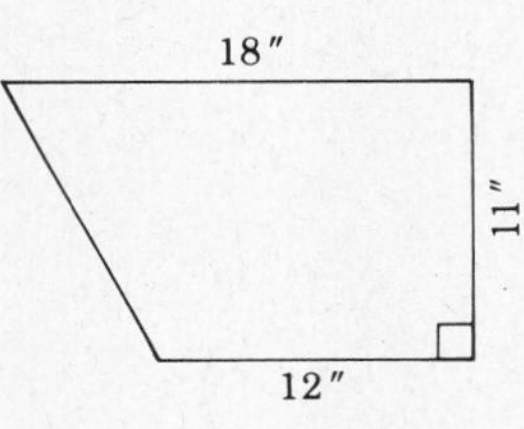

10. Find the area of the trapezoid at the right.

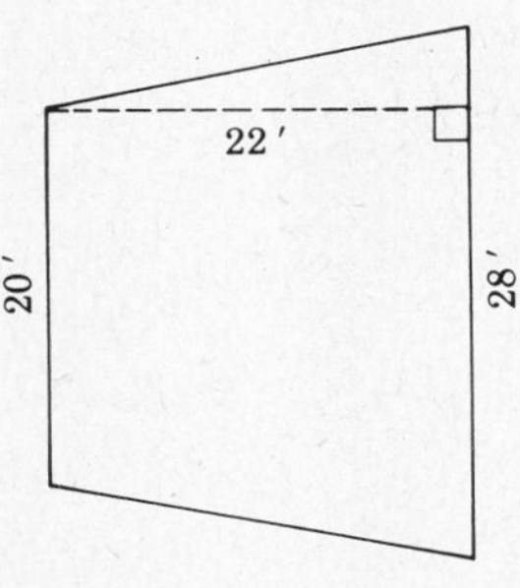

Check your answers on page 308.

CIRCLES

Up to now, you have worked only with figures that have straight sides. You have studied the ways to find perimeter and area for figures that have straight sides. Now, you are going to work with a figure that has no straight sides.

A circle is a curved line. Every part of the line is the same distance from the center of the circle. Because a circle has no straight sides, the ways to find perimeter and area of a circle are different from the ways to find the perimeter and area of sided figures.

Before you study the ways to find the perimeter and area of a circle, study the following words. You will need to know the meanings of these words to work with circles.

The **diameter** is the distance across a circle. A diameter measures the widest part of a circle. The diameter must pass through the center.

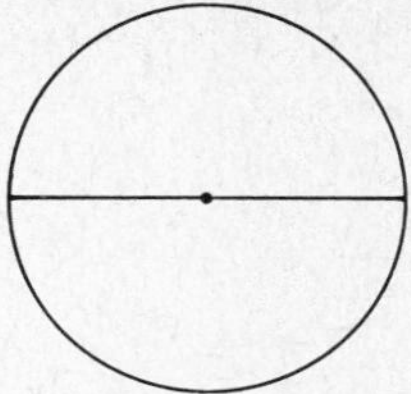

The **radius** is the distance from the center to the curved line of the circle. The radius is exactly one-half of the diameter.

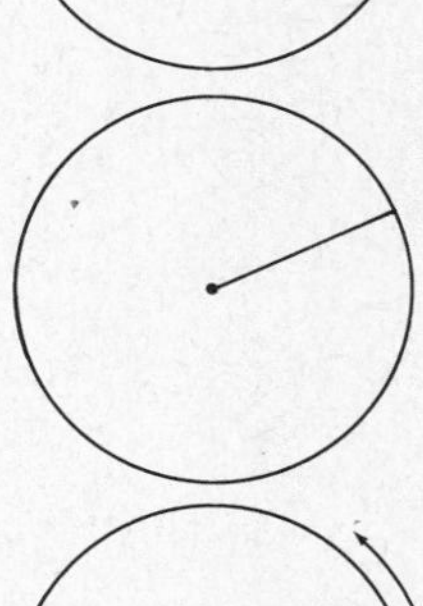

The **circumference** is the distance all the way around a circle. The circumference is the perimeter of a circle.

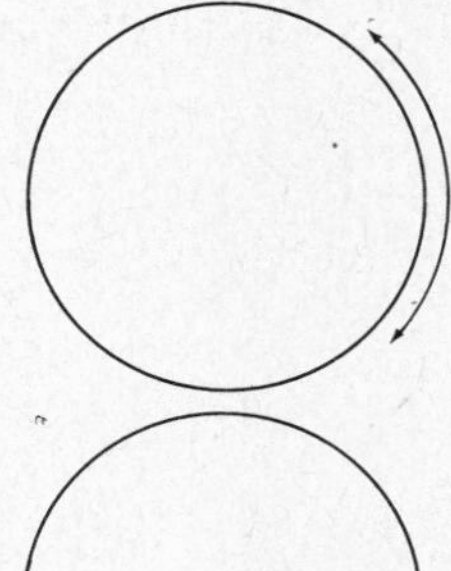

A **semicircle** is one-half of a circle. A whole circle has 360°. A semicircle has 180°.

π is a special measurement that is used with circles. π is a Greek letter that is called "pi." The value of π is about 3.14 or $\frac{22}{7}$. π is equal to the circumference of a circle divided by the diameter of the circle. π is used to figure the circumference and the area of a circle.

$$\pi = \frac{22}{7} \text{ or } 3.14$$

Study the definitions of the words about circles carefully before you try the next questions.

EXERCISE 9

DIRECTIONS: Answer each question about circles.

Use the figure at the right to answer questions 1 through 7.

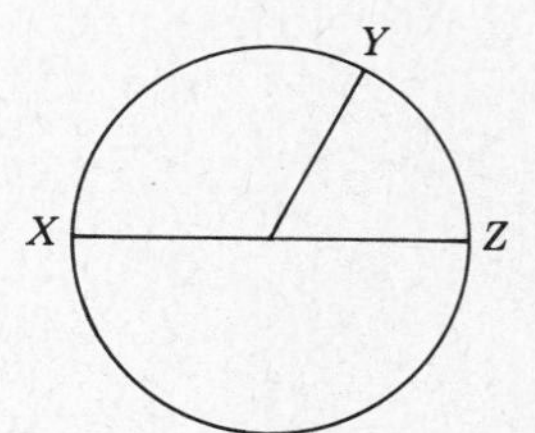

1. Line XZ is called the ______________.

2. Line OY is called the ______________.

3. What is the ratio of XZ to OY?

4. If XZ is 18 inches, how long is OY?

5. If OY is 6.5 meters, how long is XZ?

6. If XZ is 25 feet, how long is OY?

7. If OY is $\frac{3}{8}$ inch, how long is XZ?

8. A semicircle is what percent of a circle?

9. The symbol for 3.14 or $\frac{22}{7}$ is ________.

10. A measure of the distance around a circle is called the

__________.

Check your answers on page 309.

Circumference

When you measure the perimeter of a circle, you measure its circumference. Measuring the circumference is a little harder than finding the perimeter of a sided figure. With a sided figure, all you have to do is add the lengths of the sides together. With a circle, you have to use a formula to find the circumference.

The formula that you use to find the circumference is $C = \pi d$. C stands for the circumference; d stands for the diameter; and π stands for 3.14 or $\frac{22}{7}$.

EXAMPLE: Find the circumference of a circle that has a diameter of 4 inches. Use $\pi = 3.14$.

Step 1	Step 2
$C = \pi d$	$C = 3.14 \times 4$ $C =$ **12.56 inches**

Step 1: Set up the circumference formula.
Step 2: Substitute 3.14 for π and 4 for d. Solve the formula.

You can use the formula to find the circumference of a circle when the radius is given. First, you have to multiply the radius by 2, since the diameter is twice as big as the radius. Then solve the formula to find the circumference.

EXAMPLE: Find the circumference of a circle that has a radius of 7 inches. Use $\pi = \frac{22}{7}$.

Step 1	Step 2
$d = 2 \times 7 = 14$	$C = \pi d$ $C = \frac{22}{7} \times 14$ $C =$ **44 inches**

Step 1: Find the diameter. Multiply the radius by 2.
Step 2: Set up the circumference formula and substitute the given values. Solve the formula.

Sometimes, you may be given the circumference and be asked to find the diameter of a circle.

EXAMPLE: Find the diameter of a circle with a circumference of 154 feet. Use $\pi = \frac{22}{7}$.

Step 1	Step 2
$C = \pi d$	$154 = \frac{22}{7} \times d$ $\frac{7}{\cancel{22}_1} \times \cancel{154}^7 = \frac{\cancel{22}^1}{\cancel{7}_1} \times d \times \frac{\cancel{7}^1}{\cancel{22}_1}$ $\mathbf{49} = d$

Step 1: Set up the formula.
Step 2: Substitute the given values. Substitute 154 for C and $\frac{22}{7}$ for π. Then solve for d using the inverse operation.

EXERCISE 10

DIRECTIONS: Solve each problem.

1. Find the circumference of a circle with a diameter of 10 meters. Use $\pi = 3.14$.

2. Find the circumference of a circle with a diameter of 35 yards. Use $\pi = \frac{22}{7}$.

3. Find the circumference of a circle with a diameter of $\frac{1}{2}$ inch. Use $\pi = \frac{22}{7}$.

4. The radius of a circle is 20 inches. Find the circumference. Use $\pi = 3.14$.

5. The radius of a circle is $3\frac{1}{2}$ inches. Find the circumference. Use $\pi = \frac{22}{7}$.

6. The circumference of a circle is 94.2 meters. Find the diameter. Use $\pi = 3.14$.

7. The circumference of a circle is 176 feet. Find the diameter. Use $\pi = \frac{22}{7}$.

8. Find the circumference of a circle with a diameter of 1.5 inches. Use $\pi = 3.14$.

9. The distance across a circular pool is 21 feet. How many feet of sheet metal are needed to go around the edge of the pool? Use $\pi = \frac{22}{7}$.

Check your answers on page 309.

Area of a Circle

The area of a circle is the amount of space inside the circle. To find the area of a circle, you also have to use a formula. The formula for the area of a circle is $A = \pi r^2$. A stands for the area; r stands for the radius of the circle; and π stands for 3.14 or $\frac{22}{7}$.

EXAMPLE: Find the area of a circle that has a radius of 3 inches. Use $\pi = 3.14$.

Step 1	Step 2
$A = \pi r^2$	$A = \mathbf{3.14} \times \mathbf{3^2}$ $A = 3.14 \times 9$ $A = \mathbf{28.26}$ **sq. in.**

Step 1: Set up the formula to find the area of a circle.
Step 2: Substitute 3 for r and 3.14 for π. Solve the problem. Remember to express area in square units.

In some problems, you may be given the diameter and asked to find the area. Remember that the radius is $\frac{1}{2}$ of the diameter. First, multiply the diameter by $\frac{1}{2}$ to find the radius. Then find the area.

EXAMPLE: Find the area of a circle that has a diameter of 140 feet. Use $\pi = \frac{22}{7}$.

Step 1	Step 2
$r = 140 \times \frac{1}{2} = \mathbf{70}$	$A = \pi r^2$ $A = \frac{22}{7} \times 70^2$ $A = \frac{22}{7} \times 4{,}900$ $A = $ **15,400 sq. ft.**

Step 1: First, find $\frac{1}{2}$ of the diameter to get the radius.

Step 2: Set up the formula for area of a circle. Substitute 70 for r and $\frac{22}{7}$ for π. Solve the formula.

EXERCISE 11

DIRECTIONS: Solve each problem using the formula for the area of a circle.

1. The radius of a circle is 10 feet. Find the area. Use $\pi = 3.14$.
2. The radius of a circle is 7 inches. Find the area. Use $\pi = \frac{22}{7}$.
3. The radius of a circle is 3.5 yards. Find the area. Use $\pi = \frac{22}{7}$.
4. The radius of a circle is 6 meters. Find the area to the nearest square meter. Use $\pi = 3.14$.
5. The diameter of a circle is 28 inches. Find the area. Use $\pi = \frac{22}{7}$.
6. The diameter of a circle is 1.8 centimeters. Find the area to the nearest tenth of a square centimeter. Use $\pi = 3.14$.
7. The picture at the right is a diagram of a pool with a circular walk around it. The distance from the center of the pool to its edge is 10 feet. The distance from the center of the pool to the outer edge of the walk is 15 feet. Find the area of the walk. Use $\pi = 3.14$.

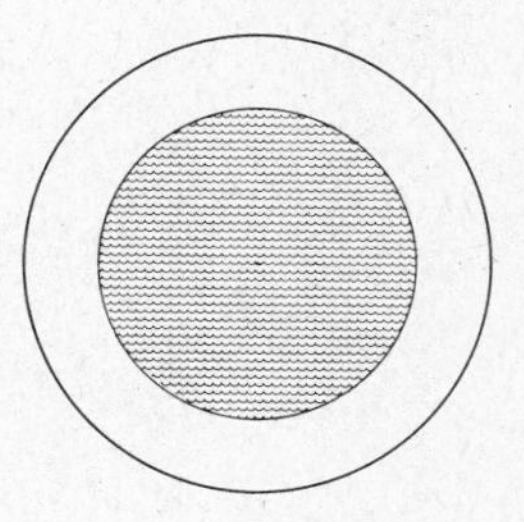

8. The diameter of the semicircle at the right is 21 feet. Find the area of the semicircle. Use $\pi = \frac{22}{7}$.

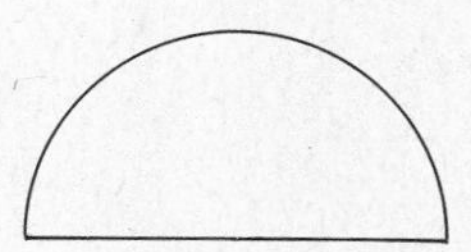

Check your answers on page 310.

VOLUME

Up to now, you have worked only with plane figures. Triangles, rectangles, squares, and circles all are plane figures. Now, you will work with solid figures. A box is a solid figure. An ice cube is a solid figure. A can is a solid figure.

On the GED Test, you will be expected to know one thing about solid figures. You will be expected to know how to find the volume of a solid figure. The volume is the amount of space inside the solid figure. For instance, look at the box on the left. The box is divided into 1-inch cubes.

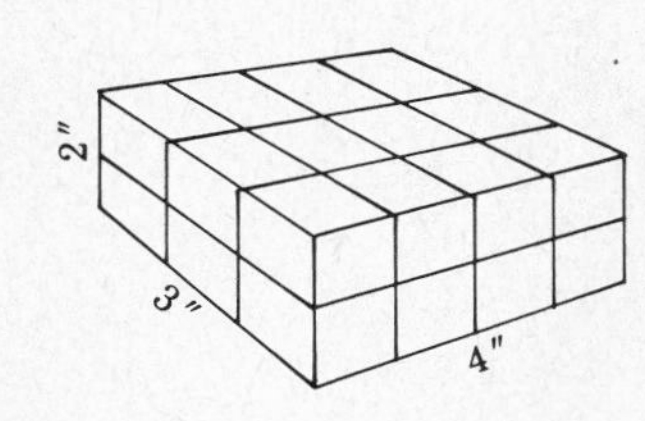

To find the volume of any solid figure, first find the area of its top surface. Then multiply the top surface by the depth of the figure. In the example above, the top surface of the box measures 4 inches by 3 inches. The depth of the box is 2 inches.

$4 \times 3 = 12$ $12 \times 2 =$ **24 cubic inches**

When you find the volume of a solid figure, always express the answer in cubic units.

EXAMPLE: The Rizzo Construction Co. dug a hole for the foundation of a house. The hole was 20 yards long, 12 yards wide, and 3 yards deep. Find the volume of the hole.

Step 1	Step 2
$20 \times 12 =$ **240**	$240 \times 3 =$ **720 cu. yds.**

Step 1: Find the area of the top surface of the hole.
Step 2: Multiply the area by the depth of the hole.

To find the volume of a cylinder, use the formula for finding the area of a circle to get the area of the flat surface. Then multiply the area by the depth of the cylinder.

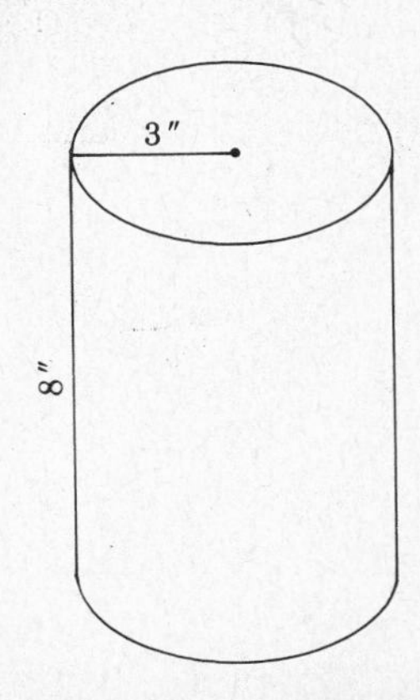

EXAMPLE: Find the volume of the cylinder at the left. Use $\pi = 3.14$.

Step 1	Step 2
$A = \pi r^2$ $A = 3.14 \times 3^2$ $A =$ **28.26 sq. in.**	28.26 $\times$ 8 **226.08 cu. in.**

Step 1: Find the area of the flat surface.
Step 2: Multiply the area by the depth of the cylinder.

EXERCISE 12

DIRECTIONS: Solve each problem.

1. Find the volume of a box that has a length of 12 inches, a width of 8 inches, and a depth of 5 inches.

2. The swimming pool at the Garfield Sports Center is 100 feet long, 24 feet wide, and 8 feet deep. Find the volume of the pool.

3. Mr. Munro poured a concrete walk at the side of his driveway. The walk is 50 feet long, 3 feet wide, and $\frac{1}{4}$ foot deep. Find the volume of the concrete in the sidewalk.

4. How many cubic feet are there in one cubic yard?

5. How many cubic inches are there in one cubic foot?

6. Find the volume of a cube of ice that measures $\frac{3}{4}$ inch on every side.

7. Find the volume of a cubic block of granite that measures 4.5 feet on every side. Round off the answer to the nearest cubic foot.

8. How many boxes, each measuring 2 feet on every side, can fit into a container that is 20 feet long, 8 feet wide, and 6 feet deep?

9. A circular pool has a radius of 7 feet and a depth of 2 feet. Find the volume of the pool. Use $\pi = \frac{22}{7}$.

10. One cubic foot of water weighs 62.4 pounds. Find the weight of the water that the pool in problem 9 can hold.

Check your answers on page 311.

Similarity

When two plane figures have the same shape, they are similar. Similar figures have the same shape but different sizes. The rectangles on the next page are similar. They are similar because the ratio of the short side to the long side for each figure is 2:3. To find the measurements of similar figures, use ratios and proportions.

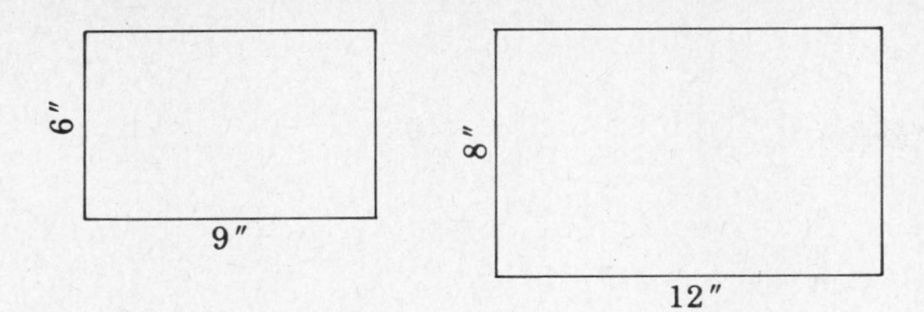

EXAMPLE: A snapshot is 3 inches wide and 5 inches long. The snapshot is enlarged to make the long side 20 inches. How wide is the enlarged snapshot?

$\frac{3}{5} = \frac{x}{20}$	$5x = 60$	$\frac{5x}{5} = \frac{60}{5}$ $x = $ **12 inches**
Step 1	Step 2	Step 3

Step 1: Set up the measurements in a proportion. Use x to stand for the missing measurement.
Step 2: Find the cross products and set up a new equation.
Step 3: Solve the equation for x.

In similar triangles, the angles are the same. The ratio of the sides of one triangle is the same as the ratio of corresponding sides of the other triangle. In the triangles below, $\angle A = \angle D$, $\angle B = \angle E$, and $\angle C = \angle F$. Side AB corresponds to side DE. Side BC corresponds to side EF. Side AC corresponds to side DF.

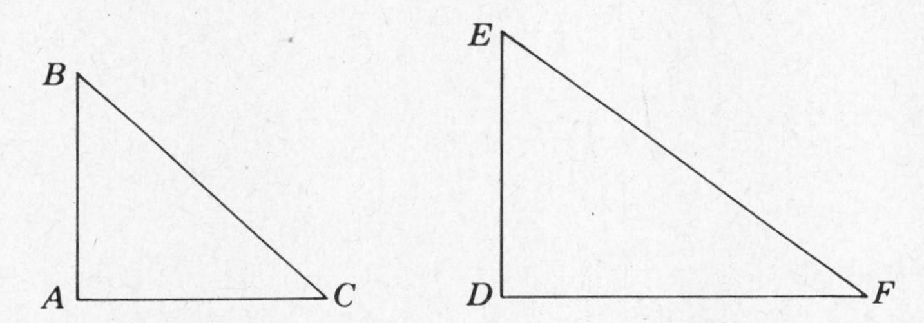

EXAMPLE: In triangle ABC above, side AB is 6″ and side BC is 10″. In triangle DEF, side DE is 9″. Find the length of side EF.

$\frac{6}{10} = \frac{9}{x}$	$6x = 90$	$\frac{6x}{6} = \frac{90}{6}$ $x = $ **15″**
Step 1	Step 2	Step 3

Step 1: Set up a proportion. Compare the short sides to the long sides. Use x for the unknown long side.
Step 2: Write a new equation with the cross products.
Step 3: Solve the equation.

In some pairs of similar triangles, it is hard to know which sides are corresponding. Remember that corresponding sides are across from equal angles.

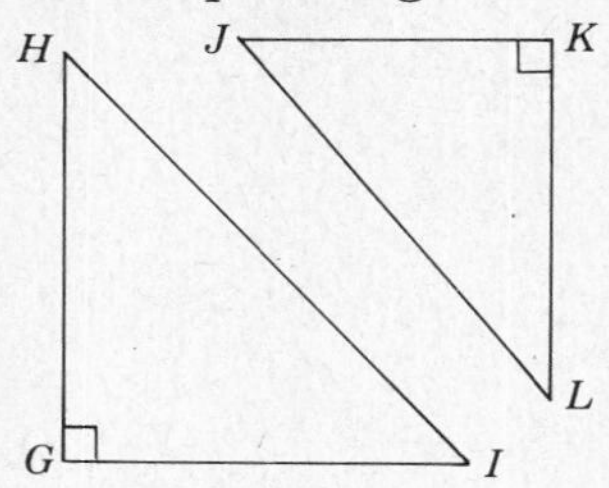

In the triangles at the left, both $\angle H$ and $\angle L$ have 50°. Side *GI* is across from $\angle H$. Side *JK* is across from $\angle L$. Side *GI* corresponds to side *JK*. Both $\angle G$ and $\angle K$ have 90°. The sides across from them, *HI* and *JL*, are corresponding.

EXERCISE 13

DIRECTIONS: Solve each problem.

1. Which side of triangle *JKL* above corresponds to side *GH* in triangle *GHI*?

2. Jill had a 12-inch wide and 18-inch long photograph reduced. The long side of the reduced picture is 6 inches. Find the short side.

In the triangles at the right, $\angle M = \angle Q$, $\angle N = \angle P$, and $\angle MON = \angle POQ$. Use these two triangles to answer problems 3 through 6.

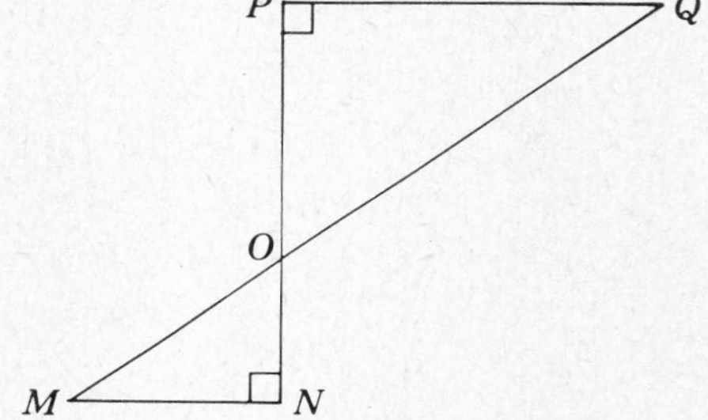

3. Which side of *OPQ* corresponds to side *MN*?

4. Which side of *OPQ* corresponds to side *NO*?

5. Which side of *MNO* corresponds to side *OQ*?

6. In triangle *MNO*, side *MN* is 12 yards and side *NO* is 8 yards. In triangle *OPQ*, side *PQ* is 21 yards. Find side *OP*.

7. Frank has a photograph 2 inches wide and 3 inches long. He wants to enlarge the photograph to make it 8 inches wide. How long will the new photograph be?

Check your answers on page 311.

Pythagorean Theorem

There is a special relationship among the sides of a right triangle. The relationship was discovered by the Greek mathematician Pythagoras. The relationship is called the **Pythagorean Theorem.**

According to the Pythagorean Theorem, the square of the hypotenuse of a right triangle equals the sum of the squares of the other two sides. (Remember that the hypotenuse is the side across from the right angle in a right triangle.)

The word **square** in this rule means a number to the second power. As a formula, the Pythagorean Theorem is: $a^2 + b^2 = c^2$; a and b stand for the legs of a right triangle, and c stands for the hypotenuse.

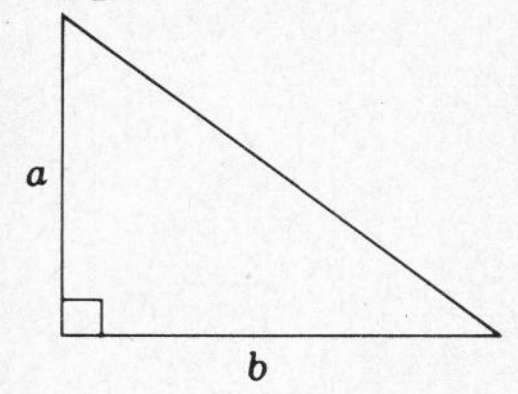

EXAMPLE: In the right triangle at the left, $a = 12$ inches and $b = 16$ inches. Find the length of the hypotenuse.

Step 1	Step 2
$a^2 + b^2 = c^2$ $\mathbf{12^2 + 16^2} = c^2$	$144 + 256 = c^2$ $400 = c^2$ $\sqrt{400} = \sqrt{c^2}$ **20 in.** $= c$

Step 1: Substitute 12 for a and 16 for b in the formula.

Step 2: Solve for c. Remember that the square root is the opposite operation for a number to the second power. To solve the equation $400 = c^2$, find the square root of both sides. (In geometry, the square root is a positive number.)

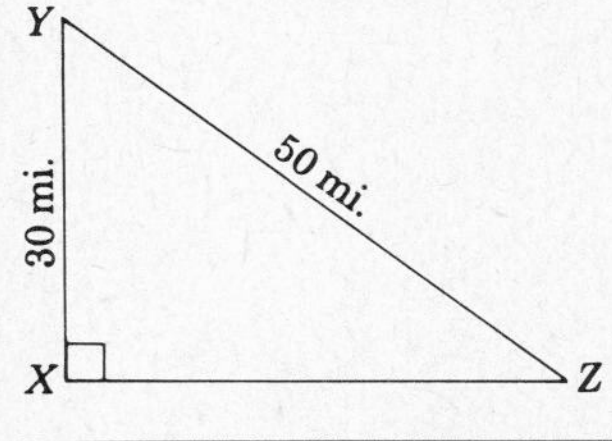

EXAMPLE: Find the distance from X to Z in the diagram at the left.

Step 1	Step 2	Step 3
$a^2 + b^2 = c^2$ $\mathbf{30^2} + b^2 = \mathbf{50^2}$	$900 + b^2 = 2{,}500$ $\mathbf{-900} \quad = \mathbf{-900}$ $b^2 = \mathbf{1{,}600}$	$b^2 = 1{,}600$ $\sqrt{b^2} = \sqrt{1{,}600}$ $b =$ **40 mi.**

Step 1: Substitute the values in the formula. In this problem, you are given the length of the hypotenuse, and you are asked to find the length of one of the legs.

Step 2: Use inverse operations to get b by itself.

Step 3: Solve for b using the square root.

Before you do the next exercises, review the list of common square roots and the ways to find square roots on page 201.

EXERCISE 14

DIRECTIONS: Solve each problem.

1. The short leg of a right triangle measures 36 feet. The long leg measures 48 feet. Find the hypotenuse.

2. One of the legs of a right triangle measures 8 inches. The hypotenuse measures 17 inches. Find the measurement of the other leg.

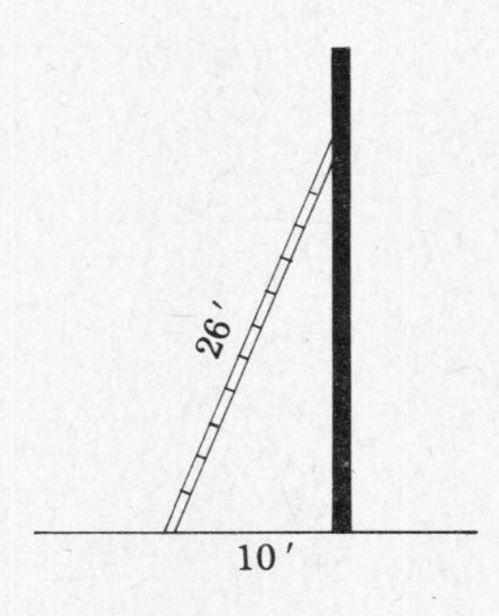

3. The picture at the left shows a 26-foot ladder leaning against a building. The bottom of the ladder is 10 feet from the building. Find the distance from the top of the ladder directly down to the ground.

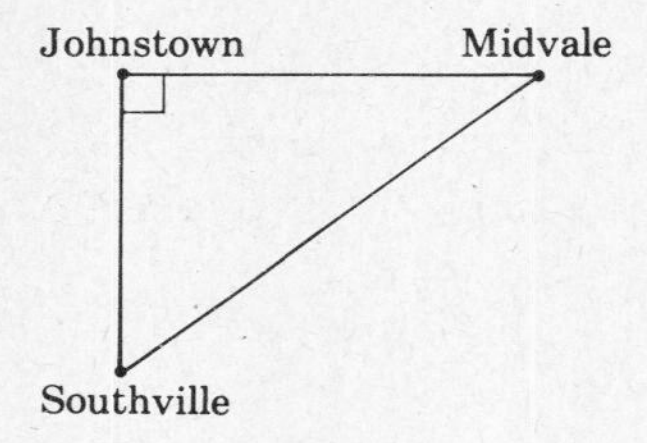

4. The distance from Johnstown to Southville on the map at the left is 24 miles. The distance from Johnstown to Midvale is 32 miles. What is the direct distance from Midvale to Southville?

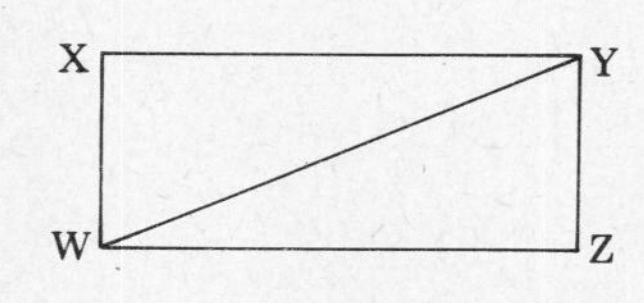

5. In rectangle *WXYZ*, side *WX* measures 5 inches. Side *XY* measures 12 inches. Find the length of the diagonal line from *W* to *Y*.

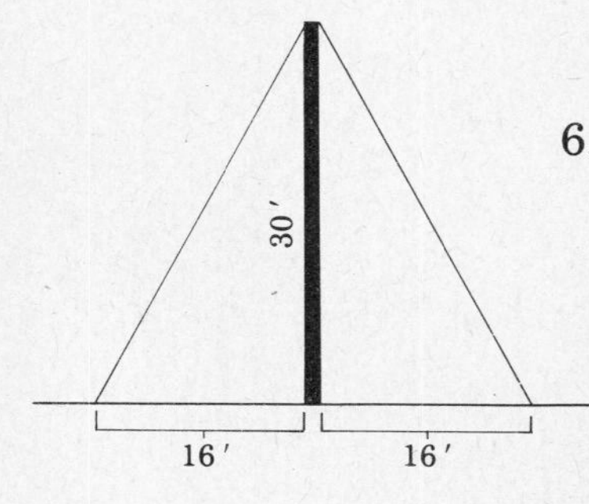

6. The diagram at the left shows two wires stretching from the top of a pole to the ground. The pole is 30 feet high. Each wire touches the ground 16 feet from the pole. How long is each wire?

7. One leg of a right triangle measures 18 feet. The hypotenuse measures 30 feet. Find the length of the other leg.

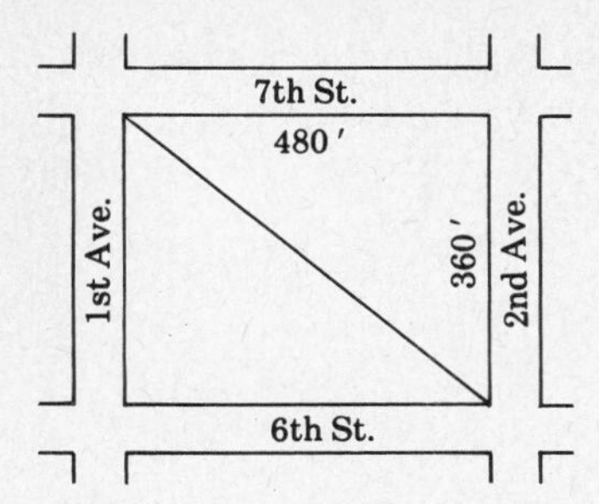

8. The diagram at the left shows a plan of a city park. The park is 480 feet long on 7th and 6th Streets. It is 360 feet wide on 1st and 2nd Avenues. What is the distance from the corner of 1st Avenue and 7th Street to the corner of 2nd Avenue and 6th Street?

9. One leg of a triangle measures 21 meters. The other leg measures 28 meters. Find the measurement of the hypotenuse.

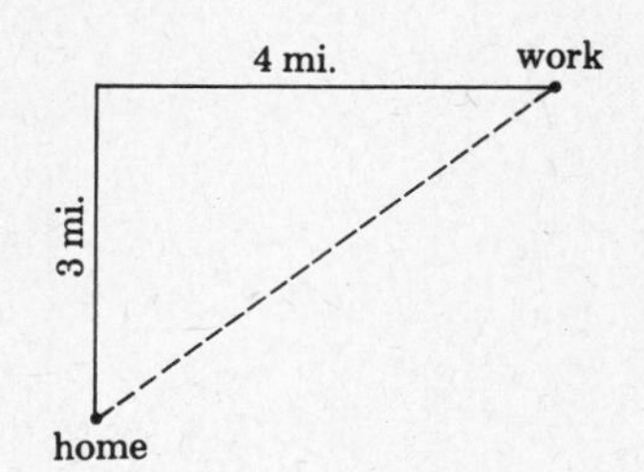

10. The solid lines on the diagram at the left show the route Pete drives to work. First he drives 3 miles north. Then he drives 4 miles east. Find the distance shown by the dotted line from Pete's house to his work.

Check your answers on page 312.

Geometry Word Problems

In the next exercise, you will have a chance to work with the types of geometry word problems that you may find on the GED Test. Work each problem slowly and carefully. Use the hints and formulas given in this section to solve the problems.

EXERCISE 15 (WORD PROBLEMS)

DIRECTIONS: Choose the correct answer to each problem.

1. An oil tank in the shape of a cylinder has a radius of 5 feet and a depth of 14 feet. The volume of the tank, in cubic feet, is

   (1) 440 (2) 880 (3) 1,100
   (4) 1,400 (5) 13,400

2. The blueprint of a floor plan for Tom's apartment is 30 inches wide and 50 inches long. Tom made a reduced copy of the blueprint. The reduced copy is 8 inches wide. How long is the reduced copy?

   (1) 10″ (2) $13\frac{1}{3}$″ (3) 16″ (4) 18″ (5) 24″

3. In triangle *AEF*, if $\angle A$ is 40° and $\angle E$ is 68°, then $\angle F$ is

   (1) 28° (2) 54° (3) 62° (4) 72° (5) 108°

4. How many square centimeters are in a square meter?

   (1) 10 (2) 100 (3) 1,000
   (4) 10,000 (5) 100,000

5. A football field is 55 yards wide and 100 yards long. In square yards, the area of a football field is

   (1) 155 (2) 310 (3) 550 (4) 5,500 (5) 55,000

6. Jed wants to put a fence around his property. If his property is a rectangular piece of land that is 40 feet long and 32 feet wide, how many feet of fencing will Jed need?

   (1) 72 (2) 144 (3) 184 (4) 288 (5) 1,280

7. One leg of a right triangle is 3 meters long. The other leg is 4 meters long. How many meters long is the hypotenuse?

   (1) 5 (2) 7 (3) 12 (4) 25 (5) 50

8. The circumference of a circle that has a radius of 14 inches is

   (1) 44″ (2) 88″ (3) 132″ (4) 176″ (5) 220″

9. A box measures 1 foot wide, 1 foot long, and 1 foot deep. In cubic inches, the volume of the box is

   (1) 1 (2) 3 (3) 12 (4) 36 (5) 1,728

10. The area, in square meters, of a triangle that has a base of 15 meters and a height of 12 meters is

    (1) 27 (2) 54 (3) 90 (4) 108 (5) 180

Check your answers on page 312.

Geometry Review

These problems will give you a chance to practice the skills you studied in the geometry section of this book. With each answer is the name of the section in which the skills needed for that problem are explained. For every problem you get wrong, look over the section in which the skills for that problem are explained. Then try the problem again.

DIRECTIONS: Solve each problem.

1. The name for an angle that measures 79° is ______________.

2. Find the complement of an angle that measures 14° 19′ 23″.

3. A triangle with three equal sides is called ______________.

4. Figure *ABCD* has four right angles. *AB* = *CD* and *BC* = *AD*. Figure *ABCD* is a ______________.

5. The distance across a circle at its widest part is called the ______________.

6. In triangle *MNO* at the right, *MN* = 9″ and *MO* = 15″. In triangle *OPQ*, *OP* = 20″. Find the length of *PQ*.

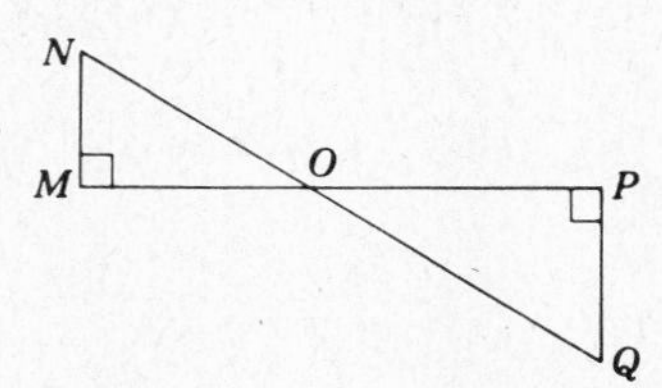

7. In rectangle *EFGH* at the right, *EH* = 24 ft. The diagonal line *EG* = 26 ft. Find the length of side *GH*.

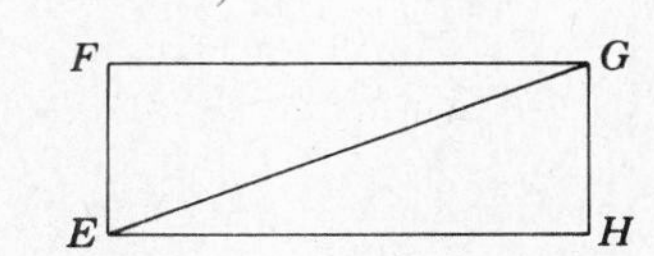

8. Find the perimeter of a square that measures 4.7 meters on a side.

9. Find the circumference of a circle that has a diameter of 2.4 centimeters. Use $\pi = 3.14$.

10. The bathroom in Fran's apartment is $5\frac{1}{2}$ feet wide and 6 feet long. Find the area of the floor of the bathroom.

11. The diameter of the semicircle at the right is 14 inches. Find the area of the semicircle. Use $\pi = \frac{22}{7}$.

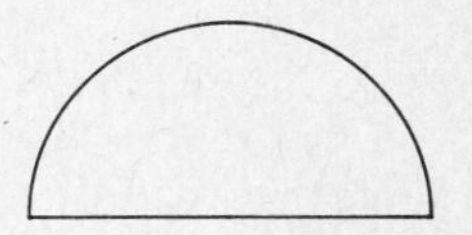

12. How many boxes each measuring 3 feet on a side can fit into a container that is 9 feet wide, 12 feet tall, and 18 feet long?

Check your answers on page 313.

ANSWERS & SOLUTIONS

EXERCISE 1

1. 30° is **acute.** 80° is **acute.** 110° is **obtuse.** 90° is **right.**

2. 270° is **reflex.** 180° is **straight.** 179° is **obtuse.** 89° is **acute.**

EXERCISE 2

1. **108°**

$$\begin{array}{r} 180^\circ \\ -\ 72^\circ \\ \hline \mathbf{108^\circ} \end{array}$$

2. **18°**

$$\begin{array}{r} 90^\circ \\ -\ 72^\circ \\ \hline \mathbf{18^\circ} \end{array}$$

3. **59° 31′ 16″**

$$\begin{array}{r} 179^\circ\ 59' \quad \\ \cancel{180}^\circ\ \cancel{60}'\ 60'' \\ -\ 120^\circ\ 28'\ 44'' \\ \hline \mathbf{59^\circ\ 31'\ 16''} \end{array}$$

4. **53° 7′ 43″**

$$\begin{array}{r} 89^\circ\ 59' \quad \\ \cancel{90}^\circ\ \cancel{60}'\ 60'' \\ -\ 36^\circ\ 52'\ 17'' \\ \hline \mathbf{53^\circ\ \ 7'\ 43''} \end{array}$$

5. **53°**

$$\begin{array}{r} 90^\circ \\ -\ 37^\circ \\ \hline \mathbf{53^\circ} \end{array}$$

6. **56°**

$$\begin{array}{r} 180^\circ \\ -\ 124^\circ \\ \hline \mathbf{56^\circ} \end{array}$$

7. **44°**

8. $\angle \boldsymbol{BOC}$

9. $\angle \boldsymbol{COD}$

10. **117°**

$$\begin{array}{r} 180^\circ \\ -\ 63^\circ \\ \hline \mathbf{117^\circ} \end{array}$$

11. **45°**

$$\begin{array}{r} 180^\circ \\ -\ 135^\circ \\ \hline \mathbf{45^\circ} \end{array}$$

12. **180°**

13. $\mathbf{37\frac{1}{2}°}$

$$\begin{array}{r} 180° \\ -\ 30° \\ \hline 150° \end{array}$$

$$\frac{4m}{4} = \frac{150}{4}$$

$$m = \mathbf{37\tfrac{1}{2}°}$$

14. **small angle = 18°**
big angle = 4(18) = **72°**

x = small angle
$4x$ = big angle

$$x + 4x = 90$$
$$\frac{5x}{5} = \frac{90}{5}$$
$$x = 18$$

15. **small angle = 76°**
big angle = 76 + 28 = **104°**

x = small angle
$x + 28$ = big angle

$$\begin{aligned} x + x + 28 &= 180 \\ 2x + 28 &= 180 \\ -\ 28 &\quad -\ 28 \\ \frac{2x}{2} &= \frac{152}{2} \\ x &= 76 \end{aligned}$$

EXERCISE 3

1. $\angle \boldsymbol{b}$ 2. $\angle \boldsymbol{e}$ 3. $\angle \boldsymbol{g}$ 4. **125°** 5. **60°**

6. **95°**

$$\begin{array}{r} 180° \\ -\ 85° \\ \hline \mathbf{95°} \end{array}$$

7. **alternate interior** 8. **corresponding** 9. **alternate exterior**

10. **vertical**

EXERCISE 4

1. **70°**

$$\begin{array}{r} 65° \\ +\ 45° \\ \hline 110° \end{array} \qquad \begin{array}{r} 180° \\ -\ 110° \\ \hline \mathbf{70°} \end{array}$$

2. $\angle \boldsymbol{A}$ **= 40°**
$\angle \boldsymbol{B}$ **=** 2(40) = **80°**
$\angle \boldsymbol{C}$ **=** 2(40) − 20 = **60°**

$x = \angle A$
$2x = \angle B$
$2x - 20 = \angle C$

$$\begin{aligned} x + 2x + 2x - 20 &= 180 \\ 5x - 20 &= 180 \\ +\ 20 &\quad +\ 20 \\ \frac{5x}{5} &= \frac{200}{5} \\ x &= 40 \end{aligned}$$

3. **130°**

$$\begin{array}{r} 25° \\ +\ 25° \\ \hline 50° \end{array} \qquad \begin{array}{r} 180° \\ -\ 50° \\ \hline \mathbf{130°} \end{array}$$

4. $\mathbf{57\frac{1}{2}°}$

$$\begin{array}{r} 180° \\ -\ 65° \\ \hline 115° \end{array} \qquad 2\overline{)115}\ \ \mathbf{57\tfrac{1}{2}°}$$

5. $\mathbf{55°}$

$$\begin{array}{r} 90° \\ +\ 35° \\ \hline 125° \end{array} \qquad \begin{array}{r} 180° \\ -\ 125° \\ \hline \mathbf{55°} \end{array}$$

6. $\angle \boldsymbol{D} = \mathbf{30°}$
$\angle \boldsymbol{E} = 30 + 20 = \mathbf{50°}$
$\angle \boldsymbol{F} = 2(30 + 20) = 2(50) = \mathbf{100°}$

$$\begin{aligned} x &= \angle D \\ x + 20 &= \angle E \\ 2(x + 20) &= \angle F \\ x + x + 20 + 2(x + 20) &= 180 \\ 2x + 20 + 2x + 40 &= 180 \\ 4x + 60 &= 180 \\ -\ 60 &\quad -\ 60 \\ \frac{4x}{4} &= \frac{120}{4} \\ x &= 30 \end{aligned}$$

7. $\angle \boldsymbol{M} = \mathbf{36°}$
$\angle \boldsymbol{O} = \mathbf{36°}$
$\angle \boldsymbol{N} = 3(36) = \mathbf{108°}$

$$\begin{aligned} x &= \angle M \\ x &= \angle O \\ 3x &= \angle N \\ x + x + 3x &= 180 \\ \frac{5x}{5} &= \frac{180}{5} \\ x &= 36 \end{aligned}$$

8. $\angle \boldsymbol{Q} = \mathbf{30°}$
$\angle \boldsymbol{R} = 2(30) = \mathbf{60°}$

$$\begin{aligned} x &= \angle Q \\ 2x &= \angle R \\ x + 2x + 90 &= 180 \\ 3x + 90 &= 180 \\ -\ 90 &\quad -\ 90 \\ \frac{3x}{3} &= \frac{90}{3} \\ x &= 30 \end{aligned}$$

9. ***TU* is the hypotenuse.**

EXERCISE 5

1. ***CD* = 12 in.**
2. ***AD* = 9 in.**
3. **90°**
4. **isosceles** because $\angle BAO$ and $\angle ABO$ are the same
5. **trapezoid**
6. **bases**
7. **parallelogram**
8. **125°**
9. **rhombus**

EXERCISE 6

1. **31 in.**

$$\begin{array}{r} 12 \\ 12 \\ +\ 7 \\ \hline \mathbf{31} \end{array}$$

2. **39 in.**

$$\begin{array}{rr} \left(8\tfrac{1}{2} \times 2\right) = & 17 \\ (11 \times 2) = & +22 \\ \hline & \mathbf{39} \end{array}$$

3. **57 ft.**

$$\begin{array}{r} 19 \\ 19 \\ +\ 19 \\ \hline \mathbf{57} \end{array}$$

4. **width = 9 ft.**
length = 4(9) = **36 ft.**

x = width $\quad x \times 2 = 2x$
$4x$ = length $\quad 4x \times 2 = 8x$

$$2x + 8x = 90$$
$$10x = 90$$
$$x = \mathbf{9}$$

5. **19 m.**

$$\begin{array}{r} 4.75 \\ 4.75 \\ 4.75 \\ +\ 4.75 \\ \hline \mathbf{19.00} \end{array}$$

6. **740 in.**

$$\left(30\tfrac{1}{2} \times 2\right) = 61$$
$$(62 \times 2) = +124$$
$$185$$

$185 \times 4 = \mathbf{740}$

7. **$61\frac{2}{3}$ ft.**

$$\begin{array}{r} \mathbf{61\tfrac{2}{3}} \\ 12\overline{)740} \\ -72 \\ \hline 20 \\ -12 \\ \hline 8 \end{array}$$

8. **$6\frac{1}{4}$ yds.**

$$\begin{array}{r} \mathbf{6\tfrac{1}{4}} \\ 4\overline{)25} \\ -24 \\ \hline 1 \end{array}$$

EXERCISE 7

1. **48 sq. ft.**

$$\begin{array}{r} 16 \\ \times 3 \\ \hline \mathbf{48} \end{array}$$

2. **126 tiles**

area of wall = 54 × 42 = 2,268 sq. in.
area of tile = 3 × 6 = 18 sq. in.

$$\begin{array}{r} \mathbf{126} \\ 18\overline{)2{,}268} \end{array}$$

3. **25 sq. yds.**

area of patio = 15 × 15 = 225 sq. ft.
area of a square yard = 3 × 3 = 9 sq. ft.

$$\begin{array}{r} \mathbf{25} \\ 9\overline{)225} \end{array}$$

You also may convert the 15 feet to yards before you multiply:

15 ft. = 5 yds. $\quad$ 5 × 5 = **25 sq. yds.**

4. **18 ft.**

Change 15 feet to yards: 15 ft. = 5 yds.
Set up to find the missing length: 30 sq. yds. = $5 \times l$

$$\frac{30}{5} = \frac{5l}{5}$$
$$6 = l$$

Convert yards back into feet: 6 yds. = **18 ft.**

5. **225 tiles**

area of table = 30 × 30 = 900 sq. in. $4\overline{)900}$ → **225**
area of tile = 2 × 2 = 4 sq. in.

6. **$572**

area of living room = 12 × 15 = 180 sq. ft. $9\overline{)180}$ → 20
area of a square yard = 3 × 3 = 9 sq. ft.

area of dining room = 9 × 12 = 108 sq. ft. $9\overline{)108}$ → 12

area of bedroom = 9 × 12 = 108 sq. ft. $9\overline{)108}$ → 12

total area = 20 + 12 + 12 = 44 sq. yds.
price = 44 × $13 = **$572**

7. **1,648 sq. ft.**

$$\begin{array}{r} 206 \\ \times\ 8 \\ \hline \mathbf{1{,}648} \end{array}$$

8. **2 bags**

area of garden = 18 × 12 = 216 sq. ft.
(1 bag covers 180 sq. ft.)

$$\begin{array}{r} 1\frac{1}{5} \\ 180\overline{)216} \\ -\ 180 \\ \hline 36 \end{array}$$

(1 bag is not enough. She will need **2** bags.)

9. **1,500 squares**

area of the space = 120 × 200 = 24,000 sq. ft. $16\overline{)24{,}000}$ → **1,500**
area of a square = 4 × 4 = 16 sq. ft.

10. **60 tiles**

area of kitchen = 54 × 90 = 4,860 sq. in. $81\overline{)4{,}860}$ → **60**
area of a tile = 9 × 9 = 81 sq. in.

EXERCISE 8

1. **$67\frac{1}{2}$ sq. in.**

$15 \times 9 \times \frac{1}{2} = \mathbf{67\frac{1}{2}}$

2. **.225 sq. m.**

$\frac{1}{2}(.75)(.6) = \mathbf{.225}$

3. **6 ft.**

$60 = b \times 20 \times \frac{1}{2}$

$60 = b \times 10$

$\frac{60}{10} = \frac{10b}{10}$

$\mathbf{6} = b$

4. **$2\frac{3}{16}$ sq. in.**

$2\frac{1}{2} \times 1\frac{3}{4} \times \frac{1}{2} =$

$\frac{5}{2} \times \frac{7}{4} \times \frac{1}{2} = \frac{35}{16} = \mathbf{2\frac{3}{16}}$

5. **400 sq. ft.**

$$\begin{array}{r} 25 \\ \times\ 16 \\ \hline \mathbf{400} \end{array}$$

6. **48 sq. in.**

$$\begin{array}{r} 8 \\ \times\ 6 \\ \hline \mathbf{48} \end{array}$$

7. **70 sq. m.**

$$\begin{array}{r} 3.5 \\ \times\ 20 \\ \hline \mathbf{70.0} \end{array}$$

8. **45 sq. ft.**

$10 + 5 = 15$

$15 \times 6 \times \frac{1}{2} = \mathbf{45}$

9. **165 sq. in.**

$12 + 18 = 30$

$30 \times 11 \times \frac{1}{2} = \mathbf{165}$

10. **528 sq. ft.**

$20 + 28 = 48$

$48 \times 22 \times \frac{1}{2} = \mathbf{528}$

EXERCISE 9

1. **diameter** 2. **radius** 3. **2:1**

4. $\frac{18}{2} = \mathbf{9\ in.}$ 5. $2(6.5) = \mathbf{13\ m.}$ 6. $\frac{25}{2} = \mathbf{12\frac{1}{2}\ ft.}$

7. $2 \times \frac{3}{8} = \frac{6}{8} = \mathbf{\frac{3}{4}\ in.}$ 8. $\frac{1}{2} = \mathbf{50\%}$ 9. $\boldsymbol{\pi}$

10. **circumference**

EXERCISE 10

1. **31.4 m.**

$$\begin{aligned} C &= \pi d \\ &= 3.14 \cdot 10 \\ &= \mathbf{31.4} \end{aligned}$$

2. **110 yds.**

$$\begin{aligned} C &= \pi d \\ &= \frac{22}{7} \cdot 35 \\ &= \frac{22}{\cancel{7}_{1}} \cdot \frac{\cancel{35}^{5}}{1} \\ &= \mathbf{110} \end{aligned}$$

3. $\mathbf{1\frac{4}{7}}$ **in.**

$$\begin{aligned} C &= \pi d \\ &= \frac{22}{7} \cdot \frac{1}{2} \\ &= \frac{\cancel{22}^{11}}{7} \cdot \frac{1}{\cancel{2}_{1}} \\ &= \frac{11}{7} = \mathbf{1\frac{4}{7}} \end{aligned}$$

4. **125.6 in.**

$$\begin{aligned} d &= 2r \\ d &= 2(20) \\ &= 40 \\ C &= \pi d \\ &= 3.14(40) \\ &= \mathbf{125.6} \end{aligned}$$

5. **22 in.**

$$\begin{aligned} d &= 2r \\ d &= 2\left(3\tfrac{1}{2}\right) \\ &= 7 \\ C &= \pi d \\ C &= \frac{22}{7} \cdot 7 \\ &= \mathbf{22} \end{aligned}$$

6. **30 m.**

$$\begin{aligned} C &= \pi d \\ 94.2 &= 3.14d \end{aligned}$$

$$d = 3.14\overline{)94.20} = \mathbf{30}$$

$$\begin{array}{r} 94\,2 \\ \hline 00 \end{array}$$

7. **56 ft.**

$$C = \pi d$$
$$176 = \frac{22}{7}d$$
$$\frac{7}{\not{22}_1} \cdot \frac{\not{176}^{8}}{1} = \frac{22}{7}d \cdot \frac{7}{22}$$
$$\mathbf{56} = d$$

8. **4.71 in.**

$$C = \pi d$$
$$= 3.14 \cdot 1.5$$
$$= \mathbf{4.71}$$

9. **66 ft.**

$$C = \pi d$$
$$= \frac{22}{7} \cdot 21$$
$$= \frac{22}{\not{7}_1} \cdot \frac{\not{21}^{3}}{1}$$
$$= \mathbf{66}$$

EXERCISE 11

1. **314 sq. ft.**

$$A = \pi r^2$$
$$= 3.14(10)^2$$
$$= 3.14 \cdot 100$$
$$= \mathbf{314}$$

2. **154 sq. in.**

$$A = \pi r^2$$
$$= \frac{22}{7} \cdot 7^2$$
$$= \frac{22}{\not{7}_1} \cdot \frac{\not{49}^{7}}{1}$$
$$= \mathbf{154}$$

3. **38.5 sq. yds.**

$$A = \pi r^2$$
$$= \frac{22}{7}(3.5)^2$$
$$= \frac{22}{\not{7}_1} \cdot \frac{\not{12.25}^{1.75}}{1}$$
$$= \mathbf{38.5}$$

4. **113 sq. m.**

$$A = \pi r^2$$
$$= 3.14 \cdot 6^2$$
$$= 3.14 \cdot 36$$
$= 113.04$ to the nearest unit
$$= \mathbf{113}$$

5. **616 sq. in.**

$$r = \frac{28}{2} = 14 \text{ in.}$$
$$A = \pi r^2$$
$$= \frac{22}{7} \cdot 14^2$$
$$= \frac{22}{\not{7}_1} \cdot \frac{\not{196}^{28}}{1}$$
$$= \mathbf{616}$$

6. **2.5 sq. cm.**

$$r = \frac{1.8}{2} = .9 \text{ cm.}$$
$$A = \pi r^2$$
$$= 3.14(.9)^2$$
$$= 3.14(.81)$$
$= 2.5434$ to the nearest tenth
$$= \mathbf{2.5}$$

7. **392.5 sq. ft.**

area of pool $= \pi r^2$
$= 3.14 \cdot 10^2$
$= 3.14 \cdot 100$
$= 314$

area of pool + walk $= \pi r^2$
$= 3.14 \cdot 15^2$
$= 3.14 \cdot 225$
$= 706.5$

area of walk $= 706.5 - 314 = \mathbf{392.5}$

8. **$173\frac{1}{4}$ sq. ft.**

$$r = \frac{21}{2} = 10\frac{1}{2}$$

area of a semicircle $\left(\frac{1}{2}\right.$ of a circle$\left.\right)$

$$A = \frac{1}{2}\pi r^2$$
$$= \frac{1}{2} \cdot \frac{22}{7} \cdot \left(10\frac{1}{2}\right)^2$$
$$= \frac{1}{2} \cdot \frac{22}{7} \cdot \left(\frac{21}{2}\right)^2$$
$$= \frac{1}{\not{2}_1} \cdot \frac{\not{22}^{11}}{\not{7}_1} \cdot \frac{\not{441}^{63}}{4}$$
$$= \frac{693}{4}$$
$$= \mathbf{173\frac{1}{4}}$$

EXERCISE 12

1. **480 cu. in.**

   $12 \times 8 \times 5 = \mathbf{480}$

2. **19,200 cu. ft.**

   $100 \times 24 \times 8 = \mathbf{19{,}200}$

3. **$37\frac{1}{2}$ cu. ft.**

   $50 \times 3 \times \frac{1}{4} = \mathbf{37\frac{1}{2}}$

4. **27 cu. ft.**

   $3 \times 3 \times 3 = \mathbf{27}$

5. **1,728 cu. in.**

   $12 \times 12 \times 12 = \mathbf{1{,}728}$

6. **$\frac{27}{64}$ cu. in.**

   $\frac{3}{4} \times \frac{3}{4} \times \frac{3}{4} = \mathbf{\frac{27}{64}}$

7. **91 cu. ft.**

   $4.5 \times 4.5 \times 4.5 = 91.125$
   to the nearest unit $= \mathbf{91}$

8. **120 boxes**

   volume of container $= 20 \times 8 \times 6 = 960$
   volume of a box $= 2 \times 2 \times 2 = 8$

   $$8\overline{)960} = \mathbf{120}$$

9. **308 cu. ft.**

   $$V = \pi r^2 h$$
   $$= \frac{22}{7} \times 7^2 \times 2$$
   $$= \frac{22}{7} \times 49 \times 2 = \mathbf{308}$$

10. **19,219.2 lbs.**

    $$\begin{array}{r} 62.4 \\ \times\ 308 \\ \hline \mathbf{19{,}219.2} \end{array}$$

EXERCISE 13

1. ***KL***

   Both *KL* and *GH* are across from 40° angles.

2. **4 in.**

   $$\frac{12}{18} = \frac{x}{6}$$
   $$\frac{18x}{18} = \frac{72}{18}$$
   $$x = \mathbf{4}$$

3. ***PQ*** 4. ***OP*** 5. ***MO***

6. **14 yds.**

   $$\frac{12}{8} = \frac{21}{x}$$
   $$\frac{12x}{12} = \frac{168}{12}$$
   $$x = \mathbf{14}$$

7. **12 in.**

   $$\frac{2}{3} = \frac{8}{x}$$
   $$\frac{2x}{2} = \frac{24}{2}$$
   $$x = \mathbf{12}$$

EXERCISE 14

1. **60 ft.**

$$\begin{aligned} a^2 + b^2 &= c^2 \\ 36^2 + 48^2 &= c^2 \\ 1{,}296 + 2{,}304 &= c^2 \\ 3{,}600 &= c^2 \\ \sqrt{3{,}600} &= c \\ \mathbf{60} &= c \end{aligned}$$

2. **15 in.**

$$\begin{aligned} a^2 + b^2 &= c^2 \\ 8^2 + b^2 &= 17^2 \\ 64 + b^2 &= 289 \\ -64 \quad & \quad -64 \\ b^2 &= 225 \\ b &= \sqrt{225} \\ b &= \mathbf{15} \end{aligned}$$

3. **24 ft.**

$$\begin{aligned} a^2 + b^2 &= c^2 \\ 10^2 + b^2 &= 26^2 \\ 100 + b^2 &= 676 \\ -100 \quad & \quad -100 \\ b^2 &= 576 \\ b &= \sqrt{576} \\ b &= \mathbf{24} \end{aligned}$$

4. **40 mi.**

$$\begin{aligned} a^2 + b^2 &= c^2 \\ 24^2 + 32^2 &= c^2 \\ 576 + 1{,}024 &= c^2 \\ 1{,}600 &= c^2 \\ \sqrt{1{,}600} &= c \\ \mathbf{40} &= c \end{aligned}$$

5. **13 in.**

$$\begin{aligned} a^2 + b^2 &= c^2 \\ 5^2 + 12^2 &= c^2 \\ 25 + 144 &= c^2 \\ 169 &= c^2 \\ \sqrt{169} &= c \\ \mathbf{13} &= c \end{aligned}$$

6. **34 ft.**

$$\begin{aligned} a^2 + b^2 &= c^2 \\ 16^2 + 30^2 &= c^2 \\ 256 + 900 &= c^2 \\ 1{,}156 &= c^2 \\ \sqrt{1{,}156} &= c \\ \mathbf{34} &= c \end{aligned}$$

7. **24 ft.**

$$\begin{aligned} a^2 + b^2 &= c^2 \\ 18^2 + b^2 &= 30^2 \\ 324 + b^2 &= 900 \\ -324 \quad &= -324 \\ b^2 &= 576 \\ b &= \sqrt{576} \\ b &= \mathbf{24} \end{aligned}$$

8. **600 ft.**

$$\begin{aligned} a^2 + b^2 &= c^2 \\ 480^2 + 360^2 &= c^2 \\ 230{,}400 + 129{,}600 &= c^2 \\ 360{,}000 &= c^2 \\ \sqrt{360{,}000} &= c \\ \mathbf{600} &= c \end{aligned}$$

9. **35 m.**

$$\begin{aligned} a^2 + b^2 &= c^2 \\ 21^2 + 28^2 &= c^2 \\ 441 + 784 &= c^2 \\ 1{,}225 &= c^2 \\ \sqrt{1{,}225} &= c \\ \mathbf{35} &= c \end{aligned}$$

10. **5 mi.**

$$\begin{aligned} a^2 + b^2 &= c^2 \\ 3^2 + 4^2 &= c^2 \\ 9 + 16 &= c^2 \\ 25 &= c^2 \\ \sqrt{25} &= c \\ \mathbf{5} &= c \end{aligned}$$

EXERCISE 15 (WORD PROBLEMS)

1. **(3) 1,100**

$$A = \pi r^2$$

$$A = 25 \times \frac{22}{7}$$

$$\frac{25}{1} \times \frac{22}{7} \times \frac{14}{1} = \mathbf{1{,}100}$$

2. **(2)** $\mathbf{13\frac{1}{3}''}$

$$\frac{30}{50} = \frac{8}{x}$$

$$\frac{30x}{30} = \frac{400}{30}$$

$$x = \mathbf{13\frac{1}{3}}$$

3. **(4) 72°**

$$\begin{aligned} 40 + 68 + x &= 180 \\ 108 + x &= 180 \\ -108 &= -108 \\ \hline x &= \mathbf{72} \end{aligned}$$

4. **(4) 10,000**

100 centimeters = 1 meter
$100 \times 100 = \mathbf{10{,}000}$

5. **(4) 5,500**

$55 \times 100 = \mathbf{5{,}500}$

6. **(2) 144**

$$\begin{aligned} (40 \times 2) &= 80 \\ (32 \times 2) &= +64 \\ \hline &\mathbf{144} \end{aligned}$$

7. **(1) 5**

$$\begin{aligned} a^2 + b^2 &= c^2 \\ 3^2 + 4^2 &= c^2 \\ 9 + 16 &= c^2 \\ \sqrt{25} &= \sqrt{c^2} \\ 5 &= c \end{aligned}$$

8. **(2) 88″**

$14 \times 2 = 28$
$C = \pi d$
$C = \frac{22}{7} \times 28 = \mathbf{88}$

9. **(5) 1,728**

12 inches = 1 foot
$12 \times 12 \times 12 = \mathbf{1{,}728}$

10. **(3) 90**

$15 \times 12 \times \frac{1}{2} = \mathbf{90}$

GEOMETRY REVIEW

1. **acute**

(Angles)

2. **75° 40′ 37″**

$$\begin{aligned} &89^\circ\ 59' \\ &\not{90}^\circ\ \not{60}'\ 60'' \\ &-14^\circ\ 19'\ 23'' \\ \hline &\mathbf{75^\circ\ 40'\ 37''} \end{aligned}$$

(Pairs of Angles)

3. **equilateral**

(Triangles)

4. **rectangle**

(Quadrilaterals)

5. **diameter**

(Circles)

6. **12 in.**

$$\frac{9}{15} = \frac{x}{20}$$

$$\begin{aligned} 15x &= 180 \\ x &= \mathbf{12} \end{aligned}$$

(Similarity)

7. **10 ft.**

$$\begin{aligned} a^2 + b^2 &= c^2 \\ 24^2 + b^2 &= 26^2 \\ 576 + b^2 &= 676 \\ -576 \quad &\quad -576 \\ \hline b^2 &= 100 \\ \sqrt{b^2} &= \sqrt{100} \\ b &= \mathbf{10} \end{aligned}$$

(Pythagorean Theorem)

8. **18.8 m**

$4 \times 4.7 = \mathbf{18.8}$
(Perimeter)

9. **7.536 cm.**

$C = \pi d$
$C = (3.14)(2.4)$
$C = \mathbf{7.536}$
(Circumference)

10. **33 sq. ft.**

$6 \times 5\frac{1}{2} = \mathbf{33}$
(Area)

11. **77 sq. in.**

$A = \frac{1}{2}\pi r^2$
(a semicircle is $\frac{1}{2}$ of a circle)

If $d = 14$, then $r = 7$

$A = \frac{\overset{1}{\cancel{1}}}{\underset{1}{\cancel{2}}} \times \frac{\overset{11}{\cancel{22}}}{\underset{1}{\cancel{7}}} \times \frac{\overset{1}{\cancel{7}}}{1} \times \frac{7}{1} = \mathbf{77}$
(Area of a Circle)

12. **72 boxes**

volume of container =
$18 \times 9 \times 12 = 1{,}944$
volume of a box =
$3 \times 3 \times 3 = 27$

$27\overline{)1{,}994}$ = **72**
(Volume)

Simulated GED Test

If you have studied all of the parts of this book, you probably have all the math skills you need to pass the GED Mathematics Test. The next two tests will give you chances to practice your math skills on the types of questions you will find on the GED.

Before you take the tests, there are a few things that you should know. First, each test has an answer sheet that matches it. Use the answer sheet to mark your answer choice for each problem. On the GED Test, you also will have to mark your choices on an answer sheet.

Second, notice that a time limit is given at the beginning of each test. The time limit for the GED Mathematics Test is 90 minutes. To get a better idea of how you will do on the actual test, try to take the tests in the time limit. If the time runs out before you finish, circle the number of the problem that you are working on. Then complete the rest of the test. This way, you can learn your score within the time limit and your total score on the test. You will be able to compare the scores to find out how much your speed in answering the questions affects your score.

If there is a question that you do not understand, try to guess the answer. On the actual GED Test, you do not get points taken away from you for guessing. Only correct answers are counted. If you guess an answer correctly, you will get a higher score than you would if you had left the answer blank.

ANSWER SHEET FOR SIMULATED TEST 1

1	①	②	③	④	⑤
2	①	②	③	④	⑤
3	①	②	③	④	⑤
4	①	②	③	④	⑤
5	①	②	③	④	⑤
6	①	②	③	④	⑤
7	①	②	③	④	⑤
8	①	②	③	④	⑤
9	①	②	③	④	⑤
10	①	②	③	④	⑤
11	①	②	③	④	⑤
12	①	②	③	④	⑤
13	①	②	③	④	⑤
14	①	②	③	④	⑤
15	①	②	③	④	⑤
16	①	②	③	④	⑤
17	①	②	③	④	⑤
18	①	②	③	④	⑤
19	①	②	③	④	⑤
20	①	②	③	④	⑤
21	①	②	③	④	⑤
22	①	②	③	④	⑤
23	①	②	③	④	⑤
24	①	②	③	④	⑤
25	①	②	③	④	⑤
26	①	②	③	④	⑤
27	①	②	③	④	⑤
28	①	②	③	④	⑤
29	①	②	③	④	⑤
30	①	②	③	④	⑤
31	①	②	③	④	⑤
32	①	②	③	④	⑤
33	①	②	③	④	⑤
34	①	②	③	④	⑤
35	①	②	③	④	⑤
36	①	②	③	④	⑤
37	①	②	③	④	⑤
38	①	②	③	④	⑤
39	①	②	③	④	⑤
40	①	②	③	④	⑤
41	①	②	③	④	⑤
42	①	②	③	④	⑤
43	①	②	③	④	⑤
44	①	②	③	④	⑤
45	①	②	③	④	⑤
46	①	②	③	④	⑤
47	①	②	③	④	⑤
48	①	②	③	④	⑤
49	①	②	③	④	⑤
50	①	②	③	④	⑤

SIMULATED TEST 1

(Time: 90 minutes)

Directions: For all problems, choose the one best answer.

1. Silvia cut a piece of material 8.75 yards long into 7 equal pieces. Find the length of each piece.

   (1) .75 yd. (2) 1 yd. (3) 1.25 yds.
   (4) 1.5 yds. (5) 2.25 yds.

2. Find the value of $4^2 + 3^3$.

   (1) 43 (2) 31 (3) 25
   (4) 17 (5) 7

3. Tom bought one board that was 7 feet 9 inches long and another that was 5 feet 11 inches long. What was the total length of the boards?

   (1) 12′ 8″ (2) 12′ 9″ (3) 13′ 6″
   (4) 13′ 8″ (5) 14′8″

4. Heather makes $680 a month. Her husband Mark makes $870 a month. Altogether, how much do Mark and Heather make in a year?

   (1) $12,000 (2) $15,500 (3) $16,600
   (4) $17,400 (5) $18,600

5. The volume of a cylinder is given in the formula $V = \pi r^2 h$, where r is the radius of the cylinder and h is the height. What is the volume in cubic feet of a cylindrical tank with a 20-foot radius and a height of 14 feet? (Use $\pi = \frac{22}{7}$.)

   (1) 1,760 (2) 8,800 (3) 15,300
   (4) 16,800 (5) 17,600

6. Bob bought a radio originally selling for $34.80. He bought it on sale for 15% off. Find the sale price of the radio.

   (1) $28.58 (2) $29.58 (3) $29.80
   (4) $33.80 (5) $40.02

7. Find the price of $\frac{3}{4}$ pound of tomatoes at 98¢ a pound.

   (1) 65¢ (2) 74¢ (3) 86¢
   (4) 90¢ (5) 98¢

8. A vertical stick 3 feet high casts a shadow 5 feet long. At the same time, a tree casts a shadow 45 feet long. How tall is the tree?

   (1) 60′ (2) 50′ (3) 45′
   (4) 36′ (5) 27′

9. On a test with 50 questions, Rita got 46 problems right. What percent of the problems did she get right?

   (1) 4% (2) 8% (3) 46%
   (4) 90% (5) 92%

GO ON TO THE NEXT PAGE

Questions 10–13 refer to the following graphs.

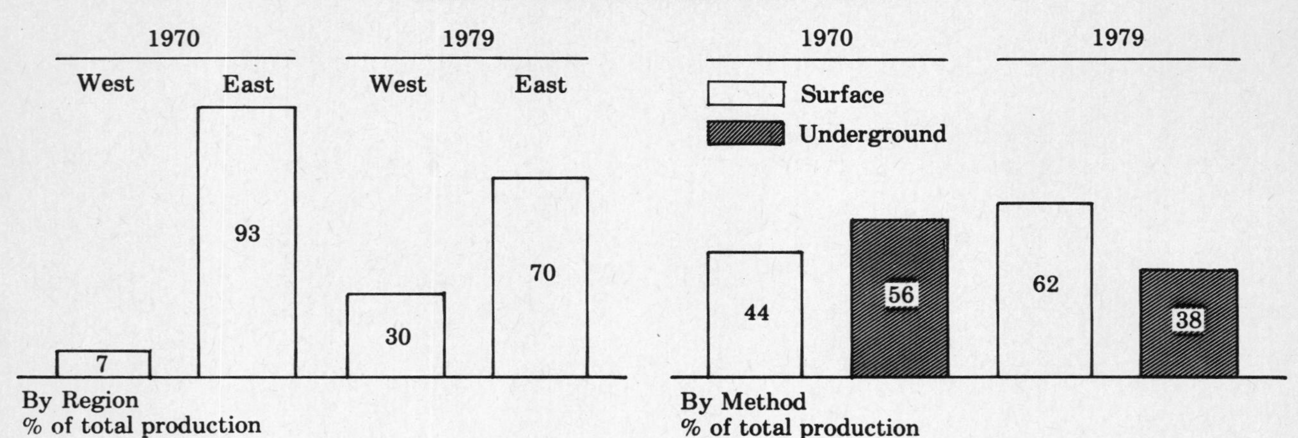

10. By what percent did soft coal production increase in the West from 1970 to 1979?

    (1) 30% (2) 23% (3) 17%
    (4) 15% (5) 7%

11. By what percent did underground production of soft coal decrease from 1970 to 1979?

    (1) 62% (2) 56% (3) 38%
    (4) 18% (5) 10%

12. In 1970 the U. S. produced about 600 million tons of soft coal. How many million tons were produced in the East?

    (1) 42 (2) 336 (3) 372
    (4) 420 (5) 558

13. Which of the following statements is false?

    (1) In 1979, more soft coal was produced in the East than in the West.
    (2) In 1970, more soft coal was produced by surface methods than by underground methods.
    (3) From 1970 to 1979, the percent of soft coal production dropped in the East.
    (4) In 1979, more soft coal was produced by surface methods than by underground methods.
    (5) From 1970 to 1979, the percentage of total coal produced in the West increased.

---

14. Find the supplement of an angle that measures 85° 14′ 19″.

    (1) 94° 85′ 81″ (2) 94° 45′ 41″
    (3) 94° 14′ 15″ (4) 14° 14′ 15″
    (5) 5° 14′ 15″

15. If $\frac{2}{3}y - 16 = 4$, then $y =$

    (1) 8 (2) 12 (3) 18
    (4) 20 (5) 30

16. From a piece of lumber 2 meters long, Jack sawed a piece .85 meter long. How long was the remaining piece?

    (1) .15 m. (2) .65 m. (3) .85 m.
    (4) 1.15 m. (5) 1.85 m.

17. Find the next term in the series 100, 88, 82, 79 . . . .

    (1) 73 (2) 75 (3) 76
    (4) $77\frac{1}{2}$ (5) $78\frac{1}{2}$

18. Find the value of $m(n + p)$ if $m = 15$, $n = 4$, and $p = 6$.

    (1) 25 (2) 66 (3) 94
    (4) 150 (5) 600

19. 1 gallon = 3.78 liters. Change 20 gallons to liters. Round off the answer to the nearest liter.

    (1) 24 l. (2) 64 l. (3) 74 l.
    (4) 76 l. (5) 80 l.

20. Ellen has a 3-inch by 5-inch snapshot. She wants to enlarge the snapshot to make it 15 inches long. How wide will the enlargement be?

(1) 6 in. (2) 9 in. (3) 12 in.
(4) 15 in. (5) 25 in.

21. If Angle A of triangle ABC measures 65°, and angle B measures 50°, then angle C measures

(1) 55° (2) 65° (3) 85°
(4) 90° (5) 245°

22. How many jars each holding $2\frac{1}{2}$ pounds of tomatoes can Brenda fill with 30 pounds of tomatoes?

(1) 75 (2) 30 (3) 20
(4) 15 (5) 12

23. The ratio of men to women working at the Edison Utility Company is 5 to 2. 70 women work for the company. How many men work there?

(1) 200 (2) 175 (3) 70
(4) 40 (5) 28

Questions 24–27 refer to the following table.

RATES FOR A 5-MINUTE PHONE CALL

	WEEKDAY		SATURDAY	
	OLD	NEW	OLD	NEW
From NYC to:				
Albany	$2.10	$1.97	$1.36	$ .75
Buffalo	2.37	2.09	1.53	.81
From Albany to:				
Buffalo	2.37	2.09	1.53	.81
Syracuse	2.10	1.97	1.36	.75
From Riverhead to:				
NYC	1.85	1.82	1.18	.70
White Plains	1.66	1.67	1.07	.65

24. By how much did the price of a 5-minute weekday call from Albany to Syracuse drop?

(1) $.27 (2) $.23 (3) $.17
(4) $.13 (5) $.10

25. A 5-minute Saturday call from NYC to Albany went down by about what percent?

(1) 75% (2) 45% (3) 20%
(4) 10% (5) 5%

26. The new price of a 5-minute Saturday call from Riverhead to NYC is how much more than a 5-minute Saturday call from Riverhead to White Plains?

(1) 5¢ (2) 9¢ (3) 10¢
(4) 15¢ (5) 19¢

27. Which sample call on the table went up in the change from the old rate to the new rate?

(1) weekday from NYC to Buffalo
(2) Saturday from NYC to Albany
(3) weekday from Albany to Syracuse
(4) Saturday from Albany to Buffalo
(5) weekday from Riverhead to White Plains

28. Find the value of $\sqrt{3{,}721}$.

(1) 186 (2) 69 (3) 61
(4) 59 (5) 51

29. The perimeter of a rectangular concrete walk is 120 feet. The walk is 4 feet wide. How long is it?

(1) 30 ft. (2) 56 ft. (3) 60 ft.
(4) 112 ft. (5) 116 ft.

30. Find the interest on $600 at $5\frac{3}{4}$% annual interest for 1 year and 6 months.

(1) $17.25 (2) $34.50 (3) $42.25
(4) $51.75 (5) $69

31. A plane took off from Central Airport and traveled due north for 30 miles. Then it traveled due east for 40 miles before it landed at Stanton Field. The shortest distance between Central Airport and Stanton Field is

(1) 70 mi. (2) 50 mi. (3) 40 mi.
(4) 35 mi. (5) 25 mi.

32. Ron drove his truck 60 miles on 4.5 gallons of gas. To the nearest whole number, how many miles did Ron drive with one gallon of gas?

(1) 17 (2) 16 (3) 15
(4) 14 (5) 13

33. The supplement of an angle that measures 175° is

(1) 5° (2) 15° (3) 25°
(4) 85° (5) 185°

34. Thursday night, 856 people went to the Royce basketball tournament. Friday night, 1,223 people went. Saturday night, 1,308 people went. Find the average attendance for each night.

(1) 987 (2) 1,019 (3) 1,129
(4) 1,187 (5) 1,214

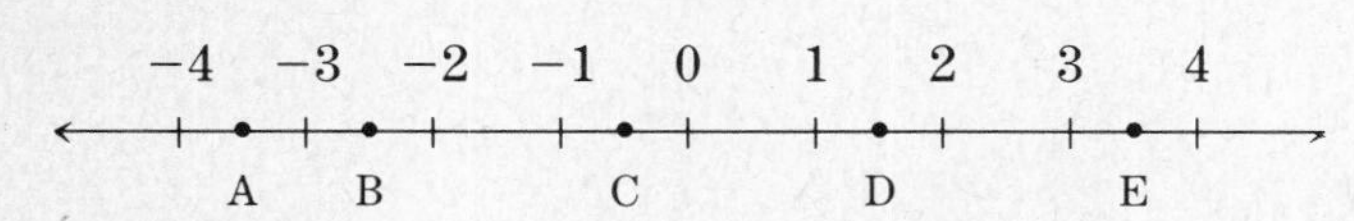

35. Which numbered letter stands for the point $-2\frac{1}{2}$?

(1) A (2) B (3) C
(4) D (5) E

36. Find the cost of $2\frac{3}{4}$ pounds of ground chuck at $1.40 a pound.

(1) $2.80 (2) $3.30 (3) $3.55
(4) $3.85 (5) $4.20

37. Eight less than seven times a number is 83. Find the number.

(1) 17 (2) 15 (3) 14
(4) 13 (5) 11

38. Charles mailed three packages each weighing 1 lb. 10 oz. Find the total weight of the packages.

(1) 4 lbs. 14 oz. (2) 4 lbs. 10 oz.
(3) 3 lbs. 14 oz. (4) 3 lbs. 10 oz.
(5) 3 lbs. 3 oz.

39. The base of a triangle measures 6.4 meters. The height measures 2.8 meters. In square meters, the area of the triangle is

(1) 17.92 (2) 12.2 (3) 9.2
(4) 8.96 (5) 2.76

40. So far, Colin has paid back $1,260 of the money he borrowed to buy a car. This is 35% of his total loan. Find the total amount of the loan.

(1) $3,500 (2) $3,600 (3) $4,410
(4) $4,800 (5) $5,000

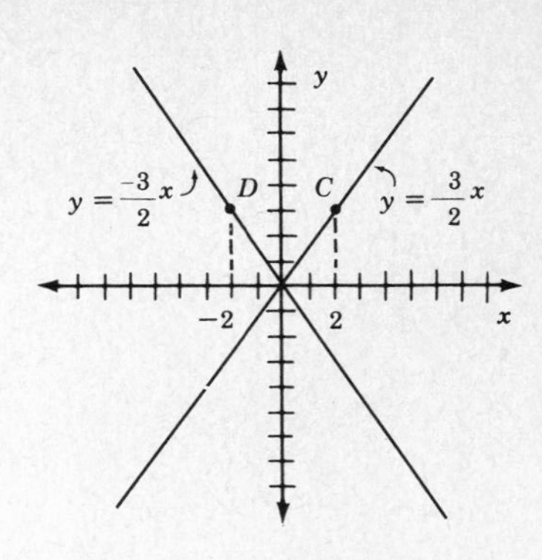

41. The distance between points C and D on the graph is

(1) 1 (2) 2 (3) 3
(4) 4 (5) 6

42. Theresa's gross income for a year was $14,260. Her employer deducted $2,978 from her pay for taxes and Social Security. What was her net income for the year?

(1) $12,282 (2) $12,182 (3) $11,282
(4) $11,188 (5) $11,182

43. How many square tiles each measuring 9 inches on a side does Shirley need to cover her kitchen floor, which is 72 inches wide and 90 inches long?

(1) 50 (2) 60 (3) 70
(4) 80 (5) 90

44. 13,000 pounds is equal to how many tons?

(1) $6\frac{1}{2}$ (2) $7\frac{1}{2}$ (3) 8
(4) 10 (5) 13

45. Find the value of $(.007)^2$.

(1) .000014 (2) .000049 (3) .0014
(4) .0049 (5) .049

46. Susan bought $2\frac{3}{4}$ pounds of oranges and $1\frac{1}{2}$ pounds of grapes. Find the total weight of her purchases.

(1) $3\frac{1}{4}$ lbs. (2) $3\frac{1}{2}$ lbs. (3) $3\frac{3}{4}$ lbs.
(4) 4 lbs. (5) $4\frac{1}{4}$ lbs.

47. If $x^2 + 4x + 4 = 0$, then $x =$

(1) 2 only (2) 0 only (3) −2 only
(4) 2 or 0 (5) 2 or −2

48. Find the difference between 6 and .473.

(1) 6.527 (2) 6.473 (3) 5.527
(4) 5.473 (5) .467

49. To the nearest foot, the circumference of a circle that has a diameter of 30 feet is

(1) 188 ft. (2) 144 ft. (3) 116 ft.
(4) 94 ft. (5) 84 ft.

50. If $x + y = 32$, and $y = 3x$, then $y =$

(1) $\sqrt{3}$ (2) $\sqrt{4} - y$ (3) $\sqrt{6}$
(4) 8 (5) 24

END OF TEST

Check your answers on page 333.

After you check your answers, look at the chart below. Circle the number of each problem you missed. Then review the section in which the skills for that problem are explained.

Section	**Problem Numbers**
Whole Numbers	4 17 34 42
Fractions	7 22 36 46
Decimals	1 16 32 48
Percents	6 9 30 40
Measurement	3 19 38 44
Tables and Graphs	10 11 12 13 24 25 26 27
Algebra	2 8 15 18 23 28 35 37 41 45 47 50
Geometry	5 14 20 21 29 31 33 39 43 49

Now, figure out your score. Count the problems that you missed. Then subtract the problems that you missed from the total number of problems on the test (50).

To find your score on the simulated test, turn to the score chart on page 332.

ANSWER SHEET FOR SIMULATED TEST 2

1	①	②	③	④	⑤	26	①	②	③	④	⑤
2	①	②	③	④	⑤	27	①	②	③	④	⑤
3	①	②	③	④	⑤	28	①	②	③	④	⑤
4	①	②	③	④	⑤	29	①	②	③	④	⑤
5	①	②	③	④	⑤	30	①	②	③	④	⑤
6	①	②	③	④	⑤	31	①	②	③	④	⑤
7	①	②	③	④	⑤	32	①	②	③	④	⑤
8	①	②	③	④	⑤	33	①	②	③	④	⑤
9	①	②	③	④	⑤	34	①	②	③	④	⑤
10	①	②	③	④	⑤	35	①	②	③	④	⑤
11	①	②	③	④	⑤	36	①	②	③	④	⑤
12	①	②	③	④	⑤	37	①	②	③	④	⑤
13	①	②	③	④	⑤	38	①	②	③	④	⑤
14	①	②	③	④	⑤	39	①	②	③	④	⑤
15	①	②	③	④	⑤	40	①	②	③	④	⑤
16	①	②	③	④	⑤	41	①	②	③	④	⑤
17	①	②	③	④	⑤	42	①	②	③	④	⑤
18	①	②	③	④	⑤	43	①	②	③	④	⑤
19	①	②	③	④	⑤	44	①	②	③	④	⑤
20	①	②	③	④	⑤	45	①	②	③	④	⑤
21	①	②	③	④	⑤	46	①	②	③	④	⑤
22	①	②	③	④	⑤	47	①	②	③	④	⑤
23	①	②	③	④	⑤	48	①	②	③	④	⑤
24	①	②	③	④	⑤	49	①	②	③	④	⑤
25	①	②	③	④	⑤	50	①	②	③	④	⑤

SIMULATED TEST 2

(Time: 90 minutes)

Directions: For all problems, choose the one best answer.

1. If $3(3a + 5) = 105$, then $a =$

   (1) 10 (2) 9 (3) 6
   (4) 5 (5) 3

2. Five people shared 18.75 pounds of fish equally. How many pounds did each person get?

   (1) 1.25 (2) 2.25 (3) 2.75
   (4) 3.25 (5) 3.75

3. Carlos built a coffee table $30\frac{1}{2}$ inches wide and 50 inches long. What is the area of the top of the table?

   (1) 1,500 sq. in. (2) 1,525 sq. in.
   (3) 1,610 sq. in. (4) 1,650 sq. in.
   (5) 1,750 sq. in.

4. The sales tax in Jane's state is 4%. She bought a toaster for $19.80. Find the total price of the toaster, including tax.

   (1) $19.01 (2) $19.76 (3) $20.
   (4) $20.20 (5) $20.59

5. Pat can type 285 words in 3 minutes. At the same rate, how many words can she type in 7 minutes?

   (1) 735 (2) 665 (3) 570
   (4) 565 (5) 407

6. Paul is 70 inches tall. His son Bill is $68\frac{1}{2}$ inches tall. How much taller is Paul than his son?

   (1) $\frac{1}{2}$ in. (2) $1\frac{1}{2}$ in. (3) 2 in.
   (4) $2\frac{1}{2}$ in. (5) $3\frac{1}{2}$ in.

7. The perimeter of a triangle that is 7″ long on each side is

   (1) 6″ (2) 15″ (3) 21″
   (4) 35″ (5) 70″

8. From a large jug holding 4 gallons of cider, Phil took 1 gallon 3 quarts. How much cider was left in the large jug?

   (1) 1 gal. 3 qts. (2) 2 gals. 1 qt.
   (3) 2 gals. 2 qts. (4) 2 gals. 3 qts.
   (5) 3 gals. 2 qts.

9. $13^2 - 5^3 =$

   (1) 294 (2) 154 (3) 144
   (4) 44 (5) 11

GO ON TO THE NEXT PAGE

Questions 10–13 refer to the following table.

PER CAPITA CONSUMPTION OF SELECTED FOODS (IN POUNDS)

Food item	1969	1979
Beef and veal	84.7	81.3
Pork	60.6	64.8
Fish	11.2	13.7
Poultry	46.7	62.0
Eggs	39.3	35.8
Milk and cream	301.0	284.2
Cheese	11.0	18.1
Fresh fruit	79.5	83.2
Fresh vegetables	98.7	104.5
Canned vegetables	53.7	55.0
Frozen vegetables	9.1	11.1
Wheat flour	112.0	112.0
Coffee	11.9	9.7
Sugar	101.0	91.6

10. How many more pounds of poultry did the average American eat in 1979 than in 1969?

(1) 3.4 (2) 4.2 (3) 8.7
(4) 12.5 (5) 15.3

11. The consumption of which food item stayed the same from 1969 to 1979?

(1) fish (2) eggs (3) cheese
(4) fresh fruit (5) wheat flour

12. The increase in cheese consumption from 1969 to 1979 was about what percent?

(1) 100% (2) 65% (3) 50%
(4) 35% (5) 7%

13. Which of the following statements is true?

(1) The average American ate less fish in 1979 than in 1969.
(2) The average American drank the same amount of coffee in 1979 as in 1969.
(3) The average American ate more sugar in 1979 than in 1969.
(4) The average American ate less beef and veal in 1979 than in 1969.
(5) The average American ate more cheese than eggs in 1979.

14. Mr. Johnson makes $296 a week. Mrs. Johnson makes $232 a week. Their son Don makes $148 a week. Their daughter Margie makes $156 a week. What is the average weekly income for each member of the Johnson family?

(1) $232 (2) $216 (3) $208
(4) $200 (5) $192

15. Mr. and Mrs. Millhouse made a down payment of $4,620 on a house. The down payment was 12% of the price of the house. Find the price of the house.

(1) $24,000 (2) $32,500 (3) $36,000
(4) $38,500 (5) $46,200

16. Find the complement of an angle that measures 41° 36′ 11″.

(1) 48° 23′ 49″ (2) 48° 36′ 11″
(3) 49° 36′ 11″ (4) 50° 23′ 49″
(5) 50° 36′ 11″

17. Janice spends $\frac{1}{4}$ of her income on rent. She makes $680 a month. What is her yearly rent?

(1) $170 (2) $204 (3) $1,680
(4) $2,040 (5) $6,800

18. Ten times a number decreased by seventeen is the same as the number increased by 55. Find the number.

(1) 4 (2) 5 (3) 6 (4) 7 (5) 8

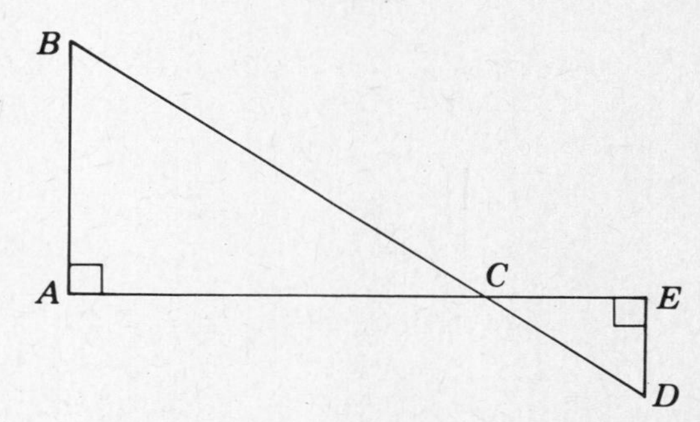

19. In triangle *CDE*, *CE* is 8 ft. and *DE* is 5 ft. In triangle *ABC*, *AC* is 20 ft. Find the measurement of *AB*.

(1) 20 ft. (2) 17 ft. (3) $12\frac{1}{2}$ ft.
(4) 8 ft. (5) $7\frac{1}{2}$ ft.

20. Find the product of 4.08 and 3.7.

(1) 0.38 (2) 1.52 (3) 7.78
(4) 11.03 (5) 15.096

21. Find the value of $cd - e^2$ if $c = 5$, $d = 8$, and $e = 4$.

(1) 56 (2) 48 (3) 36
(4) 32 (5) 24

22. The town of Ramsey has a yearly budget of $214,500 for parks and recreation. Halfway through the year, the town had spent $149,785 for parks and recreation. How much was left in the budget for the rest of the year?

(1) $65,715 (2) $65,615 (3) $64,825
(4) $64,815 (5) $64,715

23. The base of a triangle measures $3\frac{1}{2}$ inches. The height measures 4 inches. Find the area of the triangle.

(1) 14 sq. in. (2) 7 sq. in.
(3) 4 sq. in. (4) $3\frac{1}{2}$ sq. in.
(5) 1.75 sq. in.

Questions 24–27 refer to the following graphs.

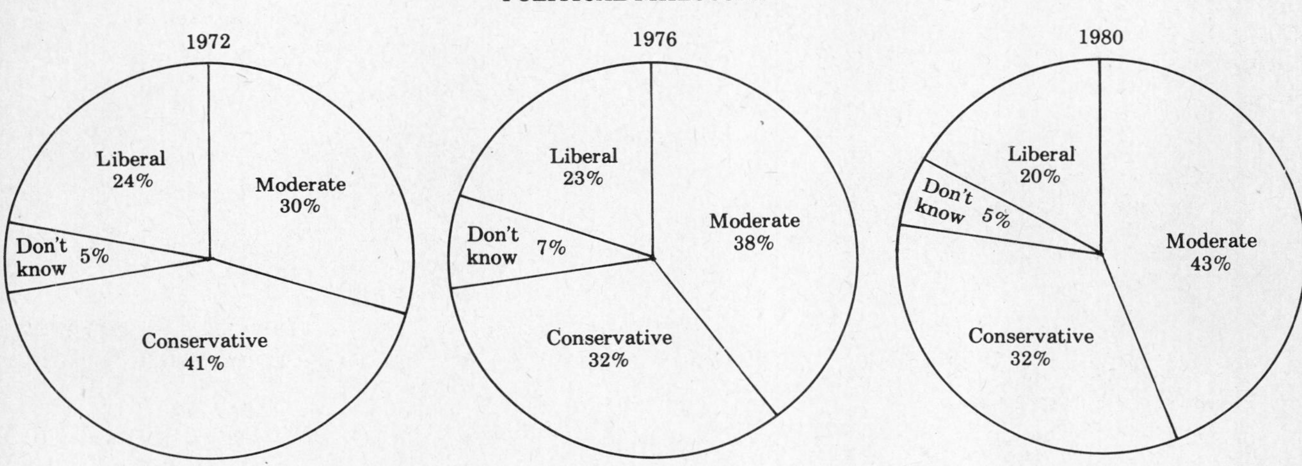

24. The percentage of Americans who said they were liberal dropped by what percent from 1972 to 1980?

(1) 1% (2) 2% (3) 3%
(4) 4% (5) 5%

25. The town of Midvale has 54,000 registered voters. Political attitudes in Midvale follow the national pattern. Find the number of Midvale voters who called themselves moderate in 1980.

(1) 23,220 (2) 20,520 (3) 17,280
(4) 16,200 (5) 10,800

26. Which political philosophy described the largest number of Americans in 1972?

(1) moderate (2) conservative
(3) liberal (4) undecided
(5) don't know

27. Which of the following statements is true?

(1) The percent of Americans who called themselves liberal grew steadily from 1972 to 1980.
(2) A greater percent of Americans was undecided in its political philosophy in 1980 than in 1972.
(3) The percent of Americans who called themselves moderate decreased from 1976 to 1980.
(4) The percent of Americans who called themselves conservative was the same in 1976 and 1980.
(5) The percent of Americans who called themselves liberal increased from 1976 to 1980.

28. 1 inch = 2.54 centimeters. To the nearest centimeter, change 36 inches to centimeters.

(1) 38 cm. (2) 81 cm. (3) 91 cm.
(4) 100 cm. (5) 103 cm.

29. Solve for a in $3a + 4(a + 2) = 43$

(1) 4 (2) 5 (3) 6
(4) $6\frac{4}{7}$ (5) $7\frac{6}{7}$

30. How many $2\frac{1}{2}$-pound bags of plaster can be filled with 75 pounds of plaster?

(1) 187 (2) 75 (3) 60
(4) 45 (5) 30

31. Find the interest on $800 at 5.6% annual interest for 6 months.

(1) $22.40 (2) $24.60 (3) $32.20
(4) $36.40 (5) $44.80

32. In triangle XYZ, if $\angle X$ is 35° and $\angle Y$ is 55°, then $\angle Z$ is

(1) 15° (2) 60° (3) 90°
(4) 180° (5) 270°

33. Find the next term in the series 2, 2, 4, 12, 48 . . . .

(1) 96 (2) 108 (3) 133
(4) 192 (5) 240

34. Pam bought 6.5 yards of material at $4.90 a yard. How much did she pay for the material?

(1) $11.40 (2) $19.40 (3) $29.40
(4) $31.85 (5) $32.50

35. The length of a rectangle is 3 times its width. The perimeter of the rectangle is 96 feet. Find the width of the rectangle.

(1) 8 ft. (2) 12 ft. (3) 18 ft.
(4) 24 ft. (5) 30 ft.

36. The diameter of a circular pond is 84 feet. The circumference of the pond is

(1) 132 ft. (2) 264 ft. (3) 396 ft.
(4) 528 ft. (5) 554 ft.

37. Sam paid $3.59 for a 16-ounce can of coffee. To the nearest cent, what was the price of one ounce of coffee?

(1) $.18 (2) $.20 (3) $.22
(4) $.25 (5) $.30

38. Which of the following expresses 12 less than 8 times a number all multiplied by 7?

(1) $8x - 12(7)$ (2) $7(8x) - 12$
(3) $8(7x - 12)$ (4) $7(8x - 12)$
(5) $8(12x - 7)$

39. In a month, the Millers make $780. They spend $156 a month on food. What fraction of their income goes for food?

(1) $\frac{1}{6}$ (2) $\frac{1}{5}$ (3) $\frac{1}{4}$ (4) $\frac{1}{3}$ (5) $\frac{1}{2}$

40. Simplify $\frac{-168}{-14}$.

(1) +154 (2) +18 (3) +12
(4) −12 (5) −154

41. Janet owed $280 on her credit card. She paid 30% of her bill. How much does she still owe?

(1) $196 (2) $224 (3) $250
(4) $283 (5) $310

42. How many square inches are there in 2 square feet?

(1) 144 (2) 216 (3) 288
(4) 576 (5) 1,728

43. The ratio of blue paint to white paint in a certain mixture is 3:2. A painter has 12 gallons of blue paint. How many gallons of white paint does he need?

(1) 6 (2) 8 (3) 12
(4) 18 (5) 24

44. A jar holds 1 lb. 3 oz. of canned tomatoes. How much do 24 jars hold?

(1) 24 lbs. (2) 24 lbs. 3 oz.
(3) 24 lbs. 15 oz. (4) 28 lbs. 8 oz.
(5) 30 lbs.

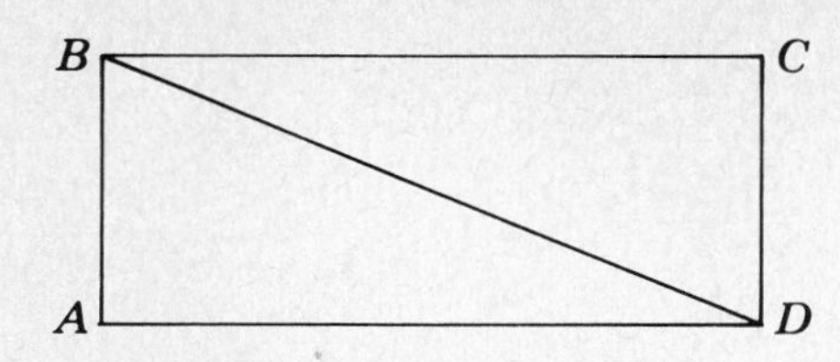

45. In rectangle $ABCD$, side $AB = 15$ inches and side $AD = 36$ inches. Find the measurement of diagonal line BD.

(1) 37″ (2) 39″ (3) 40″
(4) 42″ (5) 50″

46. How many pieces of wood each 3.5 centimeters long can be cut from a piece of wood 84 centimeters long?

(1) 12 (2) 18 (3) 24
(4) 28 (5) 30

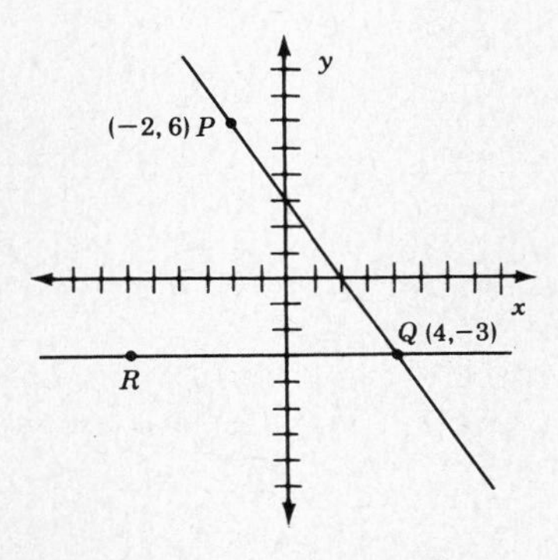

47. In the figure, line PQ intersects line QR at $(4, -3)$ and line QR is parallel to the x-axis. Find the perpendicular distance from point P to line QR.

(1) 3 (2) 6 (3) 8
(4) 9 (5) 12

48. 24,560 people paid an average of $6 each to hear a rock concert at the Smithville Arena. What was the total value of the tickets sold for the concert?

(1) $127,360 (2) $137,460
(3) $147,360 (4) $147,460
(5) $157,460

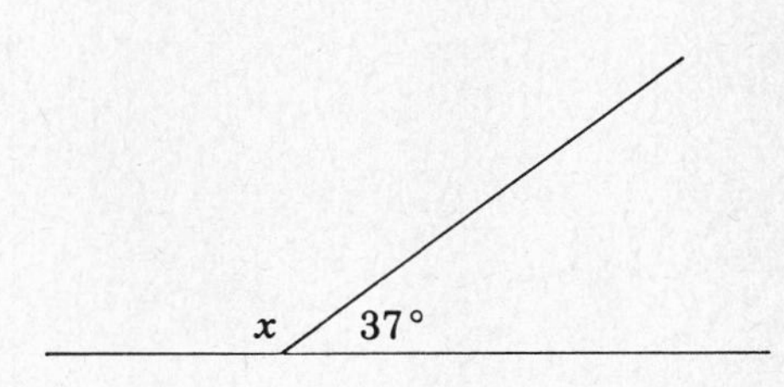

49. Find the measurement of $\angle x$ in the diagram.

(1) 143° (2) 74° (3) 53°
(4) 37° (5) $18\frac{1}{2}$°

50. Which of the following is equal to $14(73 + 26)$?

(1) $14(73) + 14(26)$
(2) $14(73) + 26$
(3) $73 + 14(26)$
(4) $(14 + 73)(26)$
(5) $73(14 + 26)$

END OF TEST

Check your answers on page 337.

After you check your answers, look at the chart below. Circle the number of each problem you missed. Then review the section in which the skills for that problem are explained.

Section	Problem Numbers											
Whole Numbers	14	22	33	48								
Fractions	6	17	30	39								
Decimals	2	20	34	46								
Percents	4	15	31	41								
Measurement	8	28	37	44								
Tables and Graphs	10	11	12	13	24	25	26	27				
Algebra	1	5	9	18	21	29	35	38	40	43	47	50
Geometry	3	7	16	19	23	32	36	42	45	49		

Now, figure out your score. Count the problems that you missed. Then subtract the problems that you missed from the total number of problems on the test. Use the score chart on the next page to find your score for the simulated test.

Score Chart

Below, you will find a chart to help you interpret the score that you got on the simulated tests. Find the number of problems that you got right in the left-hand column. The number across from the left-hand column is your score.

The scores in the chart are rough estimates of the scores you would probably get on the actual GED. Use the chart as a guide. If your score according to the chart is below 45, then you should review your math skills before you take the actual GED Test. If your score is 45 or higher, you probably have the math skills that you need to pass the GED.

Number of problems right:	Score:
less than 25	**35** or below
26–35	**45**
36–45	**55**
46–50	**65**

ANSWERS & SOLUTIONS

SIMULATED TEST 1

1. **(3) 1.25 yds.**

$$7\overline{)8.75} = \mathbf{1.25}$$

2. **(1) 43**

$4^2 + 3^3 =$
$4 \times 4 + 3 \times 3 \times 3 =$
$16 + 27 = \mathbf{43}$

3. **(4) 13′ 8″**

$$\begin{array}{r} 7'\ \ 9'' \\ +\ 5'\ 11'' \\ \hline 12'\ 20'' \end{array} = \mathbf{13'\ 8''}$$

4. **(5) $18,600**

$$\begin{array}{r} \$680 \\ +\ 870 \\ \hline \$1{,}550 \end{array} \qquad \begin{array}{r} \$1{,}550 \\ \times\ 12 \\ \hline 3\ 100 \\ 15\ 50\ \ \\ \hline \mathbf{\$18{,}600} \end{array}$$

5. **(5) 17,600**

$V = \pi r^2 h$
$= \frac{22}{7} \cdot 20^2 \cdot 14$
$= \frac{22}{\not{7}_{1}} \cdot \frac{400}{1} \cdot \frac{\not{14}^{2}}{1}$
$= \mathbf{17{,}600}$

6. **(2) $29.58**

$$\begin{array}{r} \$34.80 \\ \times\ .15 \\ \hline 174\ 00 \\ 348\ 0\ \ \ \\ \hline \$5.22\ 00 \end{array} \qquad \begin{array}{r} \$34.80 \\ -\ 5.22 \\ \hline \mathbf{\$29.58} \end{array}$$

7. **(2) 74¢**

$\frac{3}{\not{4}_{2}} \times \frac{\not{98}^{49}}{1} = \frac{147}{2} = 73\frac{1}{2}$

$73\frac{1}{2}$ to the nearest cent $=$ **74¢**

8. **(5) 27′**

$\frac{\text{height}}{\text{shadow}}\ \frac{3}{5} = \frac{x}{45}$
$5x = 135$
$x = \mathbf{27}$

9. **(5) 92%**

$\frac{46}{50} = \frac{23}{25}$

$\frac{23}{\not{25}_{1}} \times \frac{\not{100}^{4}}{1} = \mathbf{92\%}$

10. **(2) 23%**

$$\begin{array}{r} 30\% \\ -\ 7\% \\ \hline \mathbf{23\%} \end{array}$$

11. **(4) 18%**

$$\begin{array}{r} 56\% \\ -\ 38\ \ \ \\ \hline \mathbf{18\%} \end{array}$$

12. **(5) 558**

$93\% = .93$

$$\begin{array}{r} 6\ 00 \\ \times\ .93 \\ \hline 18\ 00 \\ 540\ 0\ \ \ \\ \hline \mathbf{558.00} \end{array}$$

13. (2) **In 1970, more soft coal was produced by surface methods than by underground methods.**

14. (2) **94° 45′ 41″**

$$\begin{array}{r} 179^\circ\ 59' \\ \cancel{180}^\circ\ \cancel{60}'\ 60'' \\ -\ \ 85^\circ\ 14'\ 19'' \\ \hline \mathbf{94^\circ\ 45'\ 41''} \end{array}$$

15. (5) **30**

$$\begin{aligned} \tfrac{2}{3}y - 16 &= 4 \\ +16 &\quad +16 \\ \tfrac{3}{2} \cdot \tfrac{2}{3}y &= 20 \cdot \tfrac{3}{2} \\ y &= \mathbf{30} \end{aligned}$$

16. (4) **1.15 m.**

$$\begin{array}{r} 2.00 \text{ m.} \\ -\ .85 \\ \hline \mathbf{1.15 \text{ m.}} \end{array}$$

17. (4) **$77\frac{1}{2}$**

100 88 82 79 $77\frac{1}{2}$

−12 −6 −3 $-1\frac{1}{2}$

18. (4) **150**

$$\begin{aligned} m(n + p) &= \\ 15(4 + 6) &= \\ 15(10) &= \mathbf{150} \end{aligned}$$

19. (4) **76 l.**

$$\begin{array}{r} 3.78 \\ \times\ 20 \\ \hline 75.60 \end{array}$$

to the nearest whole number = **76**

20. (2) **9 in.**

$$\frac{\text{width}}{\text{length}}\ \frac{3}{5} = \frac{x}{15}$$

$$\begin{aligned} 5x &= 45 \\ x &= \mathbf{9} \end{aligned}$$

21. (2) **65°**

$$\begin{array}{r} 65 \\ +\ 50 \\ \hline 115 \end{array} \qquad \begin{array}{r} 180 \\ -\ 115 \\ \hline \mathbf{65} \end{array}$$

22. (5) **12**

$$30 \div 2\tfrac{1}{2} =$$

$$\frac{30}{1} \div \frac{5}{2} =$$

$$\frac{\overset{6}{\cancel{30}}}{1} \times \frac{2}{\underset{1}{\cancel{5}}} = \mathbf{12}$$

23. (2) **175**

$$\frac{\text{men}}{\text{women}}\ \frac{5}{2} = \frac{x}{70}$$

$$\begin{aligned} 2x &= 350 \\ x &= \mathbf{175} \end{aligned}$$

24. (4) **$.13**

$$\begin{array}{r} \$2.10 \\ -1.97 \\ \hline \mathbf{\$\ .13} \end{array}$$

25. (2) **45%**

$$\begin{array}{r} \$1.36 \\ -\ .75 \\ \hline \$\ .61 \end{array}$$

$$\frac{.61}{1.36} = 1.36\overline{).61000}$$

$$\begin{array}{r} .448 \\ 1.36\overline{).61\,000} \\ \underline{54\ 4} \\ 6\ 60 \\ \underline{5\ 44} \\ 1\ 160 \\ \underline{1\ 088} \end{array}$$

.448 to the nearest hundredth = .45 = **45%**

26. (1) **5¢**

$$\begin{array}{r} \$.70 \\ -.65 \\ \hline \mathbf{\$.05} \end{array}$$

27. **(5) weekday from Riverhead to White Plains**

28. **(3) 61**

Guess 60.

$$60\overline{)3,721} = 62 \quad (3\,60;\ 121;\ 120)$$

$$62 + 60 = 122$$

$$2\overline{)122} = \mathbf{61}$$

29. **(2) 56 ft.**

$$120 = 2(4) + 2(x)$$
$$120 = 8 + 2x$$
$$-\ 8 = -8$$
$$\frac{112}{2} = \frac{2x}{2}$$
$$\mathbf{56} = x$$

30. **(4) $51.75**

$$1 \text{ yr. } 6 \text{ mos.} = 1\tfrac{6}{12} = 1\tfrac{1}{2} = \tfrac{3}{2} \text{ yr.}$$

$$\$600 \times 5\tfrac{3}{4}\% \times \tfrac{3}{2} \text{ yr.} =$$

$$\frac{\overset{3}{\not{6}}\not{00}}{\underset{1}{\not{1}\not{00}}} \times \frac{23}{\underset{2}{\not{4}}} \times \frac{3}{2} = \frac{207}{4} = 51\tfrac{3}{4}$$

$$= \mathbf{\$51.75}$$

31. **(2) 50 mi.**

$$a^2 + b^2 = c^2$$
$$30^2 + 40^2 = c^2$$
$$900 + 1,600 = c^2$$
$$2,500 = c^2$$
$$\sqrt{2,500} = \sqrt{c^2}$$
$$\mathbf{50} = c$$

32. **(5) 13**

$$4.5\overline{)60.0\,0} = 13.3 \quad (45;\ 15\,0;\ 13\,5;\ 1\,5\,0;\ 1\,3\,5)$$

to the nearest whole = **13**

33. **(1) 5°**

$$180° - 175° = \mathbf{5°}$$

34. **(3) 1,129**

$$856 + 1,223 + 1,308 = 3,387$$

$$3\overline{)3,387} = \mathbf{1,129}$$

35. **(2) B**

36. **(4) $3.85**

$$2\tfrac{3}{4} \times \$1.40 =$$

$$\frac{11}{\underset{1}{\not{4}}} \times \frac{\overset{.35}{\not{1}.\not{4}\not{0}}}{1} = \mathbf{\$3.85}$$

37. **(4) 13**

$$7x - 8 = 83$$
$$+\ 8 \quad +\ 8$$
$$7x = 91$$
$$x = \mathbf{13}$$

38. **(1) 4 lbs. 14 oz.**

1 lb. 10 oz.
× 3

3 lbs. 30 oz. = **4 lbs. 14 oz.**

39. **(4) 8.96**

$A = \frac{1}{2}(6.4)(2.8) = \mathbf{8.96}$

40. **(2) $3,600**

35% = .35

$$\begin{array}{r} \mathbf{\$36\ 00} \\ .35_{\wedge})\overline{\$1260.00_{\wedge}} \\ \underline{105} \\ 210 \\ \underline{210} \\ 0\ 00 \end{array}$$

41. **(4) 4**

42. **(3) $11,282**

$$\begin{array}{r} \$14,260 \\ \underline{-\ 2,978} \\ \mathbf{\$11,282} \end{array}$$

43. **(4) 80**

area of floor =
90 × 72 = 6,480
area of a tile =
9 × 9 = 81

$$\begin{array}{r} \mathbf{80} \\ 81)\overline{6,480} \end{array}$$

44. **(1) $6\frac{1}{2}$**

$$\begin{array}{r} 6\frac{1,000}{2,000} = \mathbf{6\frac{1}{2}} \\ 2000)\overline{13000} \\ \underline{12000} \\ 1000 \end{array}$$

45. **(2) .000049**

$(.007)^2 =$
$.007 \times .007 =$
.000049

46. **(5) $4\frac{1}{4}$ lbs.**

$$\begin{array}{r} 2\frac{3}{4} = 2\frac{3}{4} \\ \underline{+\ 1\frac{1}{2} = 1\frac{2}{4}} \\ 3\frac{5}{4} = \mathbf{4\frac{1}{4}} \end{array}$$

47. **(3) −2 only**

$$\begin{aligned} x^2 + 4x + 4 &= 0 \\ (\mathbf{-2})^2 + 4(\mathbf{-2}) + 4 &= 0 \\ 4 + (-8) + 4 &= 0 \\ -4 + 4 &= 0 \\ 0 &= 0 \end{aligned}$$

48. **(3) 5.527**

$$\begin{array}{r} 6.000 \\ \underline{-.473} \\ \mathbf{5.527} \end{array}$$

49. **(4) 94 ft.**

$$\begin{aligned} C &= \pi d \\ &= (3.14)(30) \\ &= 94.2 \text{ to the nearest whole number} = \mathbf{94} \end{aligned}$$

50. **(5) 24**

$$\begin{aligned} x + y &= 32 \\ x + 3x &= 32 \\ \frac{4x}{4} &= \frac{32}{4} \\ x &= 8 \\ 3x &= \mathbf{24} \end{aligned}$$

SIMULATED TEST 2

1. (1) **10**

$$\begin{aligned} 3(3a+5) &= 105 \\ 9a + 15 &= 105 \\ -15 &= -15 \\ \hline \frac{9a}{9} &= \frac{90}{9} \\ a &= \mathbf{10} \end{aligned}$$

2. (5) **3.75**

$$5\overline{)18.75} = \mathbf{3.75}$$

3. (2) **1,525 sq. in.**

$$A = 50 \cdot 30\frac{1}{2}$$

$$= \frac{\overset{25}{\cancel{50}}}{1} \times \frac{61}{\underset{1}{\cancel{2}}}$$

$$= \mathbf{1{,}525}$$

4. (5) **$20.59**

4% = .04

$19.80	$19.80
× .04	+ .79
$.79 20	**$20.59**

5. (2) **665**

$$\frac{\text{words}}{\text{time}}\ \frac{285}{3} = \frac{x}{7}$$

$$3x = 1{,}995$$

$$x = \mathbf{665}$$

6. (2) $1\frac{1}{2}$ **in.**

$$\begin{aligned} 70 &= 69\tfrac{2}{2} \\ -68\tfrac{1}{2} & \quad 68\tfrac{1}{2} \\ \hline & \quad \mathbf{1\tfrac{1}{2}} \end{aligned}$$

7. (3) **21″**

7 + 7 + 7 = **21**

8. (2) **2 gals. 1 qt.**

4 gals.		3 gals. 4 qts.
− 1 gal. 3 qts.	=	1 gal. 3 qts.
		2 gals. 1 qt.

9. (4) **44**

$13^2 - 5^3 =$

$13 \times 13 - 5 \times 5 \times 5 =$

$169 - 125 = \mathbf{44}$

10. (5) **15.3**

$$\begin{aligned} 62.0 \\ -\ 46.7 \\ \hline \mathbf{15.3} \end{aligned}$$

11. (5) **wheat flour**

12. (2) **65%**

$$\begin{aligned} 18.1 \\ -\ 11.0 \\ \hline 7.1 \end{aligned} \qquad \frac{7.1}{11.0} = \quad 11\overline{)7.100} = .645$$

.645 to the nearest hundredth = .65 = **65%**

13. (4) **The average American ate less beef and veal in 1979 than in 1969.**

14. (3) **$208**

$$\begin{array}{r} \$296 \\ 232 \\ 148 \\ \underline{156} \\ \$832 \end{array} \qquad \begin{array}{r} \mathbf{\$208} \\ 4\overline{)\$832} \end{array}$$

15. (4) **$38,500**

$12\% = .12$

$$\begin{array}{r} \mathbf{\$38{,}500} \\ .12\overline{)\$4620.00} \\ \underline{36} \\ 102 \\ \underline{96} \\ 60 \\ \underline{60} \end{array}$$

16. (1) **48° 23′ 49″**

$$\begin{array}{r} 89^\circ\ 59' \\ \not{90^\circ}\ \not{60'}\ 60'' \\ \underline{-\ 41^\circ\ 36'\ 11''} \\ \mathbf{48^\circ\ 23'\ 49''} \end{array}$$

17. (4) **$2,040**

$\frac{1}{\not{4}_1} \times \frac{\not{680}^{170}}{1} = \$170$

$$\begin{array}{r} \$170 \\ \underline{\times\ 12} \\ 340 \\ \underline{170} \\ \mathbf{\$2{,}040} \end{array}$$

18. (5) **8**

$$\begin{array}{rcl} 10x - 17 & = & x + 55 \\ \underline{-\ x\qquad} & & \underline{-x\qquad} \\ 9x - 17 & = & +\ 55 \\ \underline{+\ 17} & & \underline{+\ 17} \\ 9x & = & 72 \\ x & = & \mathbf{8} \end{array}$$

19. (3) **$12\frac{1}{2}$ ft.**

$\frac{5}{8} = \frac{x}{20}$

$8x = 100$

$x = 12\frac{4}{8} = \mathbf{12\frac{1}{2}}$

20. (5) **15.096**

$$\begin{array}{r} 4.08 \\ \underline{\times\ 3.7} \\ 2\ 8\ 56 \\ \underline{12\ 2\ 4\ \ } \\ \mathbf{15.0\ 96} \end{array}$$

21. (5) **24**

$cd - e^2 =$

$5 \cdot 8 - 4^2 =$

$40 - 16 = \mathbf{24}$

22. (5) **$64,715**

$$\begin{array}{r} \$214{,}500 \\ \underline{-\ 149{,}785} \\ \mathbf{\$\ 64{,}715} \end{array}$$

23. (2) **7 sq. in.**

$A = \frac{1}{2} \times 3\frac{1}{2} \times 4$

$= \frac{1}{\not{2}_1} \times \frac{7}{\not{2}_1} \times \frac{\not{4}^{\not{2}^1}}{1}$

$= \mathbf{7}$

24. (4) **4%**

$$\begin{array}{r} 24\% \\ \underline{-\ 20} \\ \mathbf{4\%} \end{array}$$

25. **(1) 23,220**

43% = .43

$$\begin{array}{r} 54{,}000 \\ \times\ .43 \\ \hline 1\ 620\ 00 \\ 21\ 600\ 0 \\ \hline \mathbf{23{,}220.00} \end{array}$$

26. **(2) conservative**

27. **(4) The percent of Americans who called themselves conservative was the same in 1976 and 1980.**

28. **(3) 91 cm.**

$$\begin{array}{r} 2.54 \\ \times\ 36 \\ \hline 15\ 24 \\ 76\ 2 \\ \hline 91.44 \end{array}$$

to the nearest whole = **91**

29. **(2) 5**

$$\begin{aligned} 3a + 4(a + 2) &= 43 \\ 3a + 4a + 8 &= 43 \\ 7a + 8 &= 43 \\ -8 &\quad -8 \\ 7a &= 35 \\ a &= \mathbf{5} \end{aligned}$$

30. **(5) 30**

$75 \div 2\frac{1}{2} =$

$\frac{75}{1} \div \frac{5}{2} =$

$\frac{\overset{15}{\cancel{75}}}{1} \times \frac{2}{\underset{1}{\cancel{5}}} = \mathbf{30}$

31. **(1) $22.40**

6 mos. $= \frac{6}{12} = \frac{1}{2}$ yr.

5.6% = .056

$$\begin{array}{r} .056 \\ \times\ 800 \\ \hline \$44.800 \end{array}$$

$\frac{1}{2} \times \$44.80 = \mathbf{\$22.40}$

32. **(3) 90°**

$$\begin{array}{r} 35^\circ \\ +\ 55^\circ \\ \hline 90^\circ \end{array} \qquad \begin{array}{r} 180^\circ \\ -\ 90^\circ \\ \hline \mathbf{90^\circ} \end{array}$$

33. **(5) 240**

2 2 4 12 48 **240**

×1 ×2 ×3 ×4 ×5

34. **(4) $31.85**

$$\begin{array}{r} \$4.90 \\ \times\ 6.5 \\ \hline 2\ 4\ 50 \\ 29\ 4\ 0 \\ \hline \mathbf{\$31.8\ 50} \end{array}$$

35. **(2) 12 ft.**

x = width

$3x$ = length

$$\begin{aligned} 96 &= 2(3x) + 2(x) \\ 96 &= 6x + 2x \\ 96 &= 8x \\ \mathbf{12} &= x \end{aligned}$$

36. **(2) 264 ft.**

$C = \pi d$

$= \frac{22}{\underset{1}{\cancel{7}}} \cdot \frac{\overset{12}{\cancel{84}}}{1}$

$= \mathbf{264\ ft.}$

37. **(3) $.22**

```
     $ .224     to the nearest
16)$3.590       cent = $.22
    3 2
      39
      32
       70
       64
```

38. **(4) 7(8x − 12)**

39. **(2) $\frac{1}{5}$**

$\frac{\text{food}}{\text{income}} \; \frac{156}{780} = \frac{1}{5}$

40. **(3) +12**

$\frac{-168}{-14} = +12$

Since the signs are alike, the answer is positive.

41. **(1) $196**

30% = .3

```
 $280      $280
 × .3      − 84
$84.0      $196
```

42. **(3) 288**

1 sq. ft. = 12 × 12 = 144 sq. in.

```
144
× 2
288
```

43. **(2) 8**

$\frac{\text{blue}}{\text{white}} \quad \frac{3}{2} = \frac{12}{x}$

$3x = 24$

$x = 8$

44. **(4) 28 lbs. 8 oz.**

```
   1 lb.     3 oz.
 × 24      × 24
  24 lbs.  72 oz. = 28 lbs. 8 oz.
```

45. **(2) 39″**

$$a^2 + b^2 = c^2$$
$$15^2 + 36^2 = c^2$$
$$225 + 1,296 = c^2$$
$$1,521 = c^2$$
$$\sqrt{1,521} = c$$

Guess 40.

```
    38        38       39
40)1,521    + 40     2)78
              78
```

46. **(3) 24**

```
        2 4
3.5.)84.0.
      70
      14 0
      14 0
```

47. **(4) 9**

48. **(3) $147,360**

```
  24,560
   × $6
$147,360
```

49. **(1) 143°**

```
 180°
− 37°
 143°
```

50. **(1) 14(73) + 14(26)**